Genetic Engineering in Agriculture, Horticulture and Forestry

Genetic Engineering in Agriculture, Horticulture and Forestry

Dr. Arvind Kumar

RANDOM

RANDOM PUBLICATIONS

NEW DELHI (INDIA)

Genetic Engineering in Agriculture, Horticulture and Forestry

ISBN 978-93-5111-606-6

Published in 2017 in India by
RANDOM PUBLICATIONS
4376-A/4B, Gali Murari Lal, Ansari Road
New Delhi-110 002
Phone : +9111-43580356, 011-23289044, 011-43142548
e-mail: sales@randompublications.com,
info@randompublications.com, randomexports@gmail.com

Reprinted 2026

Type Setting by : Friends Media, Delhi-110089

Preface

Genetic engineering is a radical new technology, one that breaks down fundamental genetic barriers – not only between species, but between humans, animals, and plants. By combining the genes of dissimilar and unrelated species, permanently altering their genetic codes, novel organisms are created that will pass the genetic changes onto their offspring through heredity.

Genetically engineered organisms that escape or are released from the laboratory could wreak environmental havoc. Genetically engineered "biological pollutants" have the potential to be even more destructive than chemical pollutants. Because they are alive, genetically engineered products are inherently more unpredictable than chemical products – they can reproduce, migrate, and mutate. Once released, it will be virtually impossible to recall genetically engineered organisms back to the laboratory. A report published by 100 top American scientists warned that the release of gene-spliced organisms "...could lead to irreversible, devastating damage to the ecology."

Biotechnology has offered tremendous scope and potential to conventional methods of crop improvement, crop protection, crop quality management and other horticultural traits. Biotechnology extends tremendous opportunities in fruit production by providing new genotypes for breeding purpose, supply of healthy and disease free planting material, improvement in fruit quality, enhancing shelf-life, availability of biopesticides, biofertilizers, etc.

Technological change in agriculture has been the foundation of economic growth and development. By raising yields and continually increasing the number of people who can be fed by the output of one worker in agriculture, change in agricultural technology has permitted an increasing proportion of labour and other productive resources to be devoted to non-agricultural production. Plant and animal breeding and chemical and mechanical technology plus improved husbandry have caused continual structural change in farming. These technological stimuli have arisen as a series of overlapping waves to create a process of technological change, the momentum of which looks likely to be maintained by intensified adoption of information and biotechnologies.

Highly comprehensive, erudite and lucid, it is hoped that this book proves to be extremely resourceful for the reader.

I would like to thank my team for standing beside me throughout my career and writing this book. My special thanks go to "Random Publications" who have published the book.

– Dr. Arvind Kumar

Contents

Preface .... *v-vi*

1. Technology of Genetic Engineering **1**

Introduction 1
Genetics of Chromosomes 5
The Commodification of Genetic Resources 12
The Science of Modern Genetic Engineering 16
Technological Process in Genetic Engineering 20
Useful Traits by Genetic Transformation 24

2. Genetic Engineering in Agricultural Foods **35**

Genetic Resources for Food and Agriculture 35
Genetically Modified Foods 38
Supporters and Opponents of Gmos 40
Regulation of Gmos 42
Arguments Against Gmos 46
Transfer of Dna From Gmo 48
Harmful Effects of Genetically Modified Foods 58
Food Fortification 64

3. Genetic Diversity and Agricultural Biotechnology **79**

Conservation of Plant Genetic Resources 79
Approaches of Diversity 81
Agriculture and Prospects of Cold Tolerant Genes 87
Novel Technologies for Biodiversity Assessment and Monitoring 88
Molecular Markers and the Management of Genetic Resources in Seed Genebanks 92
Regulation of Agricultural Biotechnology 94
Taxonomic Identification as Part of Germplasm Characterization 99
Facilitating the Use of Germplasm 105

4. Gene Technology in Plant Science **112**

Genes into Plants 112
Gene Mapping 117

The Gene Revolution: A Changing Paradigm for Agricultural R andD 123
Locating Genes of Major Effect 130
Theory of QTL Mapping in Doubled Haploid Lines 133
Alien Gene Transfer 137
Map-based Gene Cloning 139
New Challenge of Gene Technology 141

5. **Genetic Engineering in Plant Breeding 154**

Genetic Engineering Technology 154
Genetic Efficiency in Collecting for ex situ Conservation 157
Structural Analysis of Genome Relationships 165
Uses of Marker-aided Selection 169
Types of Genetic Markers 171
Participatory Plant Breeding 173
Reasons for Privatisation of Plant Breeding 174
Scope of Plant Breeding 177

6. **Biotechniques and the Alternative Agriculture 192**

Technological Change in Agriculture 192
Principal Directions of Biotechnol Ogical Development 193
Private-Sector Control of Biotechnology Development 196
The Neglect of Alternative Agriculture 198
Sowing the Seeds of Discontent 202
Biotechnology for Societal Development in India 208
Recognition for Farmers' Rights 216

7. **Applications of Genetic Biotechnology in Horticulture Genetic .... Engineering of Plants 232**

Tissue Culture 235
Molecular Diagnostics 236
Organic Horticulture 238
Horticulture and Anthropology 240
Applications in Horticulture and Forestry 245
Emerging Model Systems in Plant Biology 248

8. **Forest Genetics and Biotechnology 268**

The Development of Forestry 268
Transition From Exploitation to Management 269
Gene Selection and Classification of Microarray Data Using Random Forest 271
Genetic Modification in The Forestry Sector 280
Data Set Analysis in Genetic Modification 282
Global System on Conservation and Sustainable Use of Plant Genetic Resources 283
State of Genetic Modification in the Forestry Sector 286
The Great Diversity Found in Forest Trees 288

***Bibliography* 302**

***Index* 304**

1

Technology of Genetic Engineering

INTRODUCTION

Genetic engineering is a radical new technology, one that breaks down fundamental genetic barriers — not only between species, but between humans, animals, and plants. By combining the genes of dissimilar and unrelated species, permanently altering their genetic codes, novel organisms are created that will pass the genetic changes onto their offspring through heredity.

Scientists are now snipping, inserting, recombining, rearranging, editing, and programming genetic material. Animal genes and even human genes are being inserted into plants or animals creating unimagined transgenic life forms. For the first time in history, human beings are becoming the architects of life. Bio-engineers will be creating tens of thousands of novel organisms over the next few years. The prospect is frightening.

Genetic engineering poses unprecedented ethical and social concerns, as well as serious challenges to the environment, human health, animal welfare, and the future of agriculture. The following is just a sampling of concerns:

Genetically engineered organisms that escape or are released from the laboratory could wreak environmental havoc. Genetically engineered "biological pollutants" have the potential to be even more destructive than chemical pollutants. Because they are alive, genetically engineered products are inherently more unpredictable than chemical products — they can reproduce, migrate, and mutate. Once released, it will be virtually impossible to recall genetically engineered organisms back to the laboratory. A report published by 100 top American scientists warned that the release of gene-spliced organisms "...could lead to irreversible, devastating damage to the ecology."

Gene-splicing will likely result in unanticipated outcomes and dangerous surprises. Biotechnology is an imprecise science and scientists will never be able to ensure a 100 per cent success rate. Serious accidents are bound to occur. Researchers conducting experiments at Michigan State University recently found that genetically altering plants to resist viruses can cause the viruses to mutate into new, more virulent forms, or forms that can attack other plant species. Some other scary scenarios: foreign genes from genetically engineered plants could be carried by pollen, insects, wind, or rain, and flow into other crops, as well as wild and weedy relatives. Disaster would follow if genetically engineered crop traits, such as insect and virus resistance, found their way into weeds, for instance.

Genetically altered plants could produce toxins and other substances that might harm birds and other animals. Genetic engineering of plants and animals will almost certainly endanger species and reduce biological diversity. By virtue of their "superior" genes, some genetically engineered plants and animals will inevitably run amok, overpowering wild species in the same way that introduced exotic species, such as kudzu vine and Dutch elm disease which have created problems in North America.

What will happen to wild species, for example, when scientists release into the environment carp, salmon, and trout that are twice as large, and eat twice as much food, as their wild counterparts? Another danger lies in the creation of new kinds of crops and domesticated animals. Once researchers develop what is considered to be the "perfect tomato" or "perfect chicken" these will be the ones reproduced in large numbers; "less desirable" species would fall by the wayside. The "perfect" animals and plants could then be cloned (reproduced as exact genetic copies), reducing even further the pool of available genes on the planet.

Genetically engineering plants to be herbicide-tolerant will lead to increased use of chemicals in agriculture and further contamination of the environment. Biotech companies love to say that genetic engineering will end the use of dangerous chemicals in agriculture. But the leaders in biotechnology are the giant chemical companies like Monsanto, Du Pont, and Rhone-Ponlenc; they aren't interested in losing profits from the sale of chemicals. These companies are genetically engineering plants to be resistant to herbicides that they manufacture so they can sell more herbicides to farmers who, in turn, can apply more poisonous herbicides to crops to kill weeds. In fact, crops genetically engineered to be herbicide-tolerant account for nearly half of the applications for field testing submitted to the U.S. Department of Agriculture (USDA) since 1988. Even genetically engineering crops to produce their own pesticides presents dangerous problems.

Pests will eventually evolve that are resistant, then stronger chemicals will be needed to get rid of the pests. And what will happen when the pesticide gene spreads to weeds and other unwanted plants? The genetic engineering of crops and food-producing animals can produce toxic and allergic reactions in humans. Someone allergic to peanuts or shellfish, for example, would have no way of knowing if a tomato or other food had been altered with proteins from these substances, and could have a fatal reaction by eating such genetically altered foods. In addition, genetic engineers can take proteins from bacteria they find in the soil, the ocean — anywhere — and incorporate them into human food. Such substances have never been in the food supply before, so their toxic or allergenic characteristics are unknown. Genetically engineered products do not have a good track record for human safety. In 1989 and 1990, a genetically engineered brand of L-tryptophan, a common dietary supplement, killed more than 30 Americans and permanently disabled or afflicted more than 5,000 others with a potentially fatal and painful blood disorder, eosinophilia myalgia syndrome, before it was recalled by the FDA.

The manufacturer, Showa Denko K.K., Japan's third largest chemical company, had used genetically engineered bacteria to produce the over-the-counter supplement. It is believed that the bacteria somehow became contaminated during the recombinant DNA process. There were no labels on the product to identify the product as having been genetically engineered.

The patenting of genetically engineered foods, and widespread biotech food production, will eliminate farming as it has been practiced since the beginning of humankind's appearance on the planet. If the trend is not stopped, the patenting of transgenic plants and food-producing animals will soon lead to tenant farming in which farmers will lease their plants and animals from biotech conglomerates and pay royalties on seeds and offspring. Eventually, within the next few decades, agriculture will move off the soil and into biosynthetic industrial factories controlled by chemical and biotech companies. Never again will people know the joy of eating naturally produced, fresh foods. Hundreds of millions of farmers and other workers worldwide will lose their livelihoods. The hope of creating a human, sustainable agricultural system will be destroyed.

The genetic engineering and patenting of animals reduces living beings to the status of manufactured products and will result in much suffering. In January 1994, then-USDA Secretary Mike Espy announced that USDA scientists had completed genome "road maps" for cattle and pigs, a precursor to ever more experimentation on live animals. In addition to the cruelty inherent in such experimentation (the mistakes are born with painful deformities, crippled, blind, and so on), these "manufactured" creatures have no greater value to their "creators" than mechanical inventions.

Animals genetically engineered for use in laboratories, such as the infamous "Harvard mouse" which contains a human cancer-causing gene that will be passed down to all succeeding generations, were created to suffer. A purely reductionist science, biotechnology reduces all life to bits of information (genetic code) that can be arranged and rearranged at whim. Stripped of their integrity and sacred qualities, animals who are merely objects to their "inventors" will be treated as such.

Currently, more than 200 genetically engineered "freak" animals are awaiting patent approval from the federal government. No one is regulating genetically engineered organisms adequately or properly testing them for safety. In 1986, Reagan-era policymakers stitched together a patchwork of pre-existing and only marginally appropriate statutes to ease the way for new biotechnology products. But these laws were created years ago to deal with chemicals — not the unpredictable living products of genetic engineering. To date, no suitable government apparatus has been set up to deal with this radical new class of potentially overwhelming environmental and health threats.

The FDA's policy on genetically altered foods illustrates the problem. In May 1992, then Vice President Dan Quayle, and head of the Competitiveness Council, announced the U.S. Food and Drug Administration's newly developed policy on biotech foods: genetically engineered foods will not be treated differently from naturally produced foods; they will not be safety tested; they will not carry labels stating that they have been genetically engineered, nor will the government keep track of foods that have been genetically engineered. As a result, neither the government nor consumers will know which whole or processed foods have been genetically engineered.

Vegetarians and followers of religious dietary restrictions face the prospect of unwittingly eating vegetables and fruits that contain genetic material from animals — including humans. And health risks will be discovered only by trial and error — by consumers.

USDA oversight is no better. This agency has the conflicting task of both promoting and regulating agriculture, including genetically engineered plants and animals used for food. Indeed, the USDA is a primary sponsor of biotech research on plants and animals.

By patenting the genes they discover and the living organisms they create, a small corporate elite will soon own and control the genetic heritage of the plant. Scientists who "discover" genes and ways of manipulating them can patent — and thus own — not only genetic engineering techniques, but the very genes themselves. Chemical, pharmaceutical, and biotech companies such as DuPont, Upjohn, Bayer, Dow, Monsanto, Cib-Geigy, and Rhone-Poulenc, are urgently trying to identify and patent plant, animal, and human genes in order to complete their take-over of agriculture, animal husbandry, and food processing. These are some of the same companies that once promised a carefree life through pesticides and plastics. Would you trust them with the blueprints of life?

Genetic screening will likely lead to a loss of privacy and new levels of discrimination. Already, people are being denied health insurance on the basis of "faulty" genes.

Will employers require genetic screening of their employees and deny them work on the basis of the results? Will the government have access to our personal genetic profiles? One can easily imagine new levels of discrimination being directed against those whose genetic profiles reveal them to be, for example, less intelligent or predisposed to developing certain illnesses. Genetic engineering is already being used to "improve" the human race, a practice called eugenics. Genetic screening already allows us to identify and abort fetuses who carry genes for certain hereditary disorders. But within the next decade, scientists will likely have a complete map of the human genome to work with. Will we abort fetuses on the basis of non-life-threatening impairments such as myopia, because someone is predisposed towards homosexuality, or for purely cosmetic reasons? Researchers at the University of Pennsylvania have applied for a patent to genetically alter sperm cells in animals so traits passed down from one generation to the next can be changed; the application suggests that this can be done in humans too. Moving from animal eugenics to human eugenics is one small step.

Everyone wants the best for their children; but where do we stop? Inadvertently, we could soon make the efforts of the Nazis to create a "superior" race seem bumbling and inefficient.

The U.S. military is building an arsenal of genetically engineered biological weapons. Although the creation of biological weapons for offensive purposes has been outlawed by international treaty, the U.S. continues to develop such weapons for defensive purposes. However, genetically engineered biological agents are identical whether they are used for offensive or defensive purposes. Areas of investigation for such weapons include: bacteria that can resist all antibiotics; extra-hardy, more virulent bacteria and viruses that live longer and kill faster; and new organisms that can defeat vaccines or natural human or plant resistances.

Also being studied are the development of pathogens that can disrupt human hormonal balance enough to cause death, and the transformation of innocuous bacteria (such as are found in human intestines) into killers. Some experts believe that genetically engineered pathogens that can target specific racial groups are being developed as well.

GENETICS OF CHROMOSOMES

It might be best to consider the nature of the genotype at this point so that genetic principles are understood before we consider the complexities of the effects due to environmental fluctuations. As noted before, the physical bases of heredity are the genes that are borne on the chromosomes carried in cell nuclei. These chromosomes can be observed during mitosis (when the cell divides and reproduces an exact replica of its genetic content, and in meiosis (including reduction division), where the chromosome number is halved to produce the gametes (spermatozoan or ovum) involved in sexual reproduction.

The human species' 46 chromosomes are arranged in 23 pairs; in females all 23 pairs are homologous (correspond structurally and derivatively); in males, there are 22 homologous pairs plus a nonhomologous *X* and *Y* chromosome. The chromosome pairs, when properly stained, differ in length and appearance so that they can be individually recognized, much as one might recognize the familiar face of a friend. During meiosis, when the gametes are formed and the chromosome number is halved, all the chromosomes in a single gamete differ because each gamete has only one member of each chromosome pair. At fertilization, two gametes unite, each containing a set of 23 chromosomes to form the fertilized cell (zygote) with 23 pairs or 46 chromosomes once again. The number of chromosomes contained in gametes (23) is referred to as the haploid number and that of zygotes ($2 \times 23 = 46$) is the diploid number written as n and 2n, respectively. The process of meiosis affords a significant opportunity for increasing genetic variability. The most important consistent mechanism of change is crossing over, which takes place during the early phases of meiosis. Homologous chromosomes, lying next to each other, break, exchange equivalent portions of their genes, and reanneal to produce combinations that differ in arrangement from the genetic endowment provided by parents. This phenomenon is omnipresent in the normal meiotic process; when it fails to occur in at least one parent, the resulting organism is often sterile. Crossing over is the most consistent of a number of processes that provide the raw material upon which natural selection may act.

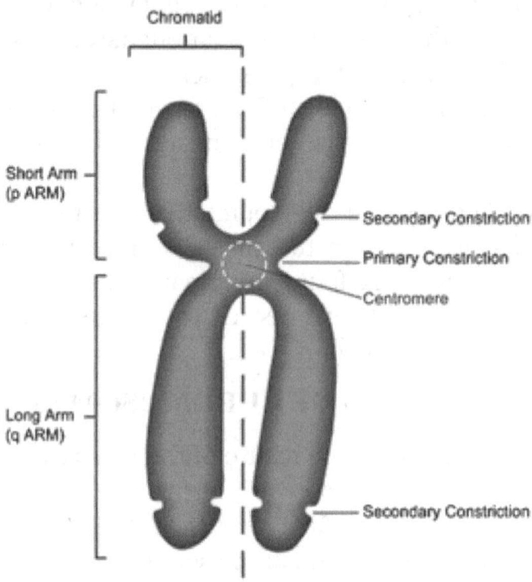

Fig. Human Chromosomes

A coarser and more sporadic mechanism is individualized chromosome rearrangement, of which there are four possible types: deletion, duplication, inversion, and translocation. Deletion is the sometimes lethal removal of a gene or sequence of genes. Death may occur because a biochemical sequence has been excised, resulting in an intolerable structural or functional gap in the organism. Duplication of a chromosome or part of a chromosome may cause an imbalance of gene activity, reducing the viability of an organism. However, because some organisms can tolerate duplications of genetic material, these duplications might play an evolutionary role if part or all of one of the duplications mutated and functioned differently from the original.

Indeed, this is one way in which evolutionary change is postulated to occur. Greater genetic variability gives rise (in minute quanta, to be sure) to physiologic and morphologic differences that endow the organism with an enlarged Behavioural repertoire so that it can more successfully cope with an ever changing environment.

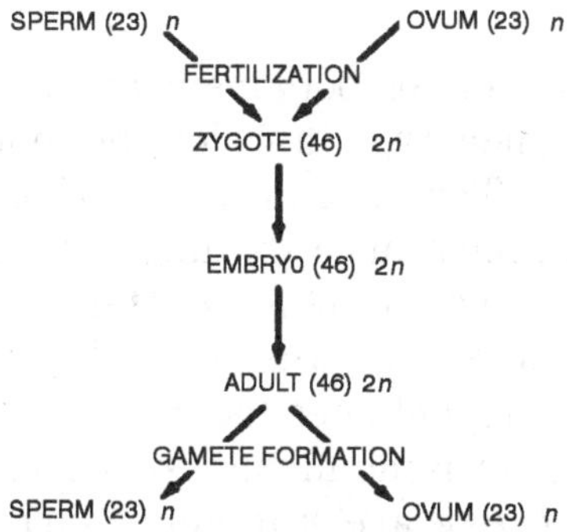

Fig. Human Chromosomes

This can be generalized for sexually reproducing organisms, which will be designated by n, where n is the haploid number and $2n$ the diploid number. An inversion occurs when a chromosome breaks in one or two places and the segment between the break rotates 180° leading to a reversal of gene order and a resulting change in biochemical product. A translocation occurs when two chromosomes from different pairs break simultaneously and exchange segments.

According to current knowledge, not all these chromo-somal and gene changes are equally important in their effects on Behaviour. To date, inversions and chromosome number changes appear to form the main categories of importance. But, crucial to the meiotic process (which reproductive systems evolved to support) is the shuffling and mingling of genetic material into infinite numbers of combinations.

MUTATION AND MUTANTS

Until now this discussion has centred on the shuffling of genetic material in sexually reproducing species. Genetic variability within a species has been discussed with emphasis on the variety of gene combinations attributable to meiosis, crossing over, duplication, and translocation. All these simply involve the reassortment of a set of variable genes indigenous to a population.

The essence of creating novelty is the process of mutation. These are alterations in the chemical structure of a gene that lead to changes in gene products that may ramify throughout the organism. Mutations at some loci occur spontaneously on the average of one in a million newly arisen gametes. The frequency is increased by radiation and by some chemicals known as mutagens.

A veritable storehouse of genetic mutations in mice is found at the Jackson Memorial Laboratory in Bar Harbour, Maine. Through selective inbreeding, when a mutant trait of interest is found, a strain of mice each of whose members possess the trait is established and bred for research uses. Many of the known mutant genes affect the nervous system during its development and produce a range of deficits in Behaviour from minor disturbances to severe defects. One of these found in the mouse is a mutation on chromosome five, an altered gene called *fidget*.

When this gene is present on both members of the fifth chromosome, it produces a Behavioural phenotype in which the mouse's head moves from side to side continuously. The gene is written *fi* and the mouse is said to have an *fift* genotype. In most mice the fidget gene is not present at this locus but the normal alternative gene, written as +, is. If the mouse is genotypically either *fi*+ or ++, the Behavioural alteration is not seen. Only if there are two *fi* genes present does the fidget phenotype become visible, as the wild type (+) gene is dominant to the recessive *fi*). However, dominance and recessivity are not necessarily complete, because heterozygotes (individuals with non-identical alleles) are often distinguishable from both homozygotes (individuals with identical alleles). Even if at first sight dominance appears to be complete, detailed biochemical tests may reveal differences between the heterozygote and the homozygote.

Untreated phenylketonuria in man provides one such example. Superficially viewed, the disease is controlled by a recessive gene *p*. with afflicted individuals being *pp* and normals *p*+ or ++. However, at the biochemical level, *p*+ and ++ may be distinguishable because the *p*+ individual tends to have more phenylalanine in blood serum. Dominance, in this case is incomplete. So, depending on which phenotype component is being examined, different conclusions as to the dominance may be drawn.

If a simple dominant—recessive relationship is assumed with two pairs of alleles at two loci (i.e., *A* and *B* are dominant to *a* and *b*), what might ensue? If a double heterozygote (*AaBb*) is crossed with a double homozygote (*aabb*) what are the statistical expectations? The double recessive homozygote produces only *ab* gametes. From the double heterozygote each separate locus yields equal numbers of *AB, Ab, aB* and *ab* gametes according to Mendel's principles of segregation and of independent assortment of alleles during gamete formation. If these gametes are fertilized with the recessive *ab* gametes, the results, assuming two independent pairs of alleles and simple dominant recessive relationships, will yield a 1:1:1:1 phenotype ratio.

The backcross to a recessive parent (one with an *aabb* genotype) is often used as a testcross in practical breeding programmes because it can determine the actual genotype of an organism whose recessive alleles may be obscured by dominant ones.

The use of this wholly recessive dihybrid parent unmasks recessive alleles in the alternate parental genotype. If we return to Mendel's garden peas and specify physical traits in a dihybrid cross (*A* is the gene for yellow and *a* the gene for green; *B* is the gene for a smooth surface and *b* the gene for a wrinkled one), we emerge with four types of gametes (*AB*, *Ab*, *aB*, *ab*) produced in approximately equal numbers, which yield the basic Mendelian ratio of 9:3:3:1.

RECOMBINATION AND GENETIC MAPS

There is an important exception to Mendel's second law. Assortment of genes is not generally independent if gene loci are on the same chromosome. The closer genes are to each other along a given chromosome, the more closely linked they are during gamete formation. Although genes on the same chromosome may recombine with each other during meiosis, the frequency of the recombination depends on the spatial distance between the relevant loci. Generally, the further apart the loci are, the greater the possibility of recombination.

From the frequency of these recombinational events, chromosome maps have been constructed for individual chromosomes. Genes located on the same chromosome are said to belong to the same linkage group. Thus, in humans we have 23 linkage groups (one for each pair of chromosomes, i.e., the haploid number of chromosomes [*n*] corresponds to the number of linkage groups).

The house mouse, possessing 40 chromosomes, has 20 linkage groups; in *Drosophila melanogaster,* an organism of historical importance in Behaviour genetics, the numbers are 8 and 4 respectively. During the formation of human eggs and sperm (gametes), the chromosome number is halved (meiosis), leaving the egg with 22 autosomes and an X chromosome, and the sperm with 22 autosomes and an X or a Y sex chromosome. Males produce four sperm from each specialized cell in the testes and X- and Y-bearing sperm occur in about equal numbers.

GAMETES AND FERTILIZATION

In ovaries, meiosis also produces four products, but three of these become polar bodies, whereas the fourth becomes the chromosome containing nucleus of the developing egg. Human females are born with their entire reproductive quotient already present and the limits are set for the number of eggs they can produce (about 400 during a reproductive life span). Males, however, continue to produce sperm from sexual maturation until senescence.

Differences in gametic size and function are extreme. The head of the sperm is about 3 to 5 microns long and 2 to 3 microns wide. The tail, which propels it, is 10 times the length of the head. Essentially, it is a pared down packet of miraculously organized DNA (genetic material) with a launching and traveling apparatus to respond to and fertilize ova.

The human egg, in contrast, is one of the largest cells in the body—a sphere about 130 microns in diametre. It moves only passively when it is essentially shoved along, taking a few days to progress from ovary to uterus. As noted by Hartl (1977, cited in Probber and Ehrman,: "All the sperm that gave rise to all the people that ever lived could be carried in a teaspoon the eggs, would require a small bucket." The egg is comparatively enormous because in addition to its complement of chromosomes, its cytoplasm provides the nourishment for, and biochemical assistance to the developing embryo for translation of the information carried in its genes.

Unlike the female gamete the function of the sperm as a cell is complete when the sperm contacts and penetrates the egg (having shed its tail), and sperm and egg fuse. With fertilization of the egg, formation of the embryo begins, but it is not until 6 weeks afterwards that sex chromosomes become operative.

At this point the *Y* chromosome, if present, will cause the inner layer of the gonads to develop into testes. If, however, there are two X chromosomes, at about 12 weeks the outer layer of the gonads will elaborate into ovaries.

X-CHROMOSOME INACTIVATION

A further complication relates to sex. Although every human embryo is predisposed to develop as a female, its eventual sex phenotype will be determined by the types of chromosomes borne by the sperm. Our species has 23 pairs of chromosomes, each of which carries thousands of genes that compose the genotype that develops and unfolds to become the visible organism, the phenotype. In humans, 22 of the total number of chromosomes are autosomes; that is, two copies of these chromosomes are present in both sexes. The 23rd pair is made up of the unique sex chromosomes; the female has two of the X chromosomes; the male has one X and one Y chromosome. These were originally called X from X the unknown factor, and Y because it follows X alphabetically.

The X chromosome was first described in 1891 from insect material as "a peculiar chromatin body" but its function was not known then or for a long time afterwards. Although the female had two doses of all the genes on the X chromosome and the male only one, there seemed to be no significant differences in X chromosome activity between the sexes. It was not until 1962 that a basis for dosage compensation emerged. This has been named the Lyon hypothesis after Mary F. Lyon, who first presented detailed studies on the subject. She hypothesized that in the normal female, one X chromosome is inactivated in each somatic body cell early in embryonic development. Whether the maternal or paternal X is rendered inert is a matter of chance and the result is that patches of cells with X-chromosomal lineages emerge.

The deactivated X becomes a tightly packed mass usually found close to the inner surface of the nuclear membrane of somatic cells in normal female mammals. It is called a Barr body after its discoverer. Accordingly, each cell in a male or a female has only one operative X chromosome. Because a female's cells are of two types (maternal X and paternal X), she is a genetic mosaic. A male, however, has only his maternal X chromosome to express.

What are the implications of this double X-chromosome dose for the female of our species? At the molecular level the X chromosome is a much more information packed and therefore biochemically active chromosome than the Y, which is much smaller and, except for male sex determination and fertility, inert. Recent work by McKusick and Ruddle cited more than 100 genes in the human X-chromosome map. The Y chromosome bears only two known genes, one controlling the TDF (Testis Determining Factor) and the other coding for the H—Y antigen.

The H—Y antigen plays an important role in the expression of male physiological and morphological characteristics, specifically for the differentiation of the testis. "Further male differentiation is imposed on the embryo by the action of testicular hormones against the inherent tendency towards the female phenotype." It is also known that males with their X and Y chromosomal complement are relatively unprotected from the expression of unpaired deleterious genes on their single X chromosome.

Every thing on their maternal X chromosome is expressed if not modified by genes on the autosomes. This is demonstrated by the much greater percentage of males afflicted by X-linked defects such as colour blindness and hemophilia. Genes on the X chromosome are said to be sex linked. In females the principles of heterozygosity and homozygosity apply to genes on the sex chromosomes just as they do for the genes on autosomes. In males, loci on the X are not matched by loci on the nearly inert Y chromosome; this condition is known as hemizygosity.

EPIGENESIS

The process of development involves constant interaction and coaction of an organism's genes and its environment. Given identical genotypes as in monozygotic human twins or highly inbred mice and *Drosophila,* phenotypes are usually very similar. Even under optimal circumstances however, phenotypic variation occurs because of perturbations in the environment. In a given situation one can speak of the average phenotype of individuals with the same genotype as a *norm of reaction*.

The *range of reaction* covers the spread of phenotypes that could potentially develop from a given fertilized ovum. Clearly, the range of reaction increases if the environment is not carefully controlled. The *epigenetic* view of development postulates that genotype and environment are involved at every stage from fertilization to birth and beyond. At fertilization, these actions begin at a molecular level.

Enzyme activities and changes in the intracellular environment are initiated. The egg provides a source of energy for the newly formed zygote and a well primed machinery for protein synthesis, which initiates the primary series of cleavages and invaginations that produce the embryo, fetus, and finally, the neonate.

The control mechanisms fostering maturation are encoded within the genes, in the internal and external environment of the DNA, and in the interrelations and interactions between them. At every level molecular, cellular, and organismic there are functional gene environment interactions and products.

THE COMMODIFICATION OF GENETIC RESOURCES

The administrative body of the CBD has been markedly influenced by the green-developmentalist perspective of Western governments and environmental organizations such as the World Conservation Union (IUCN). Interpretations of the CBD mandate prepared by the Secretariat of the CBD reflect this influence. One such document's proclamation that "modern biotechnology offers the potential to invent sustainable systems of the future, to be accompanied by a new paradigm for industry" reveals how central biotechnology has been to the interpretation of the treaty. "The private sector is the key player in benefit-sharing arrangements," the aforementioned secretariat document states flatly. Another secretariat note identifies a "policy setting" conducive to benefit sharing as one that encourages "access legislation, incentives, partnerships and contracts". This emphasis on market transactions and business partnerships illustrates how CBD administrators have promoted the market based management of genetic resources.

Case studies compiled by the secretariat interpret biodiversity benefit sharing almost exclusively in terms of North-South bio-prospecting deals. In effect, access to "biodiversity benefits" under the CBD is being made contingent on the participation of diversity-rich countries in a global genetic-resource market. But when CBD documents define biodiversity benefits as benefits to be derived from the commercialization of genetic resources by biotechnology industries, they fail to recognize that biodiversity benefits already exist—that is, that the benefits of natural resources are known and valued by people who depend on them directly for sustenance, shelter, aesthetic pleasure, and spiritual significance.

This equation of biodiversity benefits with genetic resources reduces biological variety to its purported essence as a commodity, separable from its complex interrelationships with the rest of nature and society. It is as if the values of biological variety come into existence only when its genetic information is "developed"—codified, counted, and commercialized—by biotechnology. A broader conceptualization would center on the incalculable present and future benefits of healthy ecosystems and diversity-based farming systems both to people locally and—insofar as such eco-social systems may conserve and generate biodiversity—to people in all countries.

Measuring the value of a country's biological resources on the basis of their commercial potential, or in terms of the market costs of replacing them, is compatible with the dominant discourse of environmental managerialism. However, this green-developmentalist approach privileges those aspects of nature that can be removed from their local contexts, transformed by investment, and sold. The approach fosters a view of ecosystems as warehouses of potential commodities to meet the demands of foreign consumers rather than as the bases of local and national life, as the sources of material necessities and meanings, and as the biophysical contexts of cultures. In this way, green developmentalism divorces the problems of biotechnology and genetic-resource management from the development needs of gene-rich but hard-currency-poor countries.

The globalized, market-based management of biodiversity requires clear property rights to natural resources, and the CBD secretariat has devoted what might seem, in a conservation treaty, inordinate attention to intellectual property. Conference decisions and secretariat documents display an almost schizophrenic ambivalence. The predominant assumptions in these documents support Northern intellectual-property models and their international extension, but references to IPRs are counterbalanced in nearly every statement by references to "alternative systems" or to the "concerns of indigenous and local peoples" about the impacts of IPRs.

This controversy has been kept alive by the active influence in the CBD and related international forums of social movements that oppose the privatization of genes and associated knowledge. On the basis of ethical, equity, and ecological concerns, the NGOs and indigenous peoples' and farmers' organizations that constitute this transnational movement have denounced the patenting and commoditization of life.

With the support of Sweden and a number of Southern states, these organizations have energetically resisted Northern proposals that the CBD recognize the WTO as the appropriate body for settling international disputes over property rights to organisms and knowledge about nature. In light of fears among many Southern delegations of losing sovereignty to multilateral trade bodies; growing doubts about the benefits of industrial agriculture and the safety of genetic engineering; and questions about the fairness of intellectual-property regimes, these NGOs have found a growing audience among developing-country delegations.

BIOTECHNOLOGY AND PROPERTY IN LIFE

What became known as the Cartagena Protocol on Biosafety of the CBD was finally negotiated, after significant concessions from the United States, in

January 2000. Because the protocol stipulates that countries may decline to accept exports of genetically engineered "living modified organisms", U. S.-based agribusiness interests have seen it as a threat to their export markets. Nevertheless, a united front of European and developing country delegations forced significant concessions from the United States delegation.

The United States-led Miami Group of six grain-exporting states gave up its opposition to protocol rules for the labeling of exports of genetically altered organisms meant for release into the environment, and dropped its proposed language that would have asserted the primacy of the WTO over the CBD. Included over U. S. objections was a provision allowing governments to take account of the socioeconomic impacts of transgenics in deciding whether to permit imports of particular organisms. This principle may yet collide with WTO rules against "unfair" barriers to trade, rules in which criteria of economic efficiency are pre-eminent.

Although thirty-eight states had ratified the protocol as of November 2002, it will not become international law until 50 governments also ratify it. Substantial disagreements remain over whether states or enterprises will be held liable for environmental or health damages resulting from the use of their genetically engineered products; over specific requirements for labeling, transport, and "contained use" of genetically engineered organisms; and over the interpretation of the precautionary "principle" or precautionary "approach"—both terms appear in the protocol—which contrasts with the guidelines that frame U. S. biotechnology regulatory policy.

In June 1999, a coalition of developing-country CBD members asked the CBD's scientific advisory body to call for a worldwide moratorium on the field testing and commercialization of "terminator technologies." These genetic engineering methods are being developed by the U. S. Department of Agriculture and commercial biotechnology firms to produce crops with seeds that will not germinate. By hardwiring property rights into plant genomes, these technologies would enable companies to control their privately-held crop genetic resources in cases and places where their IPR claims are not recognized. The advent of terminator technology, which confers no agronomic value to plants, has added to widespread skepticism about industry claims that transgenic crops are designed to benefit the hungry and increase the productivity of poor farmers in the developing world. Although the economic purpose of the technology—enlarging seed markets by preventing farmers from saving seed—is perfectly in keeping with the letter and spirit of the WTO, it is arguably at odds with the CBD's commitments to conserving crop biodiversity and the relevant practices of local communities.

In regional forums such as the European Union Parliament and the Organization of African Unity, governments, environmental ministries, and NGOs have called for recognition of the CBD's precedence over the WTO in matters pertaining to biodiversity. The negotiation of the International Undertaking on Plant Genetic Resources for Food and Agriculture was deadlocked for seven years over debates about IPRs, genetic-resource benefit sharing, and biopiracy.

The treaty that finally emerged from this process in November 2001 reflects an uneasy compromise that has not yet resolved issues of access to an valuation of crop genetic resources. Disputes about private property in genetic information and in biotechnology tools have also embroiled the Consultative Group on International Agricultural Research (CGIAR), which is the world's largest multilateral network of seed banks and crop-research centers, the World Intellectual Property Organization, and even the World Bank.

In August 2001, the UN Sub-Commission for the Promotion and Protection of Human Rights passed a stinging resolution against the WTO TRIPS Agreement, noting conflicts between TRIPS and "economic, social and cultural rights" in relation to the need for technology transfer and the consequences for the enjoyment of the right to food of plant variety rights and the patenting of genetically modified organisms, "bio-piracy" and the reduction of communities' (especially indigenous communities') control over their own genetic and natural resources and cultural values, and restrictions on access to patented pharmaceuticals and the implications for the enjoyment of the right to health.

The report on which the resolution was based, which described the WTO as a "nightmare" for poor countries, prompted an official complaint by top WTO officials. The repercussions of these disputes echo in the WTO. A United Statesbacked proposal to establish a WTO body on biotechnology, seen by critics as an attempt to outflank the Cartagena Protocol on Biosafety, was rejected by Europe and by developing countries at the failed 1999 WTO ministerial session in Seattle. Developing countries continue to call for reform of the TRIPS requirement that all countries enforce patents on microorganisms and property rights to animals and plants.

The African Group wants to remove references to plants and all life forms from TRIPS, whereas India wants language inserted that would require disclosure of the source of genetic material on patent applications. The Philippines Department of Agriculture has recommended that governments "get life forms (plants and animals) and biodiversity (and indigenous knowledge) out of the jurisdiction of WTO".

India, Brazil, and the African negotiating bloc, among others, have asked that the WTO's Council for TRIPS take into account biodiversity, traditional knowledge, benefit sharing, farmers' rights to save and share seeds, socioeconomic welfare, and the ethics of patenting of life forms.

The majority of council delegations have asked the council to "harmonize" TRIPS with the CBD. The U. S. delegation adamantly opposes this and objects to granting the CBD secretariat observer status in the council, insisting that the only issue to be discussed is the progress of TRIPS implementation.

Even if the majority is able to block or slow TRIPS implementation at the international level, the United States may achieve its goal of standardized IPRs by pressuring reluctant countries to include IPR promises in bilateral trade pacts or by threatening to reduce aid to countries that do not comply. Regional trade accords, including the North American Free Trade Agreement and the U. S.-proposed Free Trade Area of the Americas, include requirements that signatories maintain IPR regimes compatible with those of the United States. Bilateral trade deals, such as the recent agreement between the United States and the government of Vietnam, contain provisions requiring the new U. S. trade partner to join the UPOV (International Union for the Protection of New Varieties of Plants) convention, the international IPR agreement which the United States regards as the acceptable regime for plant-variety protection.

Nevertheless, enforcement of IPRs at the local level may be difficult in the face of growing defiance by social movements. Even where governments feel forced to comply with globalized IPR rules, many citizens are refusing to do so. In India and Bangladesh, Thailand and the Philippines, and many Latin American countries, organizations of farmers have linked their demands for land tenure and support of domestic agriculture and rural livelihoods to their opposition to crop-variety IPRs, life patenting, and food-crop biopiracy. International farmers' organizations such as Via Campesina and NGOs such as Genetic Resources Action International, the Malaysia-based Third World Network, and the United States-based Institute for Agriculture and Trade Policy stress the connections between food security (of people), food sovereignty (of nations), and preservation of agro-biodiversity and diversity-based farming systems.

Thus, a high-stakes battle continues over whether food, farming, and biotechnology will be understood and governed as a problem of corporate technoscience, economic efficiency, and universal legal standards, or whether the broader issues of who really benefits and who loses from genetic engineering of crops, privatization of research, and world-scale consolidation of agro-economic power will be addressed by emerging institutions of global governance. In the CBD and in the WTO, these international battles have brought to the surface deep discontent with persistent and widening inequalities in the postcolonial world order.

THE SCIENCE OF MODERN GENETIC ENGINEERING

Genetic engineering is one form of biotechnology and usually refers to copying a gene from one living organism—plant, animal or microbe—and adding it to another organism. In genetic engineering, a small piece of genetic material (DNA) is inserted into another organism to produce a desired effect. This is in contrast to traditional plant breeding in which all the genes desirable and undesirable contained in the male plant—pollen—are combined with all the genes of the female plant.

The progeny resulting from this cross may contain the gene for the desirable character, but it will also contain many of the undesirable genes from both parents. Genetic engineering has the advantage of being able to transfer only the gene of interest and greatly accelerate plant breeding. But genetic engineering also is more powerful than traditional breeding since it can move genes not only between similar plant species but also from distant relatives, including non-plant species. It is possible to move genes between such seemingly unrelated organisms because all living organisms share the same code for DNA and the synthesis of proteins and other basic life functions. What might seem on the surface to be very different organisms are, in fact, very similar, at least at the molecular level. All living things are more alike than different, and this is one of the reasons that genes can be moved so successfully between such seemingly different organisms as plants and bacteria. Genes are not unique to the organisms from which they came, so there really aren't "tomato genes" or "bacterial genes." It's the collection of all genes in a tomato or a bacterium that makes it a tomato or bacterium, not a single gene. As we learn more about the genetic makeup of all organisms, we see that most plant species differ by only a small percentage of their genes and those even such seemingly different organisms as tomatoes and bacteria have many common genes.

These findings suggest that in the long-term evolutionary process even tomatoes and bacteria had some common ancestor. From the discovery 50 years ago of the structure of DNA, scientists soon came to realise they could take segments of DNA that carried information for specific traits—genes—and move them into another organism. In 1972, the collaboration of Hubert Boyer and Stanley Cohen resulted in the first isolation and transfer of a gene from one organism to a single-celled bacterium where it would express the gene and manufacture a protein. Their discoveries led to the first direct use of biotechnology—the production of synthetic insulin to treat people with diabetes—and the start of what is often called modern biotechnology. Plants were first transformed through genetic engineering in the late 1970s. Mary-Dell Chilton and colleagues used a common soil-dwelling bacterium, Agrobacterium tumefaciens, that attaches itself to plants and transfers some of its DNA into the plant. Chilton and her colleagues added a gene to this bacterium, which in turn transferred the gene into a plant where it became part of the plant's DNA. This bacterium is still commonly used in genetic engineering along with another technique that uses a high-velocity mechanism to inject DNA into plant cells. The result from either technique is the same—the plant cells take up the gene and begin to express it as their own.

Benefits and Risks of Biotechnology

Plants developed through genetic engineering were first grown on 1.7 million hectares in 1996 in the United States, but by 2002 they were grown on 58.7 million hectares in 16 countries. By far the major use of the present plants is to manage pests—weeds, insects and diseases. Weed management with genetically engineered plants is accomplished because the plants have a modified enzyme (a protein) that allows them to survive an application of a specific herbicide that normally acts on that enzyme. Growers can plant the herbicide-tolerant seeds, allow the plants to emerge along with any weeds in the field and then treat the field with an herbicide.

The result is that the weeds, but not the crops, die. The advantage to growers is that they spend less time on weed management, have enhanced weed control, use safer herbicides, and in many cases use less herbicides. Additionally, this technology allows growers to use soil conservation practices such as reduced or no-tillage, thus helping to retain soil structure and moisture and reduce erosion. Herbicide tolerant crops (soybean, canola, cotton and maize) were grown on 48.6 million hectares in 2002. Insect-resistant crops developed through genetic engineering utilize the common soil bacterium, Bacillus thuringiensis (Bt), which has been commercially used for more than 50 years, as an insecticide spray.

Although safe to humans and the environment, when a susceptible insect ingests Bt, the Bt protein binds to specific molecular receptors in the gut and creates pores causing the insect to starve to death. Insecticidal products containing Bt were first commercialized in France in the late 1930s, but even in 1999 the total sales of Bt products constituted less than 2 per cent of the total value of all insecticides. Bt, which had limited use as a foliar insecticide, became a major insecticide only when genes that produce Bt toxins were engineered into major crops. The Bt crops available at present are maize and cotton. These were grown on a total of 14.5 million hectares in 2002. Virus-resistant crops were created by inserting a non-infective part of a plant virus into a plant, essentially "vaccinating" the plant to protect it from the virus.

This technique is called "pathogen-derived resistance." Squash and papaya have been engineered to resist infection by some common viruses and are approved for sale in the United States. There are fewer than 1 million hectares of these crops. The bioengineered plants available at present provide growers with better tools to manage pest problems. As with any technology, there are risks and benefits to currently available genetically engineered plants, but the present body of information indicates their use has enhanced pest management, substantially reduced the amounts of pesticides used in some crops, enabled growers to use safer pesticides, and contributed to enhanced safety for humans and the environment.

The regulatory process for managing these plants and their effects on the environment and humans has evolved with the technology and the scientific community's knowledge of these tools. Many of the more controversial issues surrounding genetic engineering of plants—such as pesticide resistance, gene flow and intellectual property issues—are not unique to this new technology but pertain to all types of agriculture. Some species of insects have developed resistance to sprays of Bt, indicating the potential for some species to become resistant to Bt plants.

However, despite Bt plants being grown on more than 62 million hectares worldwide from 1996 to 2002, there have been no documented cases of resistance development. The reasons for this lack of resistance appear to involve not only biological factors of the insects and the Bt plant, but also the fact that the regulatory agency (the Environmental Protection Agency) in the United States requires a resistance management plan for growing Bt plants. No other insecticide has such strict regulations. Still, growers, companies and federal regulatory agencies must be vigilant about resistance developing for biotech crops used to manage insects, weeds and viruses as they also must with non-biotech pest management tactics.

It will be important to consider the accrued environmental and health benefits of these biotech crops prior to the development of any resistance and how resistance can be managed if and where it occurs. In addition to pesticide resistance, gene flow from biotech to non-biotech crops may also be a concern. However, the risk of gene flow varies with each crop and each gene. Pollen flow in soybeans is very limited so the risk of a biotech soybean crop crossing with a non-biotech soybean is minimal, but this may be different for another crop. Likewise, if the gene in the biotech crop that provided a pest management trait, such as insect resistance, moved into a non-biotech plant, such as a weed, any selective advantage of the insect-protected weed in the ecosystem should be assessed. These same questions should also be answered with non-biotech crops, but these have not received the same level of attention as biotech crops because of the latter's higher profile.

Horizon of Biotechnology

In the future, the potential uses of plant biotechnology are far more wide-ranging than the pest-management biotech crops of today. Plants are being developed that serve as production "factories" for medically important drugs, sources of alternative energy, tools for cleaning toxic waste sites, and biomaterials including dyes, inks, detergents, adhesives, lubricants, plastics and the like. Consumers may see these products as more directly enhancing their quality of life than the pest-management biotech crops of today. Perhaps an even more dramatic advantage to consumers will be seen when plants are genetically engineered to have enhanced health benefits such as disease-fighting chemicals or increased amounts of essential vitamins and minerals. A healthy and well-informed discussion of the risks and benefits involved in agricultural biotechnology is needed to ensure a proper role for this new technology in our future food and health systems. No one should believe that any technology, including biotechnology, will completely solve the world's agricultural problems. Many people familiar with biotechnology, however, believe it to be an important component of the solution.

TECHNOLOGICAL PROCESS IN GENETIC ENGINEERING

In order to understand how genetic manipulation is accomplished, it is important first to understand the structure of deoxyribonucleic acid, or DNA. Within its chemical structure, DNA stores the information that determines an organism's hereditary or genetic properties. DNA is made up of a linked series of units called nucleotides (Blaese), Different nucleotide sequences determine different genes genetic information. Genetic engineering is based on this genetic information.

Genetic manipulation is carried out through a process known as recombinant-DNA formation, or gene splicing. This procedure behind genetic engineering is one whereby segments of genetic material from one organism are transferred to another. The basis of the technique lies in the use of restriction enzymes that split DNA strands wherever certain desired secjuences of nucleotides, or specific genes, occur. This desired segment of DNA is referred to as donor DNA. The process of gene splicing results in a series of fragments of DNA, each of which express the same desired gene that can then combine with plasmids (Rubenstein).

Plasmids are small, circular molecules of DNA that are found in many bacteria. The bacteria act as vectors in the process of genetic engineering. The desired gene cannot be directly inserted into the recipient organism, or host, therefore there must be an organism that can carry the donor DNA into the host. Plasmid DNA is isolated from bacteria and its circular structure is broken by restriction enzymes (Dworkin). The desired donor DNA is then inserted in the plasmid, and the circle is resealed by ligases, which are enzymes that repair breaks in DNA strands.

This reconstructed plasmid, which contains an extra gene, can be replaced in the bacteria, where it is cloned, or duplicated, in large numbers. The combined vector and donor DNA fragment constitute the recombinant-DNA molecule. Once inside a host cell, this molecule is replicated along with the host's DNA during cell division. These divisions produce a clone of identical cells, each having a copy of the recombinant-DNA molecule and thus permanently changing the genetic makeup of the host organism (Steinbrecher). Genetic engineering has been accomplished.Genetic manipulation is carried out through a process known as recombinant-DNA formation, or gene splicing. This procedure behind genetic engineering is one whereby segments of genetic material from one organism are transferred to another. The basis of the technique lies in the use of restriction enzymes that split DNA strands wherever certain desired secjuences of nucleotides, or specific genes, occur. This desired segment of DNA is referred to as donor DNA. The process of gene splicing results in a series of fragments of DNA, each of which express the same desired gene that can then combine with plasmids (Rubenstein).

Plasmids are small, circular molecules of DNA that are found in many bacteria. The bacteria act as vectors in the process of genetic engineering. The desired gene cannot be directly inserted into the recipient organism, or host, therefore there must be an organism that can carry the donor DNA into the host. Plasmid DNA is isolated from bacteria and its circular structure is broken by restriction enzymes (Dworkin). The desired donor DNA is then inserted in the plasmid, and the circle is resealed by ligases, which are enzymes that repair breaks in DNA strands. This reconstructed plasmid, which contains an extra gene, can be replaced in the bacteria, where it is cloned, or duplicated, in large numbers. The combined vector and donor DNA fragment constitute the recombinant-DNA molecule. Once inside a host cell, this molecule is replicated along with the host's DNA during cell division. These divisions produce a clone of identical cells, each having a copy of the recombinant-DNA molecule and thus permanently changing the genetic makeup of the host organism (Steinbrecher). Genetic engineering has been accomplished.

Genetic engineering is a technological process for undertaking activities which do not, and cannot, occur in nature. In this sense it is perfectly legitimate, therefore, to label genetic engineering as 'unnatural'. The European Community has incorporated this concept in its definition of genetically modified organisms, and it has been adopted by the UN Environment Programm (in the working of its Fourth Expert Panel set up under the Biodiversity Convention as part of the follow through to the United Nations Conference on the Environment and Development held in Rio de Janeiro in June 1992): 'organisms in which the genetic material has been altered in a way that does not occur naturally by mating and/or natural recombination'. Thus, hybrids and other modified organisms obtained by traditional breeding techniques are excluded from the definition of genetically modified organisms.

It is important to notc that the process as well as the products are novel. The promoters of the new technology believe that a concern for the process itself is specious, downplaying its novelty whenever confronted by discussions of regulatory oversight. However, one of the bases of the decision of the US Supreme Court in the *Chakrabarti* case (which, 5 to 4, upheld the patentability of genetically engineered microbes, and which has been unquestionably accepted by foreign governments as disposative of the issue of patentability of genetically engineered life forms) was that the element of 'novelty' required by the patent law was to be found in the process by which the microbe had been created. One reason why we cannot ignore the powerful novel aspects of the processes of genetic manipulation is because we are not omniscient as to what will, in fact, happen when we alter genomes. Genetic manipulation is not like a child's game of Legos or Tinkertoy, in which parts can be rearranged or linked up with only simple mechanical and relatively predictable consequences.

In calculating any risk from a transgenic organism, one should consider four elements: the host organism, the foreign genes, the interaction between the foreign genes and the rest of the genome, and the environment in which the organisms will be used.... [in regard to the last two elements] the literature contains many examples of genetic manipulations where inserted genes did not respond in their new environments the way they did in their old ones or where alterations with one part of the genome caused surprising activity in other parts of the genome.

There is an element of arrogance to scientific assertions which assure us that the process itself poses no risks, since scientists know so little about actual ecosystems; for example, I have been told by agronomists that over 80 poe cent of the organisms which can be identified in a soil sample from my garden are completely unknown to the scientific literature. Indeed, the Ecological Society of America itself has specifically warned about the problems presented in our lack of ecological knowledge and the resulting consequences from releasing (intentionally or accidentally) genetically altered organisms into ecological systems.

FOODS

In May 1992, the US Food and Drug Administration, responding to corporate pressures to remove the prospect of regulation of genetically altered foods, issued a set of rules which have largely left responsibility for protecting the public health and safety in the hands of the industry, permitting products to be marketed without scrutiny unless the industry indicated to the agency that it believed governmental oversight was justifiable.

However, the proposal as published by the FDA in the Federal Register perversely noted several important problem areas: creating allergens, additions of genes from sources which might violate religious and cultural norms (of vegetarians, Jews or Muslims, etc.), and implications for animal welfare and well-being (for example, the incorporation of human growth hormone-producing gene into a pig's genome produced a highly arthritic animal). Indeed, the commissioner of the FDA, David A. Kessler, and his colleagues noted the possibility that genetically engineered food might 'contain high levels of unexpected, acutely toxic substances'. In order to begin to address ethical issues presented by the genetic modification of foodstuffs, we need to have some ideas of the purposes for which these modifications are performed. (For example, if we wish to approach an analysis on utilitarian cost-benefit grounds.) Despite a great number of general statements, human nutrition and hunger do not appear to be the actual driving forces behind the development of genetically engineered foods.

Other than claiming increased shelf life (which can be indirectly related to nutrition in the sense of reducing spoilage or the consumption of foods that have begun to spoil), there is hardly any indication at all that genetic engineering is being directed to create nutritious substances out of nonnutritious ones; similarly, there are few real instances of genetic engineering reducing the cost of foodstuffs, or increasing the quantities of foods actually available to populations that are hungry because of the non-existence of consumables (as opposed to hungry because they lack money to buy sufficient foods, no matter how produced).

Indeed, early genetic manipulations seemed perversely designed to minimize the achievement of such goals: the creation of 'herbicide tolerant plants' which do not reduce the applications of dangerous chemicals to agricultural fields (which might occur if genetic manipulations of food crops were designed to make them resistant to insects, fungi, or disease) but instead permit higher levels of chemical application, or the introduction of recombinant bovine growth hormone to produce more milk at a time when developed countries suffer from milk gluts and have instituted programmes to kill cows and physically dump milk (milk as a commodity is price inelastic; quantity increases do not lead to price decreases).

The goals of genetically engineering foods seem to be more closely tied to increasing the economic gain and power of the corporations involved in food production (for example, the development of herbicide-tolerant farm crops has been led by corporations which manufacture herbicides), making production easier for corporations, reducing or altering packaging and transportation costs for the corporations, etc.

Environment

Industrial societies around the world are faced by massive pollution legacies from the previous emphases on chemical and nuclear industrial activities. The inability of the US Department of Energy to find a suitable site for a longterm nuclear waste repository illustrates our very poor track record for dealing with the scientific, technical, economic, and socio-political aspects of prior technological 'revolutions'. The environmental problems posed by genetically engineered organisms are likely to be substantially more intractable than those posed by these earlier instances of pollution because genetic wastes multiply, migrate, and mutate. A genetically engineered organism once free in the environment is impossible to recall. (Impacts which are irreversible must be considered with higher scrutiny than those which can be undone.)

Examples of environmental problems which need to be assessed include the risks of transgenic crops themselves becoming weeds; the risk of gene flow to wild relatives which might become weeds or pests; the growth of antibiotic resistance in species (particularly animals), since resistance genes are used as markers for the genetic engineering sites; the problem of exotic or non-native species taking over ecological niches (such as gypsy moth, starling, kudzu, rabbits in Australia, etc.) leading to extraordinary economic losses as well as ecological ones; the restriction (rather than increase) of biodiversity by selective advantageous breeding of transgenic organisms as opposed to natural ones, or the predatory results of expanding exotic species (such as the Dutch Elm micro-organism destroying the American Elm and severely restricting the biodiversity of certain areas in the Northeast United States); health issues (both of plant and animal species, as well as humans — worker health and safety as well as community security); and the occurrence of the completely unexpected, the inability to eliminate uncertainty (for example, the late 1993 floods in the Mississippi valley included the flooding of a field of genetically engineered corn and the dispersal of this plant material to unknown sites within the thousands of square miles of downstream flood plain).

Prospective ecological assessment is not being performed, despite the fact that its need has long since been recognized. And economic priorities, particularly those accruing directly to the promoters, as well as the nebulous spur of 'competition' with foreign countries, oftentimes overwhelms any interest in doing environmental impact analysis.

USEFUL TRAITS BY GENETIC TRANSFORMATION

NUTRITIONAL IMPROVEMENT

In addition to its basic calorific value, wheat with its high protein content is an important source of plant protein in the human diet. One of the prominent targets for the application of transgene technology for the nutritional improvement of wheat is targeted at enhancing the grain quality by, i) increasing the protein content, ii) increasing essential amino acids such as lysine, iii) increasing the high molecular weight (HMW) glutenins to improve breadmaking properties of wheat flour, iv) modifying starch composition and v) producing pharma-and neutraceuticals.

Amongst the cereals, the flour of bread wheat, Triticum aestivum, has a superior capability of forming leavened bread. This superiority stems from the structure and composition of its seed storage proteins, which upon hydration can interact to form gluten, an insoluble, but highly hydrated, visco-elastic aggregate that endows the wheat dough with its unique properties. Although the majority of wheat seed storage proteins participate in gluten formation, biochemical and genetic evidence has demonstrated that high molecular weight glutenin subunit (HMW-GS) plays a major role in determining the visco-elastic properties thereby determining bread making qualities. The HMW glutenins are necessary to create a strong dough, which is essential for making high quality, yeast-raised breads. Strong dough traps tiny bubbles of carbon dioxide gas formed naturally by yeast during mixing and subsequent raising thereby enabling the dough to rise forming leavened breads. Dough strength and the ability to contain gas bubbles is known as visco-elasticity and is an important characteristic of wheat with respect to its end product quality. Increasing understanding of the molecular basis of dough visco-elasticity would thus help in developing strategies for minimizing the effect of unfavourable environmental factors on wheat end-use quality.

Genetic transformation of wheat is a key component in a scheme proposing a complete set of approaches to apply biotechnology to improve wheat quality via direct manipulation of HMW-glutenin genes. To alter the amount and composition of these proteins by genetic engineering, a gene encoding a novel, hybrid-subunit of HMW-GS under the control of native HMW-GS regulatory sequences was inserted into wheat using the biolistic approach (Blechl and Anderson, 1996). The HMW-GS 1Ax1 gene which is known to be associated with superior bread making quality, was introduced into the cultivar (Bobwhite) lacking this gene. The introduced 1Ax1 gene under the control of HMW-GS promoter was expressed at high levels and stability was maintained for several generations.

The results demonstrate the feasibility of manipulating the composition of wheat kernels by genetic engineering and have successfully made changes in both the levels and the types of seed storage proteins. This work also demonstrated the usefulness of tissue-specific promoters for the expression of transgenic proteins in the endosperm tissue of wheat. The prospect of achieving the elusive goal of nutritional improvement of wheat brightened further with the improvement in functional properties of wheat dough due to transformation of wheat with high molecular weight subunit genes. Transformation with one or two subunit genes results in stepwise increase in dough elasticity.

The expression of a recombinant protein with x-and y-type HMW-GS subunit in transgenic wheat has resulted in altered gluten polymer assembly and composition. This study thus made it feasible to change the glutenin composition by expressing modified HMW-GS in transgenic plants.

The quality improvement studies were extended to durum wheat by He *et al.* 1999 with the introduction of one of the two subunits of HMW-GS subunit genes (1Ax1 or 1Dx5). Analysis of the expression of the additional subunits in the T2 generation using a mixograph, indicated an increase of dough strength and stability. This study demonstrated the feasibility of manipulating durum wheat for superior bread and pasta making qualities. Similarly, the overexpression of the HMW subunit 1Dx5, in transgenic wheat (T-aestivum) resulted in a four-fold increase in this proportion of component in the seed protein and also a corresponding increase in the proportions of the total HMW proteins and glutenins.

However, the overexpression of the 1Dx5 gene was found to be associated with a dramatic increase in dough strength by making it too strong, and, therefore, unsuitable for use in conventional breadmaking. Recently, Alvarez *et al.* (2000) introduced the HMW-GS genes 1Ax1 and 1Dx5 into a commercial cultivar of T-aestivum that already expresses five subunits. The overexpression of 1Dx5 gene increases the contribution of the HMW-GS to a level of 22 poe cent of the total protein content. In near future, molecular approaches including genetic transformation and marker-assisted selection will provide an opportunity for improving further the wheat processing qualities.

Modification of wheat starch is presently targeted in various laboratories to improve its potential utility. Wheat grain is predominantly composed of starch, which is a mixture of two polymers, the almost linear amylose molecules, and the heavily branched amylopectin molecules. The ratio of amylose to amylopectin in starch determines its physico-chemical characteristics and thereby its end-use. Wheat flour, low in amylose content is desirable for noodlemaking as it improves noodle texture. In wheat, scientists are thus working towards increasing the amylopectin content of starch thereby reducing amylose content leading to the formation of a value added, low amylose flour.

The starch branching enzymes catalyse formation of 1, 6-linkages in the glucan polymer and control the amount of amylopectin produced. In recent years, efforts are underway towards detailed characterization of the starch branching enzymes and starch synthases. This increasing information and characterization of the various components of starch biosynthesis will enable researchers to make rational design of novel starches and alteration in starch levels in other crop plants as well.

Wheat is widely used as an animal feed also for non-ruminants in several developed countries of the world. The phytase of Aspergillus niger is used as a supplement in animal feeds to improve the digestibility and also to improve the bioavailabilty of phosphate and minerals. The phyA gene from Aspergillus niger, encoding for the phytase enzyme has been successfully expressed in transgenic wheat lines by the microprojectile bombardment of immature embryos. The constitutively overexpressed phytase was found to accumulate at high levels in the endosperm. This work may thus open newer avenues for a wider applicability of wheat. Wheat is also an ideal system for the production of novel compounds due to its excellent storage properties and the existence of an efficient processing industry. The potential products include the high value, low-volume compounds such as biologically active proteins and peptides, as well as, high volume, low-cost raw materials for packaging and building. The production of recombinant antibodies in rice and wheat was recently reported with the expression of a medically important, single chain Fv recombinant antibody against the carcino-embryonic antigen.. The recombinant antibody was targeted to the plant cell apoplast and endoplasmic reticulum and was detected in the leaves and seeds of wheat. More significantly, the recombinant antibodies remained active after prolonged seed storage at room temperature thus opening avenues for the further exploitation of wheat for producing high-value, novel compounds.

ENGINEERING NUCLEAR MALE STERILITY

The production of hybrids is an essential component of crop breeding programmes but till date, hybrid wheat has remained elusive! The development of a suitable hybridisation system for wheat requires a high degree of male sterility in all parts of the female parent to avoid self-fertilization. De Block and coworkers have developed a nuclear male sterile system in wheat by introducing the barnase gene under the control of a tapetum specific promoter, the expression of which prevents normal pollen development at specific stages of anther development. This system employed the ribonuclease-inhibitor barstar gene to restore the fertility of male sterile plants.

To avoid complicated gene integration patterns, the target tissues were incubated on niacinamide containing medium before bombardment. The authors suggest that the enzyme poly (ADP-robose) polymerase (PARP), which plays a key role in the processes of cell division and recombination, is inhibited by niacinamide, thereby resulting in simple integration pattern of the transgene. Expressing the barnase gene at specific stages of anther development destroys the tapetum, thereby preventing normal pollen development and causes pollen sterility.

RESISTANCE TO BIOTIC STRESS

The integration of transgenic approaches with classical breeding techniques offers a potential chemical-free and environment-friendly solution for controlling pests and pathogens. Wheat is attacked by a number of viral, bacterial and fungal pathogens and also by insect and nematode pests.

Introgression of genes from the wild relatives exhibiting resistance to pests and pathogens has been successfully utilized over the years for generation of resistant varieties in wheat. With the development of plant transformation techniques, newer avenues for creating disease resistant and insect resistant crops have been created. Coupled with this is the availability of novel transgenes encoding highly potential anti-microbial peptides, defence-related proteins and enzymes for the production of anti-microbial compounds in crop plants, have greatly enhancing the possibility of engineering crop plants for resistance to pests and pathogens. Fungal pathogens of wheat cause severe crop damages by infecting the spikes, leaves and roots. Amongst the various strategies to introduce fungal resistance by transgenic approach, strengthening the host plant defence by genetic manipulation hold tremendous potential. Biochemical and structural responses against fungal attack include reinforcement of plant cell wall, accumulation of phytoalexins with microbial toxicity, ribosome-inactivating proteins (RIP) that inhibit protein synthesis, antimicrobial peptides and synthesis of other PR proteins,. Genetic engineering allows the expression of foreign genes from distant unrelated species as well as the modification of the usual pattern of expression of an already present gene.

Most of the works on genetic engineering of wheat for resistance against biotic stress have focussed on developing protection against fungal pathogens. Introduction genes encoding for chitinases from barley resulted in increased resistance against Erysiphe graminis. Bliffeld *et al.* (1999) reported the adverse effect of a ribosome inactivating protein on plant regeneration and development. Nonetheless, moderate protection against Erysiphe graminis was reported with the transformation of wheat with a gene encoding for a ribosome inactivating protein from barley.

The genes encoding for thaumatin like protein (TLP) and stilbene synthase in transgenic wheat have been shown to improve resistance of T1, T2 progeny plants against the fungal pathogens. An increase in endogenous resistance against Tilletia tritici was achieved with the introduction of virally encoded antifungal protein.

For the engineering of resistance against different pests and pathogens in wheat, genes encoding for viral coat proteins, antifungal proteins, and proteinase inhibitors have been successfully introduced. Most of the introduced genes confer increased resistance to the corresponding pests and pathogens in the transgenic plants. Introduction of canditate genes (lectin, proteinase inhibitor) for insect resistance into wheat have resulted in growth inhibition of insects on transgenic seeds, thereby decreasing the fecundity of insect population.

RESISTANCE TO ABIOTIC STRESS

Traditional approaches at transferring resistance to crop plants are limited by the complexity of stress tolerance traits, as most of these are quantitatively linked traits (QTLs). Nonetheless, the direct introduction of a small number of genes by genetic engineering offers convenient alternative and a rapid approach for the improvement of stress tolerance. Although, present engineering strategies rely on the transfer of one or several genes that encode either biochemical pathways or endpoints of signalling pathways, these gene products provide some protection, either directly, or indirectly, against environmental stresses.

Drought is a major abiotic factor that limits crop productivity, thereby causing enormous loss. The genes encoding the late embryogenesis proteins (LEA) which accumulate during seed desiccation, and in vegetative tissues when plants experience water deficiencies have recently emerged as attractive candidates for engineering of drought tolerance. Transgenic approach has been used for successfully introducing and overexpressing the barley HVA1 gene encoding for a late embryogenesis abundant (LEA) protein by Sivamani *et al.* (2000) into wheat by particle bombardment.

Most of the transgenic lines tested displayed improvement in important agronomic traits, including total dry mass and water use efficiency, shoot dry weight, root fresh and dry weights, when plants are grown under soil water deficit conditions. In general, this investigation showed that the transgenic lines expressing the HVA1 gene had improved growth characteristics including an enhanced biomass yield under water deficit conditions. The discovery of novel genes, determination of their expression pattern in response to abiotic stress and an improved understanding of their roles in stress adaptation (obtained by the use of functional genomics) will provide the basis of effective engineering strategies leading to greater stress tolerance (Cushman and Bohnert, 2000).

TRANSGENE SILENCING IN WHEAT

Success at developing improved wheat cultivars through genetic engineering depends on stable and predictable expression of the inserted gene. However, gene silencing is a common phenomenon in the production of transgenic plants and needs to be effectively controlled for the desired results. This is all the more important because gene silencing is an important phenomenon involved during natural plant defence. Gene silencing is a complicated phenomenon as it includes both transcriptional gene inactivation and post transcriptional gene inactivation. The complex and the large genome size of wheat is expected to be prone to silencing by introduced transgenes.

With the development of transformation methodologies for wheat, more information has started to accumulate regarding the inheritance and expression of the introduced transgenes. Information regarding the long term stability of transgenes is thus of immense significance for the use of genetic manipulation as a tool in wheat crop improvement and is necessary for the introduction and subsequent expression of desirable agronomic trait.

DNA methylation plays a significant role in establishing and maintaining an inactive state of the gene by rendering the chromatin structure inaccessible to the transcription machinery. Methylation of DNA is expected to result in reduced gene expression. Muller *et al.* (1996) studied the variability of transgene expression in clonal cell lines of wheat and found a negative correlation between the degree of methylation and marker gene expression. PEG-mediated approach was used to transform protoplasts isolated from suspension cultures of T-aestivum.

The integration and expression of the selectable marker gene nptII fused with different promoters, namely, alcohol dehydrogenase, shrunken from maize, and actin1 from rice, was confirmed by Southern analysis and enzyme activity test. Transgenic cell lines under selection were 'protoplasted' and clonal callus lines were cultivated from genetically identical single cells without selection pressure. A reduction/loss of marker gene expression was observed due to a reduction in the nptII transcript level and was seen to be associated with hypermethylation of the integrated DNA. The silencing effect was reversed by a 4-week culture phase on media supplemented with demethylation agent, 5-azacytidine, thus confirming the role of methylation in transgene silencing Demeke *et al.* (1999) studied the inheritance and stability of an Act1D-uidA: nptII expression cassette in T4 and T5 transgenic plants. Based on the histochemical localization of GUS activity, this study demonstrated the lack of any cytoplasmic effect on the inheritance of the transgene. The transgenes maintained a multiple integration pattern similar to that observed in the T1 generation. The transgenic plants which produced low gus and nptII activity in seeds had an intact expression cassette. Southern blot analysis of genomic DNA performed with the methylation sensitive enzyme, HpaII, showed the transgene in GUS negative plants to be highly methylated relative to the transgene in GUS positive plants. Some of the studies on gene silencing also indicate that multiple integration pattern and copy number is also associated with DNA methylation.

Cannell and coworkers studied the inheritance of gus and bar marker genes over three generations. The integration, inheritance and expression of these marker genes in the population studied were stable and predictable with a few exceptions. The inheritance of integration patterns was stable, and transmission/inheritance of the transgenes followed Mendelian ratios in a majority of lines. From this study, the authors claim that the transformation procedure, transgene integration, and marker gene expression had little effect on the transmission of transgenes to the subsequent progenies. However, over the three generations studied, a uniform, 'progressive' transgene silencing, specific to the gus gene, was observed. This gradual loss of the gus gene expression was not accompanied by a reduction in bar expression. Since in this case, the gus and the bar genes are located on the same plasmid, observations indicate a post-transcriptional gene silencing.

The silencing of HMW glutenins was also observed in transgenic wheat expressing extra HMW subunits. Characterization of six independent events involving the transformation of wheat HMW-GS genes, 1Ax1 and 1Dx5, in a cultivar expressing five subunits by particle bombardment resulted in partial or complete gene silencing. Silencing of all the HMW glutenin subunits was observed in two different events of transgenic wheat expressing the 1Ax1 subunit transgene and overexpressing the 1Dx5 subunit. Control of gene silencing thus remains a challenge for the immediate future. Meyer, (1995) advocated that the problem of gene silencing in wheat can be minimized by optimising methods for simple integration patterns, use of promoters and gene sequences isolated from cereals, use of matrix associated regions (MARs) or scaffold attachment regions (SARs), which insulate transgenes from surrounding chromatin, might help in reducing the gene silencing problem.

TRANSPOSON TAGGING IN WHEAT

Transposon mutagenesis has been widely exploited in various organisms to isolate genes that encode unidentified products. The maize activator (Ac) and Dissociation (Ds) elements are the best-studied transposable elements in heterologous host plants. Initial studies have reported the introduction and activation of the Ac/Ds elements into cultured wheat cells by using a wheat dwarf virus by particle bombardment. Takumi and coworkers has developed a transposon tagging system in wheat by introducing the Ac transposase gene under the CaMV 35S promoter into cultured wheat embryos by particle bombardment.

For the development of a transposon tagging system, embryos isolated from a stable Ac line were bombarded with a plasmid containing the maize dissociator (Ds) element located between the rice Act1 and the gus genes. The transient expression of the gus gene was observed after the excision of Ds elements. Southern and northern analysis of the T0 and T1 plants has exhibited the stable expression and inheritance of the Ac transposase gene. These results have demonstrated the precise processing of the maize Ac transposase gene and the synthesis of the transposase protein in transgenic Ac lines. The Ac transposase gene causes the transactivation and excision of the Ds in the Ac transgenic lines transformed with the maize Ds elements. Thus in near future we can expect traits of commercial importance to be tagged for a wider utility in important crop plants.

MARKER ASSISTED SELECTION IN WHEAT BREEDING

Conventionally, plant breeding depends upon morphological/phenotypic markers for the identification of agronomic traits. With the development of methodologies for the analysis of plant gene structure and function, molecular markers have been utilized for identification of traits. Molecular markers act as DNA signposts to locate the gene(s) for a trait of interest on a plant chromosome, and are widely used to study the organization of plant genomes and for the construction of genetic linkage maps. Molecular markers are independent from environmental variables and can be scored at any stage in the life cycle of a plant. Over the last several years, there has thus been marked increase in the application of molecular markers in the breeding programmes of various crop plants. Molecular markers not only facilitate the development of new varieties by reducing the time required for the detection of specific traits in progeny plants, but also fasten the identification of resistance genes and their corresponding molecular markers, thus accelerating efficient breeding of resistance traits into wheat cultivars by marker assisted selection (MAS). The availability of back-cross derived near isogenic lines (NILs) have also facilitated the analysis of various lines by using different marker systems. The introduction of alien genetic variation into wheat is a valuable and proven technique for wheat improvement. Wild relatives of wheat provide an enormous resource of new genes for wheat improvement, particularly disease and stress tolerance. These genes can be of use if recombined into lines adapted to the conditions of a particular region.

Initial studies on the application of molecular markers in wheat relied on the hybridisation based restriction fragment length polymorphism (RFLP) system-RFLP maps provided a more direct method for selecting desirable genes via their linkage to easily detectable markers thereby expediting the movement of desirable genes among varieties. The factors that had been instrumental for the use of RFLP in wheat was the limited number of polymorphisms observed among wheat lines and more significantly the availability of aneuploid stocks for the determination of chromosomal location of genes.

In wheat, RFLP's have been used to map seed storage protein loci, loci associated with flour colour, cultivar identification, vernalization (Vrn1) and frost resistance gene on chromosome 5A, intrachromosomal mapping of genes for dwarfing (Rht12) and vernalization (Vrn1), resistance to preharvest sprouting, quantitative trait loci (QTL's) controlling tissue culture response (tcr), nematode resistance, milling yield, resistance to chlorosis induction by Pyrenophora tritici-repentis.

RFLP markers are also useful in selection programmes for resistance against pests and pathogens, which is otherwise labour and time consuming, and to detect homozygous individuals and have been used for resistance to barley yellow dwarf virus, resistance to wheat spindle streak mosaic virus, resistance against powdery mildew, resistance against leaf rust, resistance against cereal cyst nematode. The use of RFLP analysis in wheat has, however, been of limited use in the intervarietal analysis due to low level of polymorphism and the high cost for screening in breeding situations.

With the development of polymerase chain reaction (PCR) methodologies, Random Amplified Polymorphic DNA (RAPD) emerged as a convenient and effective technique for tracing alien chromosome segments in translocation lines. RAPD markers provide a useful alternative to RFLP analysis for screening markers linked to a single trait within near isogenic lines and bulked segregants. He *et al.* (1992) reported the development of a DNA polymorphism detection method by combining RAPD with DGGE (denaturing gradient gel elecrophoresis) for pedegree analysis and fingerprinting of wheat cultivars. RAPD markers can be converted to more user-friendly Sequence Characterized Amplified Region (SCAR) markers, that display a less complex banding pattern.

SCAR markers linked to resistance genes against fungal pathogens have been characterized in combination with RAPD and RFLP. In recent years, RAPD and other PCR based markers like Sequence Characterized Amplified Regions (SCAR), Sequence Tagged Sites (STS) and Differential Display Reverse Transcriptase PCR (DDRT-PCR) are increasingly being used for identification of desirable traits in wheat and related genera. These markers have been used in particular for disease resistance against viral and fungal pathogens and also for insect and nematode pests and have the potential of pyramiding of resistance genes for effective breeding programmes.

PCR based markers have been extensively characterized for genes of resistance against common bunt, Tilletia tritici, ; powdery mildew, Erysiphe graminis, leaf rust, Puccinia recondita, resistance against Hessian fly, Mayetiola destructor and Russian wheat aphid, Diuraphis noxia.

Simple sequence repeats or microsatellites are more promising molecular markers for the identification and differentiation of genotypes within a species. The high level of polymorphism and easy handling has made microsatellites extremely useful for different applications in wheat breeding. Microsatellites have also been used to identify resistance genes like Pm6 from Triticum timopheevii and Yr15 from breadwheat. In near future, molecular markers can provide simultaneous and sequential selection of agronomically important genes in wheat breeding programmes allowing screening for several agronomically important traits at early stages and effectively replace time consuming bioassays in early generation screens.

GENOMICS AND FUTURE OUTLOOK

Genomics is the discovery and study of many genes simultaneously on a genome-wide scale. In the pregenomic era, gene discovery was the limiting factor for the improvement of crops by genetic engineering. For the upcoming postgenomic era, an exponential number of genes will be discovered through genome sequencing efforts on Arabidopsis and rice. Wheat genomes are large in comparision to the current plant models and are ideal targets for partial genome sequencing.

However, due to a high level of genetic colinearity at a broader genetic level and also at the gene level, comparative genome analysis studies can serve as efficient tools for transferring information and resources from well studied genomes to the related genera. DNA mapping and sequence information generated in the rice genome project would be highly useful for the characterization of larger genomes like wheat.

Molecular isolation of agronomically important genes and an understanding of their action will contribute immensely to breeding of cultivars. With the availability of detailed information regarding the location and function of gene(s) encoding for useful traits, scientists in future will be well equipped for efficiently creating varieties with exact combinations of desirable traits. However, genetic transformation will remain a significantly important tool for understanding gene functions and for testing the utility of new sequences. In near future, crop varieties could be tailor-made to meet both local consumer preferences and the demands of particular niche or environment.

The new tools of biotechnology not only have the potential for increasing the effectiveness and efficiency of wheat breeding programmes, and but will also provide insights into the genetic control of key traits to be used for genetic manipulation. Application of biotechnology will thus contribute greatly to improving yield stability by generating plants with improved resistance to biotic and abiotic stresses rather than raising the overall yield. The coming years will undoubtedly witness an increasing application of biotechnology for the genetic improvement of wheat.

2

Genetic Engineering in Agricultural Foods

GENETIC RESOURCES FOR FOOD AND AGRICULTURE

Farm animal genetic resources face a double challenge. On the one hand, the demand for animal products is increasing in developing countries: FAO has estimated that demand for meat will double by 2030; over the same thirty-year period demand for milk will more than double. On the other hand, animal genetic resources are disappearing rapidly worldwide. Over the past 15 years, 300 out of 6 000 breeds identified by FAO have become extinct. Many breeds of local importance for food security are not being improved or utilized in a sustainable manner and are in danger of being lost or diluted by crossbreeding.

Conservation and development of local breeds is important because many of them utilize lower quality feed, are more resilient to climatic stress and to local parasites and diseases, and represent a unique source of genes for improving health and performance traits of industrial breeds. It is also important to develop and utilize local breeds that are already adapted to their environments, most of which are harsh, with very limited natural and managerial input. Animals genetically adapted to these conditions are expected to be more productive at lower costs, support food, agriculture and cultural diversity, and be effective in achieving local food security objectives.

Local communities depend on these adapted genetic resources in many countries. Their disappearance or drastic modification, for example, by crossbreeding, absorption or replacement by exotic breeds, will have serious negative impacts on these human populations. Presently, most breeds at risk of extinction are not supported by any established conservation programmes or active conservation through sustainable utilization (breeding plans) and therefore breed extinction rates are increasing globally.

THE GLOBAL STRATEGY FOR THE MANAGEMENT OF FARM ANIMAL GENETIC RESOURCES

The key component of the *global strategy* is the country-based planning and implementation infrastructure, which includes five structural elements:

- The *global focal point* at FAO headquarters leads the planning, development and implementation of the overall strategy; develops and maintains the information and communication systems; oversees preparation of guidelines; coordinates regional activity; prepares reports and documents for meetings; facilitates policy discussions; identifies training, education and technology transfer needs; develops programme and project proposals; and mobilizes donor resources.
- *Regional focal points* facilitate regional communications; provide technical assistance and leadership; coordinate regional training, research and planning activities; help develop regional policies; assist in identifying project priorities and proposals, and interact with government agencies, donors, research institutions and non-governmental organizations.
- *National focal points* lead, facilitate and coordinate country activities; identify capacity-building needs; develop project proposals; assist with the development and implementation of country policies; and interface with national stakeholders, regional focal points and the global focal point.
- *Donor and stakeholder involvement* is necessary to provide financial and institutional support to the *global strategy.* In this context, the global focal point seeks to ensure stakeholder involvement in all major aspects of the *global strategy,* facilitating opportunities for governmental and non-governmental contributions.
- *DAD-IS, the Domestic Animal Diversity Information System*, is a widely available and easily accessible global database and information source. This global facility makes it possible to share data and information among countries, allowing a rapid and cost-effective distribution of guidelines, reports and meeting documents, and provides a platform to exchange views and address specific information requests, linking breeders, scientists and policy-makers. A key feature of DAD-IS is the breeds database, which provides the data for the early warning system for animal genetic resources through the World Watch List for Domestic Animal Diversity, whose third edition was released in 2001.

FIRST REPORT ON THE STATE OF THE WORLD'S ANIMAL GENETIC RESOURCES

As part of the *global strategy for the management of farm animal genetic resources,* FAO invited 188 countries to participate in the first *Report on the State of the World's Animal Genetic Resources,* which is to be completed by 2006. To date, 151 countries have accepted to submit country reports. Guidelines for preparation of country reports have been published in Animal Genetic Resources Information Bulletin (FAO) no. 30. These guidelines are used to assist countries in preparing reports as strategic policy documentation covering the state of animal genetic resources, the state of the art and national capacity to manage these resources, and country needs and priorities. Country reports will serve as the base documentation for the State of the World Reporting Process; thus the involvement of all stakeholders in the development of these reports is strongly encouraged.

The objective of the country and global assessments is to provide a comprehensive analysis of the status and trends of the world's farm animal biodiversity and of their underlying causes, as well as of local knowledge regarding its management.

The task is to go beyond description of the resources by analysing the state of these resources and the capacities to manage them, drawing lessons from past experiences and identifying problems and priorities. Country reports are policy documents covering three strategic questions: *Where are we? Where do we need to be? How do we get to where we need to be?* Country reports are intended to be used in planning and implementing priority country actions. In addition, the country report will serve as documentation for the development of the regional and global reports on strategic priorities for action and, subsequently, the first *Global Report on the State of Farm Animal Genetic Resources.*

Country reports provide an assessment in three major areas:

- the *state of diversity* to evaluate the state of conservation, erosion and utilization of farm animal agricultural biodiversity, and an analysis of the underlying processes;
- the *state of national capacity* to manage animal genetic resources, including existing policies, management plans, institutional infrastructures, human resources and equipment;
- the *state of the art* of the available methodologies and technologies to assist farmers, breeders and scientists to better understand, use, develop and conserve animal genetic resources and thereby contribute to global food security and rural development.

International organizations are also being invited to contribute to the state of the world's animal genetic resources preparatory process by providing reports. The long-term aim of the process is for countries and regions to build on the analyses contained in the country reports in order to plan and implement appropriate management of their farm animal genetic resources.

The first *Report on the State of the World's Animal Genetic Resources* will contain the *Report on Strategic Priorities for Action* and will be based on a synthesis of country reports, thematic studies and reports from international organizations.

FIELDWORK AT COUNTRY AND REGIONAL LEVELS

Countries were requested to nominate a national focal point and designate a national coordinator to facilitate the development of the country network on the management of animal genetic resources and to serve as official contact with the global focal point. Keeping in mind that the process involves both scientific and policy matters, the establishment of a National Consultative Committee is recommended to identify the primary areas and issues that need to be addressed in the preparation of the country report and to oversee its preparation. It is essential that the National Consultative Committee have wide and diverse representation and develop a broad network to ensure opportunities for all stakeholders to contribute to the country report.

The response of countries to the invitation of FAO's Director-General to participate in the first *Report on the State of the World's Animal Genetic Resources* and submit a country report has been very positive. During part of 2001 and 2002, FAO trained almost 400 professionals from 178 countries in the preparation of national reports. At the moment, FAO has a team of 15 consultants working in 14 country groupings in all regions of the world. Most countries have undertaken the organization of national stakeholder workshops to elaborate their animal genetic resources policies leading to the country reports.

FAO has organized 14 subregional workshops to discuss draft country reports and regional priorities for action. This has promoted regional cooperation and allows countries that may be experiencing delays to catch up with those in a more advanced state of country report preparation and learn from their experiences. These sessions were coordinated by the regional facilitators acting as FAO consultants.

FAO has provided technical and financial support to 115 countries with contributions from the Governments of the Netherlands and Finland and from the Nordic Gene Bank. FAO and the World Association for Animal Production (WAAP) signed an agreement to provide technical and operational support for the state of the world animal genetic resources reporting process, including training and country follow-up. FAO considers this cooperation a prime example of effective collaboration with an international non-governmental organization.

GENETICALLY MODIFIED FOODS

Increasingly, genetically modified foods are making their way to the marketplace. Since they are not labeled, consumers generally do not know that they are eating such products. Not everyone is comfortable with the present system. A few years ago, one rarely heard about genetically modified organisms (GMOs). In fact, few people knew exactly what they were. That is no longer the case. GMOs, or plants, animals, or microorganisms that have been genetically altered or engineered for a specific purpose, are everywhere.

It has been estimated that almost half of all American farmers grow some variety of genetically modified crops, also known as genetically engineered crops. The Food and Drug Administration (FDA) says that GMOs are filling the shelves of stores. "Tomatoes, potatoes, squash, corn, and soybeans have been genetically altered through the emerging science of biotechnology.

So have ingredients in everything from ketchup and cola to hamburger buns and cake mixes". In Genetically Engineered Food, Martin Teitel, executive director of the Council for Responsible Genetics, and Kimberly A. Wilson, director of the council's programme on Commercial Biotechnology and the Environment, describe the sweeping changes brought about by GMOs. "The genetic engineering of our food is the most radical transformation in our diet since the invention of agriculture 10,000 years ago. These food crops are already growing on millions of acres all around the world".

Similar sentiments were expressed in Genetically Engineered Food by Ronnie Cummins, national director of the Organic Consumers Association, and Ben Lilliston, a health and environment writer and communications coordinator for the Institute for Agriculture and Trade Policy in Minneapolis: "Genetic engineering...is a revolutionary new technology still in the early experimental stages of development.

It enables molecular biologists to permanently alter the essential characteristics or genetic codes of living organisms. This technology has the awesome power to break down fundamental genetic barriers-not only between species but between humans, animals, and plants. For the first time in history, the scientists and corporations using this technology have become, in effect, the architects, builders, and 'owners' of life". Introducing the genes of one type of food into another, thereby changing the genetic structure of an organism, gives the second plant or animal beneficial characteristics, such as resistance to disease, improved nutritional value, and better growth. These modifications enable foods to grow faster, stronger, or bigger, all while using far less pesticide. There are now tomatoes that last longer and corn that is resistant to pests.

"The acreage devoted to herbicide-resistant crops has been expanding because planting them reduces the need to plow more ground, decreases the amount of herbicidal chemicals needed, produces higher yields, and can deliver a higher grade of grain and other products". Transgenic or genetically modified crops "can be developed for pest resistance, improved yield, tolerance to biotic and abiotic stresses, use of marginalised land, improved nutritional content, and the production of vaccines and pharmaceuticals". A 2000 story in Popular Science said, "Genes from flounders can help ordinary plants like tomatoes and strawberries fight the cold. Researchers are also inserting bacterial genes into corn and soybean plants to better protect them from insects or render them immune to certain herbicides". Cross-breeding has been around for a long time. "Virtually all common fruits and vegetables look and taste the way they do because of hybridisation". However, biotechnology has added a whole new dimension to selective breeding. "This process allows for the transfer of only one or a few desirable genes, thereby permitting scientists to develop crops with specific beneficial traits and those without undesirable traits.

Current technology permits scientists to alter one plant characteristic at a time, thereby not spending years trying to develop the best tasting and hardiest plants". For example, Craig Nessler, professor and head of the Department of Plant Pathology, Physiology, and Weed Science at Virginia Tech University, inserted genes from rats into lettuce seeds. This produced leafy greens with much higher amounts of vitamin C. "Rats, it turns out, carry a gene that allows their bodies to manufacture vitamin C, which is why the little four-legged fiends never develop scurvy on long sea voyages". Years ago, such science would have been unthinkable. The differences between traditional breeding and genetic engineering. "To give a simple example, a traditional breeder interested in producing a yellow tomato must find the yellow trait in a plant that will breed with the tomato by natural mechanisms. The only plants that can breed with tomatoes are closely related ones. Unrelated plants like oak trees or cantaloupes could not breed with tomatoes, and thus could not contribute new genes. A genetic engineer, on the other hand, can consider any organism-even a butterfly or a daffodil-as a source of the yellow trait. If the gene that determines yellow colour has been identified and isolated, it can be directly transferred into tomato plants". The International Food Information Council maintains that because GMOs reduce the amounts of pesticide used by farmers, they are healthier for the environment. "Insect protected crops allow for less potential exposure of farmers and groundwater to chemical residues, while providing farmers with season-long control. Also by reducing the need for pest control, time, effort and resources spent on the land are less, thereby preserving the topsoil". It is almost impossible for the average consumer to know whether there are GMOs in his or her food. Playing on consumers' fear of the unknown, a few innovative marketers include "No GMOs" notations on their labels. However, the vast majority of products do not indicate whether they are comprised of any genetically altered ingredients. "Legal authority for food labeling rests with the Food and Drug Administration. Foods derived from biotechnology currently must be labeled only if they differ significantly from their conventional counterparts-for example, if their nutritional content or potential to cause allergic reactions is altered".

SUPPORTERS AND OPPONENTS OF GMOS

Should consumers feel comfortable eating genetically altered food? The controversy surrounding GMOs, or Franken foods, as they are called by those who oppose them, is filled with heated discussion. In general, the federal government, big business, and many researchers and numerous well-respected publications maintain that genetically altered food is perfectly safe. They contend that GMOs offer the potential to feed healthier food to far larger numbers of people. That is of particular significance, they say, when considering the vast amounts of food needed to feed the populations of developing countries. About 15 per cent of the world's population-about 800 million people-consume less than 2,000 calories per day. They are always hungry and live in a state of chronic malnourishment. Large numbers of these people are women and children. "More than 180 million children under five years of age are underweight, that is, they are more than two standard deviations below the standard weight for their age. This represents one-third of the under-fives in the developing countries."

These groups are growing at a staggering rate. "By the year 2020, there will be about an extra 1.5 billion mouths to feed. If the proportion of the population of the developing countries deprived of an adequate diet remains the same, the number undernourished 20 years from now could be well over one billion". Not everyone agrees. Sizeable numbers of equally esteemed researchers and consumer and environmental groups worry about a variety of factors. "Concerns include ethical issues related to potential long-term health effects of eating bioengineered foods, labeling, and potential environmental risks".

In Europe, particularly Great Britain, large numbers of people have supported a ban of GMOs. Grocery stores are refusing to carry foods containing GMOs. Other countries are following Britain's lead. Moreover, concern about risk of lawsuits and decline in land value has made rural land managers hesitant to authorise tenants to grow GM crops." Within the European Community, there is enormous apprehension.

"Consumers worry about the impact of any health risks that might eventually be discovered, especially if whole national populations have ingested GM food. Health effects, moreover, may stem from risk related to questions that scientists have not yet asked rather than those questions for which reassuring answers have been found".

In High-Tech Harvest, Elizabeth L. Marshall, a science writer, notes that the most serious GMO controversy has centered on food labeling. Usually, opponents of GMOs favour labeling. They want to know which foods have been changed and believe that shoppers, when presented with such information on the label, would shy away from them. "But some supporters want labels too. They believe that labels would help pave the way to consumer acceptance".

Genetic engineering food research began in the 1980s. In 1990, the enzyme chymosin was the first genetically engineered food to receive the approval of the FDA. More than half of the hard cheese currently sold in the United States is made with chymosin produced by genetically engineered fungi. The rest is made with chymosin obtained from rennet, an enzyme from the stomachs of slaughtered calves.

"Although cheese-makers can use chymosin from either source, the chymosin made from genetically engineered bacteria is easier to obtain and more pure". Two years later, in 1992, the FDA approved the first genetically engineered whole food-the Calgene Flavr Savr tomato, which became available for sale in 1994. Because of a change to a single gene, the Calgene Flavr Savr tomato ripened without getting soft. However, the Flavr Savr tomato encountered a number of problems, such as an inability to tolerate shipping. "Contrary to Calgene's expectations, the tomatoes were often so soft and bruised that they could not be sold as fresh produce, and most of the varieties did not have acceptable yields or disease resistance in tomato-growing regions". The Flavr Savr marketing campaign was unsuccessful. The tomatoes received a good deal of negative press, and consumers refused to buy them. "Calgene even tried marketing the tomatoes as a gourmet product under the friendly sounding 'MacGregor's' brand name, but despite their presence in thousands of U.S. grocery stores, consumers did not want to pay more for genetically engineered tomatoes."

In 1996, the Flavr Savr tomato was removed from the market. Teitel and Wilson noted that "the Flavr Savr grew well in the laboratory but encountered serious obstacles in the field". Nevertheless, genetic engineering moved forward. By 1996, genetically engineered seeds for corn and soybeans were available, followed by seeds for potatoes, squash, tomatoes, and rapeseed. "Each type of seed offers a characteristic that natural seeds did not. Some protect the plants against insects or disease, while others delay ripening or protect the plants from chemicals used to kill weeds.

REGULATION OF GMOS

It is important to realise that genetic engineering is a tool that may be used in a variety of ways to make many different food products. That is why, Marshall contends, the FDA decided in 1992 that genetically engineered foods would be reviewed "based on their individual safety and nutrition, rather than on the methods used to produce them." The FDA deemed them "as safe as foods developed through other agricultural technologies".

As a result, the FDA regulates GMOs as it does any other food. If a new product contains food items that are already considered safe, the manufacturer does not need to obtain special permission to sell the item. "In other words, a transgenic [another word for genetically engineered] catfish containing a trout gene would be considered safe to eat because it is already well-known that trout is harmless to humans. The only time the law requires testing of a new food is when it contains an additive that is not known and, therefore, not generally regarded as safe". In 1992, the FDA determined that there would be exceptions to this rule. If a new food contained ingredients that were likely to trigger allergic reactions in allergic individuals or if the manufacturing process dramatically altered the nutritional content, a label would be required. Unfortunately, this is not as clear-cut as it may initially appear to be. Foods are constantly combined together to form new foods. Should everything be labeled? The government and sizeable numbers of people within the food industry say that such labeling would be confusing and expensive.

In Pandora's Picnic Basket, Alan McHughen, a senior research scientist at the University of Saskatchewan, Canada, wrote that "in North America, regulators and companies agree that mandatory labels should be reserved for those products carrying a documented health risk or substantial change in nutritional composition. If the GM products are 'substantially equivalent' to conventional counterparts, the companies argue, the GM label would be misleading". Nonetheless, many consumer groups consider labeling to be an absolute necessity.

EarthSave International explains that genetic engineering "introduces new proteins into the human and animal food chains. This means that human beings are now consuming products that have never before been considered foodstuffs. There is concern that these new proteins could potentially cause toxic or allergic reactions or other health effects. Unfortunately, there is no easy way to predict the allergenic potential... since allergic reactions typically occur only after the individual consuming the food is sensitised by initial exposure to the allergen".

It is true that the government has mandated the labeling of other products such as foods that are processed and those containing sulfites. Labels must note the source of hydrolysed proteins, and they are required to state if a food-other than a spice-has been irradiated. Cigarette labels have a warning from the surgeon general. "These legal requirements are in place because large numbers of citizens want such information, and a specific fraction needs it.

An identifiable fraction of consumers actually needs information about genetic modifications-for example, regarding allergenicity and almost all consumers want it". Lawrence Kushi, an epidemiologist at Columbia University, notes that it is simply untrue to say that genetically modified food is no different from the conventional equivalents. He maintains that labeling such foods would give consumers the information they want while allowing scientists to examine whether they have any negative impacts on human health or the environment.

FAVOURABLE CONDITION FOR GMOS

Some people contend that the negative publicity that consumer groups generate about GMOs is most harmful to the world's most vulnerable people-the poor. They say that unlike residents of wealthier countries, large segments of the population of developing countries worry about dying from malnutrition and starvation. By using GMOs, farmers in developing countries may harvest more crops with less pesticide.

Playing on this theme, frequently appearing television and magazine ads have featured rice that has been genetically altered with a daffodil gene. As a result of that modification, the rice contains beta-carotene, which the body converts into vitamin A. The ads claim that by eating this form of "Golden Rice," countless numbers of people in poorer countries will be spared blindness. About 400 million people may be in danger of vitamin A deficiency. Of these, 100 to 200 million are children. Mammals do not manufacture vitamin A, so it must be obtained through the diet.

People who eat lots of fruits and vegetables and animal products obtain adequate amounts. Millions of people are not as fortunate. "1.0 to 2.5 million deaths per year of preschool children-up to 30 per cent of total deaths in that age group-could potentially be averted by bringing vitamin A deficiency under control worldwide. Combined with expanded vitamin A supplementation programmes-which will continue to be important-Golden Rice is expected to make a major contribution to improving the health of millions of the world's poorest children".

For the unspoken challenge here is that if we don't get over our queasiness about eating genetically modified food, kids in the third world will go blind." But Pollan asserts that there is far more to the Golden Rice issue than is initially evident. For example, in order to meet the minimum daily requirement for vitamin A, it would be necessary for an average 11-year-old to eat 15 pounds of cooked Golden Rice. Moreover, in order to convert beta-carotene into vitamin A, fat and protein must be present, and these are lacking in the diet of malnourished children. Pollan also wonders whether Asians will accept rice that is golden in colour.

It is already well known that brown rice is healthier than white rice, yet Asians consistently prefer white rice. Apparently, at some point, Confucius "extolled the pure whiteness of rice as the ideal backdrop for green vegetables". Pollan further questions the ethics of an industry that is using the suffering of children to sell its food. In "Vitamin A Deficiency Disorders: Origins of the Problem and Approaches to Its Control," Alfred Sommer writes that there are some "hurdles" to be overcome before Golden Rice may have an impact on society. "The strains must be able to grow under the varied conditions in countries with vitamin A deficient populations. The yield and the cost must be attractive to the farmer (or benefit from public sector subsidisation). The organoleptic [sensory] qualities of the rice must be acceptable to the target population (women and children). The beta-carotene needs to be bioavailable, the degree dependent on its concentration in the rice, the matrices to which it is bound, the effect of traditional cooking methods and the amount consumed". Sommer warns that Golden Rice will never be a complete solution to the problem of vitamin A deficiency. "Many deficient populations do not consume rice, and even within traditional rice-consuming countries, some high-risk groups will not be able to afford it". Regardless, it is impossible to deny that the population of the world is growing at a sometimes mind-boggling rate. Presently, the world has about 5 billion people. By the year 2050, that figure will double to 10 billion. Large numbers of these people will live in the developing world, where hunger is rampant.

The International Food Information Council states that "by increasing a crop's ability to withstand environmental factors, growers will be able to farm in parts of the world currently unsuitable for crop production. Along with additional food, this could provide the economies of developing nations with much-needed jobs and greater productivity". Feeding so many people takes a staggering toll on the environment. "Erosion can claim precious topsoil, farm chemicals sometimes reach streams, rivers and ground water supplies, and livestock can deplete grazing lands.

Wetlands and other sensitive habitats sometimes get plowed under for use as farmlands. And, in the world's tropical forests where an estimated 90 per cent of the world's species exist, poor farmers clear trees in order to provide food and a living for their families." The International Food Information Council believes that GMOs may play a pivotal role in mitigating this burden. Crops that have been genetically modified are often more resistant to disease and require a reduced level of insect control.

Therefore, fewer acres need to be planted to reap the same amount of food. One of the key players in the effort to increase the production of rice in Asia has been the Rockefeller Foundation. The foundation believes that GMOs will enable Asian farmers to obtain more rice from the same amount of land. "By the year 2005, the Foundation's hope is to increase rice production in Asia by 20 per cent through the use of biotechnology without degrading the environment or reducing farm incomes."

The Rockefeller Foundation has also been conducting similar work in Africa. Former President Jimmy Carter, who has become well known for his humanitarian works, has observed that "responsible biotechnology is not the enemy; starvation is. Without adequate food supplies at affordable prices, we cannot expect world health or peace". Earth Save International disagrees with the notion that genetic engineering may play an important role in feeding the world's growing population.

"World hunger is not a problem of technology or insufficient production, but primarily one of unequal distribution and economic inequality. As farmers lose their land and move to the cities, they also lose their food-independence and begin to rely on money, often in dramatically short supply for many in the third world, in order to buy food that they formerly grew themselves". In The Ecological Risks of Engineered Crops, Jane Rissler and Margaret Mellon agree. "Even if research is done on the right crops and result in increased agricultural productivity, increased production is only one factor in the complicated equation of world hunger. Poverty, trade policies, subsidies, soil erosion, and water shortages are also important causes of hunger.

Increases in productivity obtained through genetic engineering will have mixed effects if not developed with due regard to the other important aspects of the hunger problem". In The Future of Food, Brian J. Ford, an English scientist, writer, and television host, says that a number of everyday farm animals and crops-such as pigs, cattle, wheat, oats, barley, and rye-are the result of human intervention. "Traditional farmers have been producing new animals and plants by cross-breeding for ten thousand years." While genetic engineering "offers a more radical way of manipulating characteristics," it is fast becoming a part of daily life. "Genetic modification is inevitable. Like electric power, road transport, or computers, it is a facet of the future and the public will gain little by campaigning to ban this potentially rewarding technology. Properly applied, it could offer us so much".

All the same, Ford does not believe that the public should simply accept whatever the bioengineering companies do. Rather, he calls for a careful monitoring of the industry and the introduction of safeguards. "This is a huge new industry, and it will have pronounced effects on us all. We will need to control it". Ford suggests the following controls:

- Approval for new experiments should always be sought from a regulatory authority well versed in the subject.
- None of the members of such a committee should be in a position to benefit commercially from approval.
- No genes conferring problematical properties (*e.g.*, antibiotic resistance) shall be used, whether as markers or otherwise, outside enclosed laboratories.
- Agents capable of transfer to wild plants, like pollen-carrying genes conferring resistance, shall not be liberated into the environment.
- The public shall be consulted about the siting and the benefits of experiments.
- All GMO products should be properly labeled and the source declared.
- Records of possible unwanted side effects should be meticulously maintained. All such events should be investigated by an organisation unconnected with the source of the product.

ARGUMENTS AGAINST GMOS

The controversy shows no sign of diminishing. GMO opponents present some compelling arguments. The Organic Consumers Association contends that genetic engineering deconstructs basic, fundamental genetic barriers between species. "By combining the genes of dissimilar and unrelated species, permanently altering their genetic codes, novel organisms are created that will pass the genetic changes onto their offspring through heredity.

Animal genes and even human genes are being inserted into plants or animals creating unimagined transgenic life". Completely new organisms will be created. The association believes that these have the potential to compromise human health and well-being, animal welfare, and the environment. The following are a few examples of ways in which this might occur:

- Genetically engineered organisms can reproduce, migrate, and mutate. If they escape or are released from the laboratory, they could wreak environmental havoc.
- Gene splicing can have unpredictable and dangerous results.
- Genetic engineering of crops and food-producing animals can produce toxic and allergic reactions in humans.
- Safety testing and regulation of genetically engineered organisms are inadequate or lacking.
- Patenting genes they discover and living organisms they create will enable a small corporate elite to own and control the genetic heritage of the planet.

The poll found that only about one-third of the Americans surveyed view GMOs as safe to eat. A little more than half found them unsafe for human consumption. Women, who tend to do the family shopping, were even more against GMOs than men. While 40 per cent of men think that GMOs are unsafe, 62 per cent of women feel that way. A striking percentage of those surveyed-93 per cent-want the federal government to require food labels to indicate the presence of GMOs.

Fifty-seven per cent said that they were less likely to purchase food that contains GMOs. There are even age differences. "People under 45 are about 10 points more likely than their elders to think genetically modified foods are safe to eat. But a majority of young adults still calls genetically modified foods unsafe". In Brave New Seeds, Robert Ali Brac de la Perrière, an international consultant, and Franck Seuret, a journalist, contend that throughout the world, GMOs will weaken the role of small-scale farmers who lack the wherewithal to pay higher prices for genetically modified seeds and lack the expertise that more technically oriented farming demands.

Increasingly, large farms will absorb the smaller ones. Further, farmers are losing control over their products. "With transgenic plants, the freedom of agricultural practice and liberty of choice is reduced as the farmer works with a patented product which is subject to very specific conditions of use. For example, contracts for the use of a transgenic soya variety that is resistant to the herbicide RoundUp forbids, amongst other things, the cultivation of other varieties, the use of herbicides other than those allowed by Monsanto and the exchange of seeds with neighbours".

Crop diversification reduces the amount of herbicide and synthetic nitrogen fertilizer that may be needed, improves the quality of soil and water, controls insect-pest and pathogen populations, increases crop yield, and reduces yield variance. Insect-resistant crops may have unintended victims and disrupt other forms of pest control. Also, toxins from these products may remain in the soil for extended periods of time and reduce the level of soil fertility. As if this were not sufficiently problematic, there is the issue of "terminator" seeds, genetically altered plants that are programmed to kill their own seeds. Unlike most plants, these do not regenerate. "After the seeds are planted and the crop matures, the [seed-killing] toxin is produced, killing the new seeds the plants carry".

In the previously noted Genetically Engineered Food, Martin Teitel and Kimberly A. Wilson say that terminator seeds end the long-held practice of saving seeds for future plantings. "Farmers who use terminator seeds will be forced to buy new seeds for every planting, ending an age-old tradition of seed saving and creating a perpetual cycle of dependence on big seed companies". In Brave New Seeds, Brac de la Perrière and Seuret observe that terminator seeds "achieved their supreme goal-enslaving their clients". There is an even more alarming possibility. "Some doomsday scenarios suggest pollen from Terminator plants could drift with the wind like a toxic cloud, cross with ordinary crops or wild plants, and spread from species to species until flora all around the world had been suddenly and irreversibly sterilised". That is probably a huge exaggeration. Gene drift does not occur that easily. A large number of crops pollinate themselves. As for those that do not, border fields may be established to contain genes. In "Genetically Modified Foods: Are They a Risk to Human/Animal Health?," Generally, when testing products, the industry has compared a GM food to a non-GM food.

"When they are not significantly different the two are regarded as 'substantially equivalent,' and therefore the GM food crop is regarded as safe as its conventional counterpart." Pusztai said that "substantial equivalence is an unscientific concept that has never been properly defined and there are no legally binding rules on how to establish it." Additionally, there are currently no reliable methods to test new genetically engineered crops for food allergies. "It is at present impossible to definitely establish whether a new GM crop is allergenic or not before its release into the human/animal food/feed chain".

TRANSFER OF DNA FROM GMO

There is potential for the presence of live GMOs in foodstuffs such as yoghurt and fermented sausage produced with starter cultures containing genetically modified organisms or perishable foods treated with protective cultures of GM organisms. The environment within living food may permit gene transfer by conjugation or transduction and this has been seen at high frequency in complex food matrices.

Free recombinant DNA may also be present in food products resulting from the killing of GMMs during food processing. The human gut microflora is extremely complex, being made up of more than 500 different species of bacteria and climax microbial populations reaching up to 1011 CFU/g luminal contents in the colon. The high density and species diversity of bacteria present in the colon provides ideal conditions for the transfer of DNA between bacteria.

However, the diversity of the human gut microflora also presents significant problems in monitoring the DNA transfer events between bacteria and, in particular, monitoring the fate of recombinant DNA, which may be ingested in food. It has been determined that only a fraction of the bacteria present in the human gut microflora may be cultured under labouratory conditions and characterized.

Many species found in the human gut microflora defy culture by standard microbiological techniques, and selective agars and growth conditions exist for only a small proportion of the predominant species present. Much progress as been made in recent years in the development of culture-independent methods for identifying bacteria present in natural environments.

Using the phylogenetic information present in 16S rRNA and fluorescent in situ hybridization techniques, it is now possible to identify bacteria in situ in natural environmental samples including human faeces. However, such techniques tell us little about the genetic capabilities of bacteria and by themselves do not allow us to monitor the distribution or indeed dissemination of specific genetic determinants amongst natural assemblages of bacteria.

Thus, studies on the transfer of DNA between members of the human gut microflora and from GMOs to members of the human gut microflora have employed specific marker systems borne on the mobile genetic elements or on recombinant DNA. Traditional marker systems used in pure culture microbiology such as the lac operon or X-gal are of little use when employed to monitor DNA transfer events in complex assemblages of gut microorganisms, because of the widespread occurrence of such genes amongst the mammalian gut microflora.

Antibiotic-resistance determinants, when shown to be absent from background bacteria, have been widely employed in labouratory-based studies. However, such marker systems cannot be used in studies with human volunteers because of the emergence of antibiotic-resistance determinants in pathogenic strains. Novel marker systems such as green fluorescent protein are now being employed in labouratory studies to monitor the persistence of GMOs in the human gut microflora and DNA transfer events between bacteria. Another hurdle to monitoring the dissemination of recombinant DNA in complex microbial consortia and investigating the factors governing DNA transfer in natural environments is the non-expression of marker genes in novel bacterial hosts.

Marker genes transferred between bacteria may not always be expressed in the recipient cell. Thus not all DNA transfer events will be detected by existing marker systems. One way around this problem is to use a specific plasmid transfer system comprising a donor and recipient bacterium in which the marker gene is known to be expressed, and then examining the factors governing DNA transfer between these two strains in models of the ecosystem under investigation.

Recent advances in molecular biology offer some hope that more reliable means of monitoring the dissemination of genetic determinants amongst diverse bacteria may be developed. Fluorescent in situ hybridization is now being employed to enumerate bacteria in natural assemblages by hybridizing 16S rRNA targets to fluorescently labelled DNA probes and visualization with fluorescent microscopy. Similarly, it is now possible to visualize PCR products within bacterial cells fixed on microscope slides.

By combining these techniques it may soon be possible to visualize concomitantly, fluorescently labelled PCR products of single-copy genes and labelled 16S rRNA sequences within a single cell in environmental samples. To enable us to ask specific questions about the human gut microflora, in vitro and in vivo models of the human gut have been developed.

In vitro models generally employ anaerobic chemostats of varying degrees of complexity or short-term batch cultures of human faeces or luminal samples extracted from animals. They allow generation of data on the effect of specific biotic and abiotic environmental parameters on the gut microflora, which may otherwise be unavailable due to inaccessibility of the target environment in the gastrointestinal tract or for ethical reasons.

Freter *et al.* studied the transfer of plasmid DNA between E. coli strains using a single-stage continuous flow culture model of the mouse gut microflora. This in vitro model of the mouse caecal microflora simulated microbial interactions observed in the mouse gut. Data generated from the in vitro model also allowed validation of mathematical models of plasmid transfer between E. coli strains in the presence of the complex mouse gut microflora. Rang *et al.* employed short-term batch cultures of luminal extracts from the mouse gastrointestinal tract to monitor transfer of plasmid RP1 between E. coli strains. More complex in vitro models of the human gastrointestinal tract developed recently have yet to be employed to monitoring DNA transfer events in the human gut microflora. In vitro models allow us to examine interactions between specific members of the gut microflora or between the whole human faecal microflora and introduced GMOs and non-GMOs under the physiological conditions of the human gut.

They provide a vital first stage in the biosafety assessment of GMOs intended for use in human food. However, such models do not take into consideration the biological complexity of the mammalian gut ecosystem. In vitro models may not simulate the biological surfaces or the input of endogenous substrates and secretions of the human immune system, which may play a vital role both in the maintenance and establishment of the gut microflora, and the frequency of DNA transfer events.

Gnotobiotic and germ-free animals have been employed as in vivo models of the human gastrointestinal ecosystem for some years. Gnotobiotic technology provides a useful tool in studying the interactions between specific members of the human gut microflora, e.g. the role played by specific members of the microflora in colonization, resistance to pathogenic bacteria, bacterial pathogenesis, and metabolism of dietary constituents or xenobiotic compounds.

Transfer of plasmid DNA between specific members of the human gut microflora has also been studied using di- or poly-associated animals. Conventional animals may not provide information of direct relevance to the human gut ecosystem because of the major differences in the composition of the gut microflora between animals and man. To circumvent these problems, germ-free animals have been associated with groups of bacteria from the human gut microflora or the whole human gut microflora.

Such human flora-associated (HFA) animals provide an essential tool in studies of human gut microbiology in that they reflect the complexity of the human gut ecosystem in terms of microbial species and numbers, and they provide some of the biotic parameters provided by the mammalian mucosa, immune system and digestive functions.

Such in vivo models provide an extremely useful tool in the generation of data on the safety of GMOs in food until a greater degree of confidence in the safety of such products is attained and human volunteer studies can ethically be considered.

Factors Affecting Transfer of Plasmid DNA from Bacteria in vivo

Very little is known about the extent to which transformation and transduction contribute to DNA transfer between bacteria in the human gastrointestinal microflora. Only one study on bacterial transformation in fresh human saliva samples has been presented in the literature. No reports of transformation or transduction in lower regions of the human gastrointestinal tract were found after exhaustive search of the literature.

This is probably due to the fact that conjugation has been viewed as the dominant contributor to DNA transfer between bacteria in natural environments, because of the ubiquitous occurrence of conjugative elements amongst diverse bacterial species. Thus most studies looking at DNA transfer in the gut have concentrated on conjugation.

Earlier, it was proposed that the frequency of plasmid transfer in a given environment will be a function of the interaction between environmental factors and the bacterial populations involved, characteristics of the plasmid itself and the mode of conjugation employed by the plasmid. Some of these parameters have been studied in vivo in animal models.

Effect of Antibiotics on Plasmid Transfer

A number of early investigators observed the transfer of plasmid DNA between bacteria in the gastrointestinal tract of human volunteers. However, many of these studies were conducted with antibiotic-resistance determinants in E. coli strains and in the presence of antibiotic selective pressure. In the light of the recent emergence of pathogenic bacteria with multiple antibiotic resistance determinants it is no longer considered ethically justifiable to carry out such investigations in human volunteers.

Ingestion of antibiotics leads to perturbation of the natural homeostasis of the human gut microflora and presents a selective pressure for the evolution of antibiotic resistance strains through mutation and transfer of resistance determinants. The human gut microflora is commonly exposed to antibiotics, both through clinical practice and use of antibiotics in animal husbandry. Similarly, antibiotic-resistant bacteria have been isolated from a wide range of human foodstuffs.

Duval-Iflah *et al.* observed the transfer of an R plasmid from Serratia liquifaciens to an E. coli recipient strain, naturally present in the human microflora of the HFA mice. The donor S. liquifaciens strain became established in the HFA mice at between 106 and 107 CFU/g faeces. Transfer was observed 12 days after inoculation with the donor strain and their numbers ranged from between 103 and 105 CFU/g faeces. The number of transconjugants greatly increased after addition of selective antibiotic in drinking water. Morelli *et al.* showed that a subtherapeutic dose of selective antibiotic (erythromycin at 10 μg/ml drinking water) increased the level of pAMâ1 transfer from L. reuteri to Ent. faecalis from 1 × 10-7 to 1 × 10-4 transconjugants per donor in anexic mice (i.e. gnotobiotic mice associated with donor and recipient strains).

Although selective antibiotic pressure previously has been employed to demonstrate transfer of genetic elements between bacteria in animal models and in humans, the observation that subtherapeutic doses of selective antibiotic increase the rate of plasmid transfer has important implications for both the design in GMOs and the use or misuse of antibiotics in clinical practice and agriculture.

Persistence of GMOs and Their Recombinant DNA in vivo

The ability of GMOs, which may be ingested in a viable form in fermented foods, to persist in the human gastrointestinal tract will greatly determine the extent to which recombinant DNA may be transferred to the human gut microflora.

Similarly the stability of recombinant DNA within GMOs in vivo is of major interest when designing GMOs for use in food or in the human gut itself. Stable maintenance of recombinant DNA will greatly determine the ability of a GMO to effectively carry out the task for which it was designed. On the other hand, a certain degree of recombinant DNA instability may limit the risks posed by recombinant DNA persisting in non-target ecosystems.

Gruzza *et al*. investigated the colonization potential and recombinant DNA stability of genetically modified L. lactis strains in germ-free mice. All L. lactis constructs reverted to plasmid-free derivatives in the digestive tract of anexic mice and plasmid-free derivatives became dominant over plasmid-bearing parental strains. Two high copy number, non-self transmissible plasmids exerted strong ecological disadvantage on plasmid-bearing L. lactis strains and were quickly lost from the digestive tracts of the anexic mice.

Transfer of the conjugative plasmid pIL205 was observed between an Ent. faecalis strain colonizing the digestive tract of anexic mice and an L. lactis donor strain. Both the L. lactis donor strain and Ent. faecalis (pIL205) transconjugants were lost from the anexic mice 10 days after dosing with the donor strain. Thus it appears that carriage of recombinant DNA on non-conjugative plasmids and indeed conjugative plasmid DNA, imposed considerable ecological disadvantages on plasmid-bearing strains in anexic animals. Schlundt *et al*. investigated transfer of pAMâ1 between L. lactis donor and recipient strains in gnotobiotic and conventional rats. In the conventional animals, both donor and recipient strain were quickly lost from the system and no transconjugants were recovered.

In gnotobiotic animals both donor and recipient strain colonized the gastrointestinal tract. Tranconjugants were recovered within a few days of dosing and maintained at between 103 and 105 CFU less than the recipient strain in faecal samples. Upon sampling of luminal contents of the jejunum, caecum and colon it was observed that carriage of the plasmid seemed to endow a competitive advantage to transconjugants in the small intestine but not in the caecum.

The population of transconjugants was maintained at about 104 CFU/g throughout the intestine but numbers of recipient strain varied from between 104 and 105 CFU/g jejunum contents to 108-109 CFU/g caecal and colonic contents. Clearly the ability of plasmid DNA either to impose ecological advantage or disadvantage on the bacterial host will depend on the plasmid and bacterium involved, and may very well vary considerably between different environmental habitats or microhabitats.

Effect of Microbial Interactions on DNA Transfer

The human gut microflora exerts considerable colonization resistance on extraneous microorganisms entering the colonic environment. Limiting the persistence of a GMO in a given environment will also limit its ability to undergo DNA transfer with the microflora. Gruzza *et al*. investigated the ability of members of the human gut microflora to limit the colonization ability of and DNA transfer from, genetically modified Lactococcus lactis strains, similar to those likely to be used in the food industry.

L. lactis strains bearing the conjugative plasmid pIL205 and/or a non-conjugative but a pIL205 mobilizable plasmid pIL253 were used. Gnotobiotic mice were associated with four strictly anaerobic bacteria commonly found in the human gut microflora, i.e. Bacteroides sp., Bifidobacterium sp., Peptostreptococcus sp. and Ent. faecalis.

No transfer of plasmid DNA was observed from L. lactis donor strain to Bacteroides sp., Bifidobacterium sp. or the Peptostreptococcus sp. However, upon addition of Ent. faecalis to the polyassociated mice, Ent. faecalis transconjugants were observed despite the competitive exclusion imposed on the L. lactis donor strain by the colonizing Ent. faecalis strain. Gene transfer from the L. lactis strains was also investigated in HFA mice. HFA mice were dosed with L. lactis donor strain, bearing pIL205 and pIL253, seven days after introduction of the human faecal microflora.

From an initial inoculum of 8 × 108 CFU, no L. lactis donor cells were recovered from the faecal samples 24 hours after dosing (detection limit 102 CFU/g faeces). However, 21 hours after introduction of the donor strain, transconjugant facultatively anaerobic streptococci were recovered resistant to antibiotic markers borne on each of the two plasmids.

The number of transconjugants increased until day 15 after dosing. Thus, although rather quickly eliminated from the HFA mice, the L. lactis donor strain was able to transfer its plasmid DNA (both the conjugative pIL205 and the non-conjugative but mobilizable pIL253) to members of the human gut microflora in HFA mice.

Brockmann *et al.* investigated the transfer of plasmid pAMâ1 and pLMP1 (bearing a genetically modified proteinase gene) from L. lactis donor strains to a proteinase-deficient recipient strain in gnotobiotic and conventional rats.

No transfer of the proteinase gene to the recipient L. lactis strain was observed in anexic or conventional rats. Plasmid pAMâ1 was transferred between L. lactis donor and recipient strains in anexic animals but not in conventional rats where both donor and recipient were quickly lost from faecal samples. Transfer of pAMâ1 from the L. lactis donor strain to Ent. faecalis (or a close relative) was observed in one conventional rat.

Using an in vitro single stage continuous flow culture (CFC) of human faecal microflora and in vivo human flora-associated (HFA) rat model of the human colonic microflora we have shown that the human gut microflora exerts a considerable barrier to persistence of L. lactis strains in the colonic ecosystem. A plasmid-free L. lactis strain MG1614 was washed out from the in vitro model at a rate greater than expected for non-growing cells.

Similarly, L. lactis MG1614 introduced into HFA rats was eliminated from the HFA rats within eight days. By continuously dosing HFA rats with L. lactis MG1614 in the drinking water we succeeded in maintaining a stable population of L. lactis MG1614 in the rats. It was hoped that this strain implanted in the microflora of HFA rats would act as a recipient in plasmid transfer studies with a closely related L. lactis strain.

Transconjugants were maintained at between 104 and 105 CFU/g faeces for 27 days. Thus despite the fact the human gut microflora exerts such a drastic barrier effect on the persistence of L. lactis strains in models of the human colonic microflora, transfer of plasmid DNA from an ingested L. lactis donor strain to Enterococcus spp. present in the microflora was observed.

Design of GMOs and Colonization of Gut by GMOs

GMOs considered for release into the environment in a viable form will employ several measures to limit the ability of recombinant DNA to be transferred to indigenous bacteria in the target environment. Recombinant DNA may be borne on non-conjugative vectors, on the chromosome of the host strain or by the use of suicidal genetic elements.

However, inbuilt safeguards which prevent the transfer of recombinant DNA under labouratory conditions cannot be relied upon to prevent transfer of recombinant DNA in the natural environment. Thus studies employing in vivo models of the human gastrointestinal tract have been conducted with such GMOs in order to determine their ability to persist in the digestive tract and transfer DNA. Gruzza *et al.* investigated the ability of genetically modified L. lactis strains to colonize gnotobiotic mice.

The genetically modified L. lactis strains were constructed with antibiotic-resistance marker genes inserted on different replicons, chloramphenicol resistance was inserted into a self-transmissible plasmid pIL205, erythromycin resistance was inserted into non-self transmissible plasmids pIL252 (low copy number) and pIL253 (high copy number) and a second erythromycin-resistance determinant was implanted into the L. lactis chromosome.

All strains carried a naturally occurring pIL9 plasmid encoding the machinery for lactose fermentation. It was observed that anexic mice were colonized with the recombinant strains or their plasmid-free derivatives, with plasmid-free derivatives becoming dominant in all cases. Plasmids pIL9 and pIL205 were lost from donor strains but plasmid-bearing parental strains and plasmid-free derivatives were maintained at co-dominant levels, suggesting equilibrium between plasmid loss and conjugation in the mouse gastrointestinal tract.

L. lactis strains bearing the low-copy number non-self transmissible plasmid pIL252 rapidly lost their plasmid complement in the gut of anexic mice, probably due to high segregational instability of this plasmid observed in vitro. L. lactis strains bearing pIL253, the high-copy number, non-self transmissible plasmid colonized the digestive tract of anexic mice but only became dominant when plasmid pIL205 was not present in the same cell, indicating a degree of incompatibility between the two plasmids both plasmids belong to the same incompatibility group. The erythromycin-resistance determinant borne chromosomally was maintained stably in the digestive tract of anexic mice. The authors concluded that in vitro determinations of recombinant DNA stability may highlight if a particular insert is extremely unstable but does not give reliable data on the in vivo stability of the construct.

This, highlights the need for in vivo studies both on the stability of recombinant DNA for efficient GMO function and the effect instability has on the persistence of recombinant DNA in the digestive tract, which is a major factor governing the extent to which DNA transfer between ingested GMOs and members of the human gut microflora occurs.

Duval-Iflah *et al.* investigated the mobilization of pBR-derived recombinant plasmids from donor E. coli K12 strains to E. coli strains of human origin in vitro and in the digestive tract of anexic and gnotobiotic mice associated with the human gut microflora. The genetically modified plasmids used were designed to be non-conjugative (mob-, tra-) and either mobilizable or non-mobilizable. E. coli strains isolated from human faeces were screened for their ability to mobilize the oriT- and oriT+ plasmids.

In anexic mice, associated with donor and recipient E. coli strains, mobilization of an oriT+ pBR derivative by trans-complementation with the gene products of tra (encoded by conjugative plasmid R388) and mob was observed. Such mobilization was only observed sporadically with one E. coli strain of human origin in di-associated mice and not at all in HFA mice. It was determined that about 50 poe cent of E. coli strains of human origin were able to mobilize oriT- pBR derivatives in in vitro matings but not in anexic mice or HFA animals.

Dietary Factors Affecting DNA Transfer

The ecological significance of the metabolic burden imposed on a bacterium by plasmid carriage will be greatly determined by the availability of substrates and nutrients to the cell. Similarly, densities of donor and recipient cells have been shown to affect conjugation frequencies in vitro. In our studies, transfer of plasmid pAMâ1 from an L. lactis donor strain to an Enterococcus spp. recipient strain naturally present in the microflora of HFA rats was found to be strongly affected by dose of donor strain employed.

Significant numbers of Enterococcus spp. (pAMâ1) transconjugants were recovered only at a donor cell concentration of 109 CFU/ml or above. The diet fed to the HFA rats also had an effect on transfer of pAMâ1 from the L. lactis donor strain to the Enterococcus spp. in vivo. A high-fat, low-fibre synthetic diet (similar to human high-fat diets) reduced the numbers of Enterococcus spp. (pAMâ1) transconjugants recovered from the faeces of HFA rats compared with the commercial rodent chow (higher levels of complex carbohydrate and fibre).

The presence of increased substrate availability in the caecum of HFA rats fed the rodent chow or increased surfaces provided by the higher fibre content of the rodent chow may be involved in the higher numbers of Enterococcus spp. (pAMâ1) transconjugants recovered from HFA rats fed this diet. Duval-Iflah *et al.* looked at the effect of milk fermented with Lactobacillus bulgaricus and Streptococcus thermophilus on the transfer and mobilization of plasmids between E. coli strains in the gut of anexic mice.

Plasmid transfer studies were conducted in gnotobiotic mice associated with E. coli K12 donor strain and an E. coli recipient strain PG1 of human origin. The affect of milk fermented with L. bulgaricus, S. thermophilus or with both lactic acid bacteria on transfer of plasmid between the E. coli strains was determined.

Fermented milk consumption had little effect on the populations of R388 or pUB2380 transconjugants in the anexic mice but the milk fermented with both LAB resulted in a slight increase in E. coli PG1 (R388) transconjugants. Long-term consumption of milk fermented with both LAB strains resulted in inhibition of occurrence and maintenance of E. coli PG1 (pCE325) transconjugants in the anexic mice. L. bulgaricus-fermented milk resulted in a reversible reduction in the numbers of E. coli PG1 (pCE325) transconjugants in mouse faecal samples.

When L. bulgaricus and S. thermophilus were grown in BHI broth, the transconjugant lowering capabilities observed with the LAB plus fermented milk were reversed. Administration of L. bulgaricus and the combination of both strains grown in BHI broth increased the population of E. coli PG1 (pCE325) transconjugants recovered from mouse faecal samples. Thus, the authors concluded that probiotic LAB have an effect on the formation and subsequent maintenance of transconjugants in the gastrointestinal tract of anexic mice, and that this effect was dependent on whether the LAB were administered in fermented milk or labouratory growth media, indicating that bacterial metabolic end products as well as viable LAB may be involved.

Clearly much work remains to be done to elucidate the role played by specific environmental parameters, such as diet, ratio of donor and recipient cells, activity of donor and recipient and antagonistic effects of the human gut microflora on plasmid transfer and persistence of GMOs which may be ingested in a viable form in fermented foods. Genetic engineering is now being employed by the food industry to improve the performance of microorganisms used in the production of fermented foods. Concern has been expressed over the exposure of consumers to genetically modified microorganisms which may be ingested in a viable form in future food products. Of chief concern is the risk of recombinant DNA transfer from ingested GMOs to members of the human gut microflora. A range of natural DNA transfer mechanisms are available to microorganisms in the environment. Chief among these are conjugation, transformation and transduction. Apart from one case of transformation in the oral cavity, most studies examining DNA transfer in the human gastrointestinal microflora have been concerned with conjugation. In vitro and in vivo models of varying degrees of complexity have been employed to simulate the prevailing ecological conditions in the human gastrointestinal microflora. Such studies have demonstrated the transfer of plasmid DNA, both naturally occurring and genetically modified, in the digestive tract.

Some of the factors thought to affect conjugation between bacteria in the gut microflora have also been studied in vitro and in vivo, e.g. metabolic burden of plasmid maintenance, recombinant DNA stability, colonisation ability of GMOs and antagonistic activities of the microflora, presence or absence of antibiotics, cell concentrations of bacteria involved in conjugation and the effect of dietary supplements on DNA transfer events in the human gut microflora.

In order to provide information of direct relevance to risk assessment of the use of genetically modified microorganisms in food, further scientific investigations employing in vivo and in vitro models.

HARMFUL EFFECTS OF GENETICALLY MODIFIED FOODS

Largely between 1997 and 1999, genetically modified (GM) food ingredients suddenly appeared in *2/3rds of all US processed foods*. This food alteration was fuelled by a single Supreme Court ruling. It allowed, for the first time, the patenting of life forms for commercialization. Since then thousands of applications for experimental genetically-modified (GM) organisms, including quite bizarre GMOs, have been filed with the US Patent Office alone, and many more abroad. Furthermore an economic war broke out to own equity in firms that legally claimed such patent rights or the means to control not only genetically modified organisms but vast reaches of human food supplies. This has been the behind-the-scenes and key factor for some of the largest and rapid agri-chemical firm mergers in history.

The merger of Pioneer Hi-Bed and Dupont (1997), Novartis AG and AstraZeneca PLC (2000), plus Dow's merger with Rohm and Haas (2001) are three prominent examples, Few consumers are aware this has been going on and is ever continuing. Yet if you recently ate soya sauce in a Chinese restaurant, munched popcorn in a movie theatre, or indulged in an occasional candy bar-you've undoubtedly ingested this new type of food. You may have, at the time, known exactly how much salt, fat and carbohydrates were in each of these foods because regulations mandate their labelling for dietary purposes. But you would not know if the bulk of these foods, and literally every cell had been genetically altered!

In just those three years, as much as 1/4 of all American agricultural lands or 70-80 million acres were quickly converted to raise genetically-modified (GM) food and crops. And in the race to increase GM crop production verses organics, the former is winning.

CORE PHILOSOPHICAL ISSUES

When Gandhi confronted British rule and Martin Luther King addressed those who disenfranchised Afro-Americans, each brought forth issues of morality and spirituality. They both challenged others to live up to the highest principles of humanity. With the issue of GM food technology, we should naturally do the same, and with great respect for both sides. It is not enough to list fifty or more harmful effects but we need to also address moral, spiritual and especially worldview issues. Here the stakes are incredibly huge. The philosophical issues involving GMOs, why this technology represents the impregnation of a mechanical worldview, a death-centred vision of nature that is greatlyt accelerating the death of species on earth.

FROM HYBRIDIZATION TO GMOS

Another challenging phenomenon to face in our modern world is that of hybridization. It seems to have worked so very successfully in some commercial realms, and as a major application of Gregor Mendel's revolutionary Gene Theory. Mendel offered a logical extension of the larger mechanical worldview. Just as we create factory assembly lines for manufacturing inanimate products, why can't we also manufacture living organisms, and using the same or similar principles? Why not take this assembly-line process to the next logical and progressive level?

What's wrong then with the "advance" of genetic engineering? No doubt, with hybridizations conscious life is manipulated. But living organisms continue to make some primary genetic decisions amid limited selections. We can understand this with an analogy. There is an immense difference between being a matchmaker and inviting two people to a dinner party, to meet and see if they are compatible. This differs essentially from forcing their meeting and union or a violent date rape. The former act may be divine, and the latter considered criminal. The implication is that biotechnology involves vital moral issues in regard to the whole of life in nature.

With biotechnology, roses are no longer crossed with just roses. They are mated with pigs, tomatoes with oak trees, fish with asses, butterflies with worms, orchids with snakes. The technology that makes this all possible is called biolistics-a gunshot-like violence that pierces the nuclear membrane of cells. This essentially violates not just the core chambers of life (physically crossing nuclear membranes) but the conscious-choice principle that is part of living nature's essence. Some also compare it to the violent crossing of territorial borders of countries, subduing inhabitants against their will.

What will happen if this technology is allowed to spread? Fifty years ago few predicted that chemical pollution would cause so much vast environmental harm. Now nearly 1/3rd of all species are threatened with extinction (and up to half of all plant species and half of all mammals). Few also knew that cancer rates would skyrocket during this same period. Nowadays approximately 41 poe cent on average of Americans can expect cancer in their lifetime.

ALARM SIGNALS

No one has a crystal ball to see future consequences of the overall GMO technology. Nevertheless, there are silent alarm signals like the early death of canaries in a mine shaft. There is, for example, growing evidence that the wholesale disappearance of bees relates directly to the appearance of ever more GM pollen. If we understand certain philosophical issues about the 17th century's worldview, the potential harm of GMOs actually can potentially far outweigh that of chemical pollution.

This is because chemistry deals mostly with things altered by fire (and then no longer alive, isolated in laboratories-and not infecting living terrains in self-reproducible ways). Thus a farmer may use a chemical for many decades, and then let the land lie fallow to convert it back to organic farming. This is because the chemicals tend to break down into natural substances over time, Genetic pollution, however, can alter the oil's life forever!

Farmers who view their land as their primary financial asset have reason to heed this warning. They need to be alarmed by evidence that genetically-modified soil bacteria contamination can arise. This is more than just possible, given the numerous (1600 or more) distinct microorganisms that can be found in a single teaspoon of soil.

If that soil contamination remains permanently, the consequences can be catastrophic. Someday the public may blacklist precisely those farms that have once planted GM crops. No one has put up any warning signs on product packaging for farmers, including those who now own 1/4 of all agricultural tracks in the US. Furthermore, the spreading potential impact on all ecosystems is profound.

"Our way of life is likely to be more fundamentally transformed in the next several decades than in the previous one thousand years...Tens of thousands of novel transgenic bacteria, viruses, plants and animals could be released into the Earth's ecosystems...Some of those releases, however, could wreak havoc with the planet's biospheres."

In short these processes involve unparalleled risks. Voices from many sides echo this view. Contradicting safety claims, no major insurance company has been willing to limit risks, or insure bio-engineered agricultural products. The reason given is the high level of unpredictable consequences. Over eight hundred scientists from 84 countries have signed The World Scientist open letter to all governments calling for a ban on the patenting of life-forms and emphasizing the very grave hazards of GMOs, genetically-modified seeds and GM foods. This was submitted to the UN, World Trade Organization and US Congress. The Union of Concerned Scientists (a 1000 plus member organization with many Nobel Laureates) has similarly expressed its scientific reservations. The prestigious medical journal, Lancet, published an article on the research of Arpad Pusztai showing potentially significant harms, and to instill debate. Britain's Medical Association (the equivalent of the AMA and with over a 100,000 physicians) called for an outright banning of genetically-modified foods and labelling the same in countries where they still exist.

In a gathering of political representatives from over 130 nations, drafting the Cartagena Protocol on Biosafety, approximately 95 poe cent insisted on new precautionary approaches. The National Academy of Science report on genetically-modified products urged greater scrutiny and assessments. Prominent FDA scientists have repeatedly expressed profound fears and reservations but their voices were muted not due to cogent scientific reasons but intense political pressure from the Bush administration in its efforts to buttress and promote the profit-potentials of a nascent biotech industry.

To counterbalance this, industry-employed scientists have signed a statement in favour of genetically-modified foods. But are any of these scientists impartial? Writes the New York Times (about a similar crisis involving genetic engineering and medical applications).

"Academic scientists who lack industry ties have become as rare as giant pandas in the wild...lawmakers, bioethics experts and federal regulators are troubled that so many researchers have a financial stake [via stock options or patent participation] ...The fear is that the lure of profit could colour scientific integrity, promoting researchers to withhold information about potentially dangerous side-effects."

Looked at from outside of commercial interests, perils of genetically modified foods and organisms are multi-dimensional. They include the creation of new "transgenic" life forms-organisms that cross unnatural gene lines (such as tomato seed genes crossed with fish genes)-and that have unpredictable behaviour or replicate themselves out of control in the wild. This can happen, without warning, *inside of our bodies* creating an unpredictable chain reaction. A four-year study at the University of Jena in Germany conducted by Hans-Hinrich Kaatz revealed that bees ingesting pollen from transgenic rapeseed had bacteria in their gut with modified genes. This is called a "horizontal gene transfer." Commonly found bacteria and microorganisms in the human gut help maintain a healthy intestinal flora. These, however, can be mutated.

Mutations may also be able to travel internally to other cells, tissue systems and organs throughout the human body. Not to be underestimated, the potential domino effect of internal and external genetic pollution can make the substance of science-fiction horror movies become terrible realities in the future. The same is true for the bacteria that maintain the health of our soil-

and are vitally necessary for all forms of farming-in fact for human sustenance and survival.

Without factoring in biotechnology, milder forms of controlling nature have gravitated towards restrictive monocropping. In the past 50 years, this underlies the disappearance of approximately 95 poe cent of many native grains, beans, nuts, fruits, and vegetable varieties in the United States, India, and Argentina among other nations (and on average 75 poe cent worldwide). Genetically-modified monoculture, however, can lead to yet greater harm. Monsanto, for example, had set a goal of converting *100 poe cent* of all US soy crops to Roundup Ready strains by the year 2000. If this plan were effected, it would have threatened the biodiversity and resilience of all future soy farming practices. Monsanto laid out similar strategies for corn, cotton, wheat and rice. This represents a deepest misunderstanding of how seeds interact, adapt and change with the *living* world of nature.

One need only look at agricultural history-at the havoc created by the Irish potato blight, the Mediterranean fruit fly epidemic in California, the regional citrus canker attacks in the Southeast, and the 1970's US corn leaf blight. In the latter case, 15 poe cent of US corn production was quickly destroyed. Had weather changes not quickly ensued, most all crops would have been laid waste because a fungus attached their cytoplasm universally. The deeper reason this happened was that approximately 80 poe cent of US corn had been standardized (devitalized/mechanized) to help farmers crossbreed-and by a method akin to those used in current genetic engineering. The uniformity of plants then allowed a single fungus to spread, and within four months to destroy crops in 581 counties and 28 states in the US. According to J. Browning of Iowa State University: "Such an extensive, homogeneous acreage of plants... is like a tinder-dry prairie waiting for a spark to ignite it."

The homogeneity is unnatural, a byproduct again of deadening nature's creativity in the attempt to mechanize, to grasp absolute control, and of what ultimately yields not control but wholesale disaster. Europeans seem more sensitive than Americans to such approaches, given the analogous metaphor of German eugenics.

HISTORICAL SYNOPSIS

Overall the "biotech revolution" that is presently trying to overturn 12,000 years of traditional and sustainable agriculture was launched in the summer of 1980 in the US. This was the result of a little-known US Supreme Court decision *Diamond vs. Chakrabarty* where the highest court decided that biological life could be legally patentable.

Ananda Mohan Chakrabarty, a microbiologist and employee of General Electric (GE), developed at the time a type of bacteria that could ingest oil. GE rushed to apply for a patent in 1971. After several years of review, the US

Patent and Trademark Office (PTO) turned down the request under the traditional doctrine that life forms are not patentable. Jeremy Rifkin's organization, the Peoples Business Commission, filed the only brief in support of the ruling. GE later sued and won an overturning of the PTO ruling. This gave the go ahead to further bacterial gmo research throughout the 1970's. Then in 1983 the first genetically-modified plant, an anti-biotic resistant tobacco was introduced. Field trials then began in 1985, and the EPA approved the very first release of a GMO crop in 1986. This was a herbicide-resistant tobacco. All of this went forward due to a regulatory green light as in 1985 the PTO also decided the Chakrabarty ruling could be further extended to all plants and seeds, or the entire plant kingdom.

It then took another decade before the first genetically-altered crop was commercially introduced. This was the famous delayed-ripening "Flavr-savr" tomato approved by the FDA on May 18, 1994. The tomato was fed in laboratory trials to mice who, normally relishing tomatoes, refused to eat these lab-creations and had to be force-fed by tubes. Several developed stomach lesions and seven of the forty mice died within two weeks. Without further safety testing the tomato was FDA approved for commercialization. Fortunately, it ended up as a production and commercial failure, and was ultimately abandoned in 1996. This was the same year Calgene, the producer, began to be bought out by Monsanto. During this period also, and scouring the world for valuable genetic materials, W.R. Grace applied for and was granted fifty US patents on the neem tree in India. It even patented the indigenous knowledge of how to medicinally use the tree (what has since been called biopiracy). Also by the close of the 20th century, about a dozen of the major US crops-including corn, soy, potato, beets, papaya, squash, tomato and cotton-were approved for genetic modification.

Going a step further, on April 12, 1988, PTO issued its first patent on animal life forms (known as oncomice) to Harvard Professor Philip Leder and Timothy A. Stewart. This involved the creation of a transgenic mouse containing chicken and human genes. Since 1991 the PTO has controversially granted other patent rights involving human stem cells, and later human genes. A United States company, Biocyte was awarded a European patent on all umbilical cord cells from fetuses and newborn babies. The patent extended exclusive rights to use the cells without the permission of the donors. Finally the European Patent Office (EPO) received applications from Baylor University for the patenting of women who had been genetically altered to produce proteins in their mammary glands. Baylor essentially sought monopoly rights over the use of human mammary glands to manufacture pharmaceuticals. Other attempts have been made to patent cells of indigenous peoples in Panama, the Solomon Islands, and Papua New Guinea, among others. Thus the groundbreaking Chakrabarty ruling evolved, and within little more than two decades from the patenting of tiny, almost invisible microbes, to allow the genetic modification of virtually all terrains of life on Earth.

Certain biotech companies then quickly, again with lightening speed, moved to utilize such patenting for the control of first and primarily seed stock, including buying up small seed companies and destroying their non-patented seeds. In the past few years, this has led to a near monopoly control of certain genetically modified commodities, especially soy, corn, and cotton (the latter used in processed foods when making cottonseed oil). As a result, between 70-75 poe cent of processed grocery products, as estimated by the Grocery Manufacturers of America, soon showed genetically-modified ingredients. Yet again without labelling, few consumers in the US were aware that any of this was pervasively occurring. Industry marketers found out that the more the public knew, the less they wanted to purchase GM foods. Thus a concerted effort was organized to convince regulators (or bribe them with revolving-door employment arrangements) not to require such labelling.

FOOD FORTIFICATION

Food fortification has been defined as the addition of one or more essential nutrients to a food, whether or not it is normally contained in the food, for the purpose of preventing or correcting a demonstrated deficiency of one or more nutrients in the population or specific population groups. Other terminology exists for the addition of nutrients to foods. Restoration means the addition to a food of essential nutrients which are lost during the course of Good Manufacturing Process (GMP), or during normal storage and handling procedures, in amounts which will result in the presence in the food of the levels of the nutrients present in the edible portion of the food before processing, storage or handling. Enrichment has been used interchangeably with fortification, but elsewhere it has been defined as the restoration of vitamins and minerals lost during processing.

IMPACT OF OTHER TECHNOLOGIES ON FORTIFICATION PROCEDURES

In order to achieve the required level of nutrients in fortified products reaching the consumer, manufacturers have to estimate processing and storage losses and add the necessary excess during production. The introduction of new processes, equipment and packaging materials can affect processing and storage losses and hence fortification procedures.

THERMAL PROCESSES

Thermal treatments in food processing serve multiple functions: destruction of microorganisms, inactivation of enzymes and toxic factors, modification of flavour and texture. In many cases the heat treatment is selected based on consideration of the required lethality of the process so as to render the product safe under stated conditions of storage.

The lethality of a process is a function of both temperature and time of exposure at each temperature. The Arhenius activation energy of spore inactivation is high as compared with that for vitamin destruction. For this reason, at high temperatures the microbiological requirements of the thermal process can be met with relatively low losses of vitamin being incurred. Heat treatments based on the principle of 'High Temperature Short Time' (HTST) or Ultra High Temperature (UHT) are less destructive to the vitamin content and require the addition of lower levels of excess vitamins to account for processing losses.

As the lethality of a process is determined by sum of lethal effects at the 'cold spot' of the product unit (can, pouch, etc.), innovations in equipment design which lead to the faster attainment of thermal equilibrium will also affect the thermal process which must be delivered and hence vitamin loss during processing. The introduction of agitating retorts and retorts which can be used in the sterilisation of materials in flexible packages as well as the development of flame sterilisation processes also can have an affect on fortification procedures.

PACKAGING MATERIALS

The use of retort pouches and other deformable containers in thermal processing influence processing procedures and the extent of thermal degradation of nutrients. The important role of packaging in the application of aseptic processing technologies has important implications for fortification procedures. This is not only in terms of the UHT treatments which can be applied to thin liquid films prior to packaging, but also to the impact on storage losses. The shelf life of aseptically processed foods can exceed 1 year and storage losses over this entire period must be taken into consideration in the calculation of required overages. During aseptic processes the product is cooled prior to packaging into sterile containers, headspace oxygen and dissolved oxygen levels are therefore higher than for hot-filled products or for traditional thermal processes which involve exhausting or vacuum sealing prior to processing.

Problems of vitamin loss in aseptically processed foods have been attributed to dissolved and headspace oxygen, oxygen and light permeability of the packaging material and to commodity-specific reaction which are only dependent on storage temperature. In aseptically processed milk the dissolved and headspace oxygen lead to rapid initial losses of vitamin C activity. This in turn causes substantial losses in folate as vitamin C plays an important role in the protection of folate. Oamen also reported heavy losses of vitamins B_6, and B_{12} during storage of aseptically processed milk. Improvements in the barrier properties of plastics and laminates also impact on the potential for nutrient loss on storage of products other than those aseptically processed.

IRRADIATION AND MICROWAVES

Irradiation is used to a limited extent in food processing and as such cannot be expected to impact upon fortification procedures to a large extent. One common use of food irradiation is in the prevention of insect infestation in grains. Losses of vitamin B_1 on irradiation of whole grains are small, but increase in the irradiation of milled grains. Losses of this vitamin can be reduced by exclusion of oxygen during irradiation and storage. The increased use of microwave cooking on sous-vide and other cook-chill foods does not have a major effect on vitamin retention.

OTHER CONSIDERATIONS IN THE USE OF FORTIFICANT COMPOUNDS

BIOAVAILABILITY OF COMMERCIAL PREPARATIONS

While the practice of coating many of the water soluble vitamins does offer protection against interaction with the food medium and development of organoleptic problems, it can create other difficulties. Slow dissolution of the coating in the gastrointestinal tract in order to free the active component can cause reduced bioavailability of the micronutrient. Care should therefore be taken in the use of such preparations that the nutritional objectives are being met.

NUTRIENT-NUTRIENT INTERACTIONS

When more than one fortificant is being added to a particular vehicle, consideration must be given to the interactions, both positive and negative, which could take place. An example of this is seen with vitamin C and iron. Vitamin C has been shown to improve the absorption of iron. Other studies have indicated that iron and other trace metals increase the rate of ascorbic acid destruction. This deteriorative reaction was dependent on their being sufficient moisture to facilitate the mobility of the metal ions and was therefore demonstrated not to be a factor at water activity levels below 0.4. The presence of vitamin E has been shown to increase the bioavailability of vitamin A, one explanation of this is that tocopherol as a lipid phase antioxidant, stabilises vitamin A in the GI tract. Calcium has also demonstrated an inhibitory effect on iron absorption. Phosphorous negatively influences calcium absorption and due to this fact, calcium/osteoporosis claims can only be made in the United States if the calcium: phosphorous ratio is greater than 1:1.

In solution, ascorbic acid has been shown to destabilise folic acid and cyanocobalamin. Ryley and Kajda reported that vitamin C in UHT milk played an important role in the protection of the folate content. Degradation products of thiamine have also been shown to accelerate the rates of degradation of folic acid and cyanocobalamine, whereas the presences of iron salts in solution reportedly have a stabilising effect on cyanocobalamin. In solutions of the B-vitamins, riboflavin can cause the oxidation and consequent loss of thiamine.

If ascorbic acid is included into the solution the reaction does not occur. This observation is of practical significance in cases where solutions of water soluble vitamins are sprayed onto foods during the fortification process. Riboflavin also promotes oxidative degradation of folic acid, but this effect can be minimised by deaeration. Riboflavin has also been implicated in oxidative loss of vitamin C. Light-induced vitamin C loss from milk is reduced in the absence of riboflavin.

NUTRIENT-MATRIX INTERACTIONS

Besides nutrient - nutrient interactions, other components of the food matrix may also affect the functionality of the fortificant. Selection of the vehicle in fortification programmes much be such as to avoid reduced bioavailability of nutrients due to the presence of anti-nutritional compounds. Some interactions of added nutrients with the food matrix have been noted but are poorly understood. For example, the absorption of iron from sugar fortified with ferric orthophosphate and ascorbic acid was shown to be greatly improved when added to maize porridge before cooking as compared with after cooking.

Ascorbic acid is involved in many browning reactions. Vitamin B_6 B can take part in the maillard browning reactions under suitable conditions and thiamine can react with polyunsaturated fatty acids to produce a typical 'meaty' flavour and odour. Owing to such interactions, most of the water soluble vitamins are commercially available in coated form. The use of these coatings may be associated with problems of bioavailability. lodisation of salt and the compounds used for this have been widely studied in relation to the stability of the fortified salt. More research may be required if iodisation of processed foods through the use of iodised salt is to be widely practised.

The available information on this topic has shown no problems with off-flavour and off-colour development in many common processing applications such as vegetable fermentation, production of fresh pack pickles and canning of vegetables. In some specific cases, however, the use of iodised salt might be problematic. Off-flavour development in a cake mix was shown to be due to the use of iodised salt in its manufacture. There also needs to be more information on the stability of iodised salt under conditions experienced in vegetable fermentations.

NUTRIENT ADDITIONS TO FOOD

The following is not an exhaustive review of all food products which are fortified. Food products have been selected for discussion on the basis of their nutritional impact or their applicability to fortification programmes in developing countries.

CEREALS AND CEREAL BASED PRODUCTS

This is one of the most important areas for consideration of food fortification technologies. Foods from this category form the major component of diets around the world. This is especially true in the case of developing countries where dietary diversity is limited. On average, cereals provide 52 per cent of caloric intake globally.

For Africans and Asians they represent 60-75 per cent of the caloric intake; for Latin Americans, 50 per cent; and in the United States of America, 26 per cent. In developing countries 95 per cent of the population consume cereals as a dietary staple which also provide about 47 per cent of the per capita protein intake. Milling of cereal grains prior to their consumption is a common practice. During the milling process a substantial proportion of the nutrients are lost from the refined product. The fortification/enrichment of cereal grains can therefore be rationalised in more than one way. One valid reason is to restore to refined products, nutrients which have been removed during the milling process. Another reason is to improve the nutrient intake levels of target populations which are at risk of micronutrient deficiency.

RICE AND OTHER WHOLE CEREAL GRAINS

Among the cereals, rice presents unique problems in fortification. These are due to the fact that it is most commonly consumed as a whole grain and also in many countries, extensive washing of the grain prior to cooking is the normal practice. The earliest methods of rice enrichment involved the production of parboiled and converted rice. By this means nutrients from the bran layer were transferred to the starchy endosperm. The parboiling process involved the soaking of the rough rice, the application of heat followed by drying and milling. It was demonstrated that in this way 50-90 per cent of the thiamine was retained. The process for converted rice developed by Huzenlaub was similar to that for parboiling, but also employed pressure differences to facilitate transfer of nutrients. Acid parboiling, described by Kondo was similar to the parboiling except that it was carried out in the presence of acetic acid.

After parboiling and converting, the next methods of enrichment involved the actual addition of nutrients to the milled products. Techniques used for this have been classified into two main groups 'powder type' and 'grain type' enrichment. In powder type enrichment, a powdered pre-blended mixture of vitamins and minerals has been added at a rate of 1, 0.5. or 0.25 oz. per 100 lbs of rice (a w/w ratio of 1:1600, 1:3200 or 1:6400). For white parboiled rice, the normal practice has been to add the premix soon after milling as the heat and moisture at the grain surface at this point facilitates adherence of the powder.

A major disadvantage of this method of nutrient addition has been that 20-100 per cent of the nutrients are lost on washing. In the USA, rice enriched in this way must bear a label stating 'to retain vitamins do not rinse before or drain after cooking'. In the second major type of enrichment, a powdered nutrient mixture has been applied to the milled rice grains followed by coating with a water insoluble substance. A fortified rice premix produced in this way has then been added to milled rice at a rate of 0.5 per cent, to yield an enriched product conforming to the required standard of identity. In the United States of America, this standard requires between 2.0 - 4.0 mg of thiamine, 1.2 - 2.4 mg of riboflavin, 16 - 32 mg of niacin or niacinamide and 13 - 26 mg of iron per 100 lbs. of rice.

It may contain 250 -1000 USP units of vitamin D and 500 -1000 mg of calcium. The requirement regarding riboflavin has been stayed pending final action. The objections regarding the use of this vitamin in the premix have been due to its role in the colour deterioration of the enriched grain.

'Grain type' or 'coated grain' enrichment have also been carried out by spraying the premix solution onto the rice which is contained in a rotary cylinder, followed by hot air drying of the rice, application of a water insoluble sealant, addition of the iron compound and finally a second application of the water insoluble compound.

Water insoluble coatings which have been reported include an ethanol or isopropanol solution of zein, palmitic or stearic acid and abeitic acid. Other coating materials have contained ethyl cellulose. They dissolve at the elevated temperatures employed during cooking. Using this 'coated grain' procedure, Cort *et al* and Rubin *et al.* were able to successfully enrich rice with niacin, thiamine, pyridoxine, vitamin A, vitamin E, folic acid, iron, and zinc using expanded versions of the procedures already described. The water insoluble vitamins and minerals were added in different layers with intermittent coatings of shellac. Stability to rinsing treatments were reportedly high, with vitamin losses in the range of 0.2-1.1 per cent. Ricegrowers Co-operative Ltd. (RCL) in Australia revised the HLR rice fortification procedure for the enrichment of rice with thiamine, niacin and iron. The HLR process involved dissolution of vitamins in dilute sulphuric acid, application of vitamin solution to rice, application of a water insoluble substance, application of ferric pyrophosphate and talc followed by reapplication of a water insoluble layer and iron mixture before screening and packaging.

Due to increased cost of raw materials, low factory output, use of hazardous materials, reported problems with browning of enriched grains, RCL modified the procedure to include the application of the nutrient mixture to the rice followed by drying, screening and packing. The nutrient mixture was composed of ferric pyrophosphate suspended in an acid solution of the vitamins. They reported that the browning found in the HLR premix on storage was due to the formation of ferric sulphates and that this problem was eliminated with the use of an alternative acid. Acid hydrolysis at the surface of the rice grain provided a sufficiently sticky surface to ensure adhesion of the iron compound. Cost of the process was reduced and the factory output was doubled. There was no mention, however, of the stability of premix to nutrient loss through rinsing.

In the Philippines, attempts at fortifying rice by application of vitamin A followed by a water insoluble layer were abandoned since washing losses of between 10 - 20 per cent were recorded. Rice fortification with iron has also been tried in the Philippines using the coating method. The fortificant used in this programme was anhydrous ferrous sulphate. Some discolouration was found after 20 weeks of storage at room temperature and washing losses were 9 per cent.

Ferric orthophosphate (white iron) is a recommended form of iron for use in the fortification of rice. This iron compound is almost water insoluble and has been preferred for mixing with milled rice due to its white colour. When it is oxidised or contains excessive moisture it may become tan, yellow, purple and or black. Hurrel reported that the bioavailability of ferric orthophosphate varied widely from batch to batch and was highly negatively correlated with particle size as was also found with elemental iron. This fortificant compound is more expensive than anhydrous ferrous sulphate. According to Hurrel the cost of ferric orthophosphate was about six times that of anhydrous ferrous sulphate, for the same level of total iron.

In Japan, a multinutrient enriched rice has been on the market since 1981. The first step in this procedure was acid parboiling in the presence of thiamine, riboflavin, niacin, pantothenic acid and pyridoxine. The second step involved coating the grain with separate layers of vitamin E, calcium and iron in separate layers and finally a protective coating material. Modified atmosphere packaging, utilising aluminium laminate and carbon dioxide, ensured the stability of vitamin E during storage. The yellow or brown colour of the premix was not a problem in this case as it was used in the fortification of 'brown' rice. Apart from these well established procedures there have been other innovations regarding alternative methods for rice fortification. Joseph *et al.* described an enrichment procedure using a premix containing thiamine, riboflavin, niacin and pyridoxine. This procedure involved soaking the milled rice in an acid medium containing the water soluble vitamins followed by the cross linking of starch granules in the enriched grains. The cross linking procedure itself was demonstrated to have caused significant vitamin loss, but the added vitamins were highly cook and wash stable. It is possible that this method could have some utility in the future.

Use of fortified simulated grains has featured prominently in attempts at rice fortification in the developing countries. Murphy *et al.* described the production of such a fortified rice product for use in the Philippines. The synthetic rice grains were produced by extrusion of rice flour in a pasta machine. The best formulation contained vitamin A stabilised by a mixture of tocopherol, ascorbate and lipids with a low level of unsaturation.

Retinyl palmitate stabilised in an acacia matrix, type 250 SD was the fortificant used. Retention of vitamin A after washing was reported to be 100 per cent. Vitamin retention after cooking, however, ranged from 60 - 94 per cent. The formulations which demonstrated the better storage stability, particularly at high humidity, suffered greater cooking losses. Drawbacks with this technology have been reported to exist with respect to blending with the natural product and in the consistency of the simulated grains after cooking. Field trials in Brazil showed no problems with acceptability of rice enriched in this way. The enrichment of whole wheat grains with vitamin A has been attempted. The fortificant used was a premix comprised of concentrated vitamin A attached to wheat grains, for mixing at a level of 0.25 per cent with wheat grains. The feasibility of this procedure was not determined. The fortification of whole grain cereals with soluble iron compounds is difficult because they promote oxidation of the lipid component of the grain, thus reducing the shelf life.

FLOURS, CORNMEAL AND BREAD AND PASTA

In the fortification of flour the required nutrient mixture is mixed with an appropriate diluent to produce a premix, which is then accurately metered into the flour. The addition of vitamins B_1 and B_2, niacin, iron and calcium to wheat flour is a common practice in many developed countries. It is technologically feasible to add other vitamins and minerals as well. Cort *et al* successfully used two premixes to fortify wheat flour; the first comprised vitamin A, pyridoxine, folic acid, tocopherol acetate, thiamine, riboflavin, niacin and iron while the second contained calcium, magnesium and zinc. The vitamin/iron premix demonstrated excellent stability on storage. The flour enriched with both premixes also demonstrated excellent stability on storage at room temperature. Under conditions of accelerated storage at elevated temperature (45 °C), however, there was substantial loss of vitamin A beyond 4 weeks of storage. Parrish *et al.* also reported good stability of enriched wheat flour stored at room temperature, but about 50 per cent losses in flour stored at 40 °C for 6 months.

The form of vitamin A most commonly used in the fortification of flour was dry stabilised vitamin A palmitate (type 250-sd) powder form. The water soluble vitamins (thiamin, riboflavin, niacin, pyridoxine, folate and calcium pantothenate) are used in pure crystalline form. The mononitrate salt of thiamine is preferred for this use. Iron is normally used in its reduced elemental form. In the work of Parrish *et al.* ferrous sulphate was also used as the source of iron and there was no significant difference demonstrated between the flour enriched with ferrous sulphate and that enriched with elemental iron. It has been recommended that ferrous fumarate be used in the fortification of wheat flour within the P.L. 480 programme, due to its greater bioavailability. When mixing dry fortificants with dry foods, careful selection of the physical characteristics of the fortificant compound is important so as to ensure adequate mixing and to minimise segregation on storage.

Rubin *et al.* investigated the stability of bread made from flour enriched by the 6 vitamins and 4 minerals described by Cort *et al.* They found that the inclusion of calcium and magnesium adversely affected vitamin A retention during the baking process.

There was also an off-flavour detected in bread which included magnesium after 5 days of storage. Crumb colour and grain were also negatively affected by the nutrient multimixes tested by Rubin *et al.*, with the greater affect occurring with the nutrient mix which included magnesium. Fortification of bread with zinc salts exerted no adverse effect on loaf volume, flavour or any other index of bread quality.

Iodisation of bread has been carried out in the Netherlands and Tasmania by the addition of 2-4 ppm KIO_3 to the bread improver which was already in general use. KIO_3 has been used in bread production in the past, not as a fortificant but as an oxidising agent to improve dough quality. The fortification of corn meal with 6+4 vitamin/iron mixture was shown to be technically feasible. Losses of vitamin A on storage at room temperature for up to six months were greater than with wheat flour but still remained below 20 per cent. In many countries pasta or noodles are commonly eaten and these can therefore be important vehicles for fortification.

The manufacture of such products involves the production of a dough which is then extruded and dried. Enrichment can be through the use of enriched flour or alternatively, wet addition of a dispersion of the required vitamins can be carried out at the dough-making stage. Vitamin losses during production depend largely on the drying conditions employed. Dexter *et al.* enriched durum wheat flour with a vitamin mixture containing riboflavin, thiamine mononitrate and niacin for the production of spaghetti. They reported that high temperature drying treatments resulted in significant losses of riboflavin, whereas the other vitamins were stable to the processing conditions. In all cases cooking losses, estimated between 40 - 50 per cent, exceeded those experienced during processing.

BREAKFAST CEREALS

The fortification of ready to eat breakfast cereals is a wide-spread practice. In the United States the general practice has been to provide 25 per cent of the RDA of thiamine, riboflavin, niacin, pyridoxine, folate, ascorbic acid and vitamin A per 1 ounce. 10 or 25 per cent iron, up to 20 per cent RDA for calcium have also been added; vitamin D has also sometimes been included. Surveys have indicated that fortified breakfast cereals play an important role in ensuring nutritional adequacy of the diets of the U.S. population.

The minerals and the more heat stable vitamins like niacin and riboflavin have been added to the basic formula mix prior to processing. The heat labile vitamins such as vitamins A, C and thiamine are usually sprayed onto the cereals after the high temperature processes such as oven or extruder. According to Johnson *et al.* two important considerations in the design of the spray system are:

- To ensure uniformity in spray coverage
- To optimise protection of both vitamins A and D.

As the hydrophobic vitamins cannot form true aqueous solutions, they have been sprayed on in the form of an emulsion. The composition of the emulsion has been shown to greatly affect the vitamin stability. The inclusion of 15-20 per cent sucrose into the spraying emulsion stabilised the vitamin A. It acted by forming a protective coating around the vitamin on drying. This fact explained the greater stability of vitamin A included in raisin bran as compared with cornflakes. In the latter case a no-sugar spray was used. Vitamin E can be added to the spray emulsion, it was at least as stable when added under these conditions.

Good storage stability of the fortified breakfast cereals was reported. This was partly due to the fact that packaging materials with appropriate barrier properties have to be selected to ensure the maintenance of equilibrium moisture contents below about 5 per cent. Above this level there was textural deterioration of the product which was perceived as 'staleness'. Lund reported almost 60 per cent loss of vitamin C from raisin bran on storage for a 12 month period.

This was ascribed to the fact that the moisture levels in this cereal were maintained at a higher level to prevent hardening of the raisins. Improved procedures for the coating of raisins with a moisture barrier material prior to mixing with the material can minimise this problem. The addition of ascorbic acid was found to stabilise vitamin A to a greater extent than did phenolic antioxidants. The addition of tocopherols did not provide any additional stability, possibly because they were already present at optimal levels.

FUTURE DIRECTIONS OF FORTIFICATION

TECHNICAL PROBLEMS REQUIRING ATTENTION

Loss of vitamin A on storage was found to be significant in a number of fortification processes. The solution to this problem might lie in the design of a new fortificant preparation in which vitamin activity is better protected. This approach has, however, in the past led to new problems regarding cost and bioavailability. Alternatively more studies might be carried out to determine critical factors pertaining to packaging materials or storage conditions which affect vitamin content. Problems persist with iron fortification as well. Discolouration of the food vehicle due to the presence of iron hinders consumer acceptance in many cases. With the existing iron fortificants available commercially, trade-offs with respect to reactivity, bioavailability and cost must be accepted.

Sodium ferric EDTA has several useful properties in respect of fortification, but its status as a food additive is only now being considered by Codex Alimentarius Commission for use as a dietary supplement for use in supervised food fortification programmes in populations where iron-deficiency anaemia is endemic. JECFA has requested that additional studies be carried out to assess the site of deposition of iron administered in this form.

Loss of vitamin C during storage, particularly due to the presence of oxygen, is a problem with many fortified foods. Possible solutions to this problem include the development of new, more highly protected vitamin fortificants or investigation of new packaging materials or storage conditions which promote increased vitamin stability.

The use of iodised salt in food processing is another area which requires attention. There are at least two distinct issues to be considered: the first is the role of iodine compounds in deteriorative reactions such as off-flavour development and discolouration, the second is the effect of processing operations on the iodine content of the fortified food.

Some relevant data for selected food processes have been reported but further research needs to be done. If, for example, the iodine fortification of fermented vegetable products was to be considered through the use of iodised salt in their production, it would be essential to know the effect of the iodine compound on product quality as well as the stability of the iodine compound under fermentation conditions.

ANALYTICAL CONSIDERATIONS

Many advances have been made with respect to the analysis of micronutrients in foods. According to Walter problems still remain with the analysis of vitamins which occur in very small concentrations such as vitamins D_3, B_{12}, folate and biotin. Research is required into the development of appropriate analytical methods for use in the field. Such testing is a required component of quality assurance in food fortification programmes. 'Appropriate' is determined by such factors as availability of equipment and materials for testing, accuracy and precision of the testing method, simplicity and rapidity of the experimental procedure.

As analytical methodologies continue to be developed or modified there is a need to verify agreement of results obtained using different procedures. There may also be a need to standardise some aspects of experimental procedures.

HEALTH IMPLICATIONS OF MICRONUTRIENTS

Fortification practices are predicated on our understanding of our physiological requirement for the micronutrients. This understanding has continued to be expanded. The possible preventative effect of antioxidative vitamins on cardiovascular disease and cancer is currently the subject of much investigation. Based on the findings of these investigations, an increase in the recommended daily allowances for these vitamins may be indicated.

Since fortification procedures are based upon delivering a predetermined proportion of the RDA for the nutrients in question, such research will impact upon food fortification.

Up to the present time, due to high variability between and within individuals, it has not been possible to establish generally accepted cut-off values for the health protective potential of B-carotene, vitamin E and vitamin C. RDA's and National Reference Values (NRV) for an expanding number of essential trace minerals are being determined. Continued international discussion and agreement on these is essential.

NUTRITIONAL IMPLICATIONS OF NEW DIETARY TRENDS

Consumer demand for low fat products has given rise to concern that the loss of fat soluble vitamins associated with the removal of fat could lead to inadequate intake of these vitamins among certain sectors of the population.

The finding of the paper prepared by Germany for the Codex Committee on Nutrition and Foods for Special Dietary Uses was that it was premature to determine fortification requirements for lower tat products. Their recommendation was for intensive study of consumer behaviour before other measures be taken.

STANDARDISATION AND REGULATION

A thorough review of existing legislation pertaining to fortified foods needs to be carried out. This should provide the basis for suggested changes where it is found necessary. An example of the effect of legislation on fortification procedures was the use of a mixture of table salt and curing salt in the production of certain iodised sausages in Germany since iodine was allowed in table salt but not curing salt.

The General Agreement on Tariffs and Trade (GATT) Uruguay Round of Multilateral Trade Negotiations have had significant implications for the work of the Codex Alimentarius Commission. The GATT Agreements on the Application of Sanitary and Phytosanitary Measures and on Technical Barriers to Trade emphasize the need for international standards, guidelines and recommendations to facilitate international trade.

The Codex Alimentarius Commission needs to work further on the development of standards for fortified foods. Such standards must of course be consistent with the general principles of the Commission regarding fortification. International standards must be sufficiently flexible to allow national governments to meet the specific nutritional needs of target populations while still providing guidance with respect to acceptable ranges of fortification and unambiguous definition of terms.

International standards for fortified foods used in food aid programmes need to be considered. This would facilitate the interchangeable use of foods coming from different sources in meeting the needs of target groups, and could provide the standard which quality assurance programmes should demonstrate all food aid products to meet.

FOOD FORTIFICATION IN DEVELOPED COUNTRIES

Food fortification continues to be a widely used mechanism in many developed countries. In this context, it is the rapidly changing lifestyles and increasing reliance on more highly processed foods has been used to justify the addition of nutrients to an expanding range of foods in order to ensure nutritional adequacy of the diet.

It is important to note that there is no general consensus regarding the extent to which food fortification should be practised and the prevailing attitude towards it varies in the developed world.

France, the Netherlands, Norway and Finland have restrictive legislation regarding the addition of micronutrients to foods. Nutrification has, however, continued in these countries for certain processed foods. In France restoration of vitamins to compensate for processing losses to the extent of 80 - 200 per cent of naturally occurring levels, is allowed. It is also permitted to add nutrients to foods for special dietary uses.

In the Netherlands, addition of nutrients to certain foods for special dietary uses is allowed as well. Vitamin A fortification of margarine and iodine addition to salt are also practised. In Norway and Finland, special permission for nutrient addition to specific foods may be granted. In his recommendations concerning the micronutrient enrichment of foods in the Netherlands, Anderson stated that the increase in processed ready to use foods had not lowered nutrient intakes in that country and that there was no basis for extending the authorisation for food enrichment. In Germany there is currently an effort to promote the fortification of foods with iodine in order to combat iodine deficiency in that country.

In North America, fortification of foods is viewed positively. Many studies underline the role of such nutrified foods in assuring nutritional adequacy in the North American diet.

The expanding range of fortified foods can be justified by the fact that RDA's for many nutrients are commonly not met. In a study carried out on a group of women in New York, it was shown that the mean intakes of Ca, Fe and vitamin

D were all below the recommended daily allowance. Other factors can complicate food fortification policy, however, as was shown by the USA National Dairy Council's request for monitoring of Ca fortification on the basis that the required element is readily available in dairy products. In developed countries, regulation of fortification is currently receiving more attention than the technologies involved.

This is because there is a legitimate fear of over-fortification as manufacturers seek to use fortification as a marketing tool. Whatever are the issues that now surround food fortification in developed countries, there can be no question of the role which this practice has played in the past. The virtual elimination of micronutrient deficiencies from these countries has been attributed in large part to this.

Food scientists have risen to the challenges posed by micronutrient fortification. Evidence of their success is clearly demonstrated on the supermarket shelves in most developed countries.

Technical problems such as the interaction of the added nutrients with the food carrier matrix due to the pro-oxidant and catalytic properties of many of the essential minerals as well as the susceptibility of many vitamins to destruction by heat, light and oxygen have been overcome for an ever widening range of applications.

FOOD FORTIFICATION IN DEVELOPING COUNTRIES

The 1992 ICN conference held in Rome emphasised the importance of food based activities in their plan of action geared at addressing the issue of micronutrient malnutrition.

Food fortification is one of the relevant modes of action. Advantages of food fortification relative to other modes of intervention have been widely noted and a result of these is that fortification programmes can be implemented and yield results within a short period. The focus of the international community has so far been on the three most prevalent deficiencies: vitamin A, iodine and iron. There are different considerations involved in the establishment of food fortification programmes in developing countries as opposed to developed ones.

There are several steps in the implementation of a food fortification programme in developing countries. Having identified the need for nutritional intervention on behalf of a population or a sub-group thereof and the required levels of fortification, then a suitable carrier must be identified, appropriate fortificants selected, the technologies used in the fortification process determined and some mechanism put in place to determine whether the nutritional objectives of the programme are being met. Selection of the carrier is a critical step and several required characteristics of the carrier have been noted.

The identified vehicle must be consumed in roughly constant quantities throughout the year by the majority of the population. The food must pass through a centralised point to facilitate a rigidly controlled fortification process. The addition of fortificants at the required levels must not affect the organoleptic qualities of the food.

It is important to emphasise that micronutrient fortification is not universally applicable. In some cases the high degree of decentralisation of food processing activities makes it impossible to regulate or even establish a viable fortification programme. In cases where subsistence farming is practised, fortification is not likely to be a practicable solution to nutrient deficiencies. It is imperative, before embarking on food fortification activity, to carefully define the context in which it is indicated.

3

Genetic Diversity and Agricultural Biotechnology

CONSERVATION OF PLANT GENETIC RESOURCES

The conservation of plant genetic resources has, in recent years, attracted growing public and scientific interest and political support. There is an increasing awareness of the relevance of biological diversity and its conservation to the health of the biosphere. Particularly urgent is the need to raise agricultural output to meet the basic nutritional needs of increasing populations.

The dilemma is that the required increase in food production must be obtained through 'sustainable' forms of agriculture which are less dependent on the use of modern high-yielding varieties bred for intensive, high-input systems. It seems unlikely that increased food production on this scale will come from increasing the area under cultivation. Rather, major solutions will depend upon innovations which exploit sophisticated methods to manipulate plant genomes and the totality of natural genetic variation.

The area of disease resistance will serve to illustrate this argument. Many authors have emphasized the importance of innovation in the area of crop protection since crop losses due to pests and disease may account for between 20 and 40 poe cent of productivity worldwide.

Indeed, over 50 poe cent of the research and development (R andD) spent in the plant breeding/seed industry is focused on the identification and incorporation of resistance traits. In a situation where there is some evidence that the pool of available variation for disease resistance genes within commercial breeding lines is becoming limited (e.g. for yellow rust in cereals) the search is on for novel sources of resistance.

There are now many examples of biotechnology being put to the service of pest control, through the incorporation into plants of transgenes for toxins, antifeedants, enzyme inhibitors, etc. But, as Woolhouse (1992) has pointed out, such is the long coevolutionary history of pests and pathogens with their hosts that present strategies for interfering with these relationships represent only a small proportion of the likely defence mechanisms used by plants in resisting foreign organisms. This enormous diversity of mechanisms represents an arsenal of ammunition which can be used in the fight to minimize disease and predation.

Are there any hard facts which support these assertions concerning the value of genetic resources to agricultural productivity? Can we put a cost on the loss of diversity? The sources of variation used by the plant breeder range from current breeding lines to wild species and the products of direct gene manipulation. In a recent analysis of the plant breeding/seed industry, Swanson (1996) showed that, over a five-year period, 6.5 poe cent of all genetic research within this industry, resulting in a marketed innovation, was concerned with germplasm from wild species and landraces.

This compares with only 2.2 poe cent of new germplasm arising from technological approaches of induced mutation. The vast majority of germplasm successfully used by the industry still arises from 'conventional' sources, i.e. the heavily exploited commercial cultivars, which at first sight might be taken to suggest that wild resources are relatively unimportant in R andD.

In fact Swanson (1996) argues just the opposite, suggesting that the stock of existing commercial varieties should be seen as the information base from which bioindustries develop innovations, whereas the sources of new diversity (wild species, landraces, induced mutation) should be seen as supplying increments to the information base. In other words, R andD requires an annual injection of 'new' genetic material from natural sources, estimated as amounting to approximately 7 poe cent of the stock of material currently within the system.

Biodiversity is valuable to industries other than those that seek novel genes for crop plant improvement. Biodiversity has traditionally provided a source of compounds to the pharmaceutical, foodstuffs, crop protection and other industries and here it is possible to give more precise evaluations of its importance. As outlined in Chapter of this volume, drug sales based on natural products from plants were estimated at $US43 billion in 1985 alone and the value of yet undiscovered pharmaceuticals in tropical rainforests is estimated to be as much as $US147 billion to society as a whole. Vast markets exist for the replacement of synthetic biocides, many of which have associated toxicological and environmental problems, by 'natural' compounds, which may have more specific modes of action and fewer side-effects.

Even greater may be the use of genetic resources in more fundamental research where their value in the longer term may be unknown or impossible to estimate accurately at the present time. Diverse genetic material is needed and now regularly utilized to create mapping populations and to study inheritance of a wide range of traits which may or may not be of immediate practical value. Very large numbers of germplasm samples are regularly screened and evaluated for different characteristics so that the underlying molecular biology can be investigated. The intention here is that a greater understanding of fundamental biology will give rise to wealth creation in the future.

APPROACHES OF DIVERSITY

The criticisms levelled at those who believe that biotechniques herald a new 'sustainable agriculture' is that existing problems of uniformity will be exacerbated. Can the transition from BCM to BML mode re-introduce diversity into a system based on uniformity? From a purely technical view, biotechniques' capacity to draw upon genetic material from the tertiary genepool would suggest that additional genetic variation might be introduced. Thus, Bassett argues that new varieties 'will simply coexist with the old varieties: diversity will have been increased, not decreased'.

But this approach has two major shortcomings. Firstly, the basic research and the application are inextricably linked, so much so that possible applications are driving the direction of basic research. Thus, a growing proportion of public sector research is supporting commercially oriented research. Secondly, what actually happens is a subset of what could happen, as the earlier case-studies have argued. Kloppenburg bluntly states: 'the baby of biotechnology is not so easily separated from the corporate bath water,' which is exactly why, as Lesser points out, trends towards uniformity predate the existence of PBR and patents.

Strengthening IPR as applied to living material has encouraged private industry to engage in biotechnical research at levels above those that would have existed in their absence. Furthermore, because Governments now see biotechnical prowess as important for maintaining competitiveness, public research is beginning to resemble privately sponsored research, either through its increasingly subservient position to private industry, or through more overt aims to generate revenue from patentable research outputs. Erstwhile President of Harvard University, Derek Bok, has expressed concern that Universities will increasingly 'differ from corporations only because there are no shareholders and no dividends'.

An element of historical contingency is at work here, since many governments are cutting state spending on education and research where it is perceived to be of low market value. To the extent that environmental issues are intimately related to issues of social welfare, and because many environmental costs are not captured in market transactions, one assumes that private organisations are, notwithstanding their own public relations, less likely to integrate environmental issues into their research programmes. Indeed, one survey, aimed at eliciting the ranking of breeding companies' priorities, placed the environment at the bottom of seven criteria.

However, the same is increasingly true of publicly funded research. Institutional changes are making the public / private distinction irrelevant in predicting the social welfare goals which will be pursued by one or other form. This is one reason why, notwithstanding some of the claims made for the efficacy of biotechnologies (most of which, incidentally, are made with the BCM technologies as the implied baseline), IPR will if not increase, then maintain, the vulnerability of agriculture in the field. Other reasons include the following:

Distinctness of Varieties

Already, anecdotal evidence suggests that seed companies rely on a few elite cultivars in their research programmes and new varieties are developed through minor modifications to these. Very little hard evidence is available, but it is clear that the number of varieties available:

- Does not reflect genetic diversity,
- Masks the concentration in varietal use out in the field.

The future strategy for breeders will be structured by the IPR legislation in place. The combination of UPOV 1991 and patents ensures that the work of the breeder, and the value of an identified gene, is both recognised. The 'essentially derived' clause was introduced to deal with the problem of genetic distancing. Biotechniques make it possible to reduce the genetic distancing required to discriminate between varieties, in which case, the genetic variation existing in the field would become even further divorced from consideration of the number of varieties grown.

For example, since distinctness could now be measured at the level of the gene, a superfluous (in agronomic terms) gene could be spliced into a variety thus making it, potentially, distinct. As Smith points out, the practice of reverse engineering of varieties is increasingly common and would make the aforementioned practice more likely, rendering the granting of PBR meaningless. The 'essentially derived' clause, though of theoretical value, raises important questions of definition. For both Espenhain and Smith, who, in his excellent account, notes that numerous controversies in this regard are in no one's interest, case law will provide the answers.

Because UPOV 1991 extends the breeder's right to the commercialisation of essentially derived varieties (the principle of dependence), companies using genetic transformation techniques will either work with their own PBR-protected varieties, or license genes of interest to other companies for incorporation into their varieties. Strategic alliances between those specialising in biotechniques and those with greater specialism in traditional breeding seem likely.

Hence, Pioneer markets both soyabeans containing Monsanto's Roundup Ready gene (tolerant of Monsanto's glyphosate herbicide) and soyabeans containing the DuPont-owned sulfonylurea tolerant gene. DEKALB also has cross-licensing agreements with both Monsanto and DuPont. Those licensing technologies will seek to gain from the technology premium which biotechnically developed products aim to attract. To a significant degree, varietal make-up will remain as before, but with genes spliced into a particular variety's background, and with plants themselves being made more uniform. However, since industrial structure will be affected by the evolving IPR framework, as well as the techniques themselves, and developments in individual sectors, there may be implications for the diversity of what is offered to farmers. This is considered in what follows.

Research Concentration

Patent enforcement prevents companies from carrying out research on a patented genetic sequence or process without paying royalties to the patent holder, whilst essentially derived varieties are subject to PBR under UPOV 1991. The strategic importance of patents for any company depends to a considerable degree on the extent of exclusion implied by the text of the patent. In this context, the current trend towards granting broad patents to companies is worrying indeed.

For example, Agracetus, a subsidiary of W. R. Grace (and recently taken over by Monsanto), was granted patent rights in the US over any genetically manipulated variety of cotton, and by the European Patent Office over all genetically transformed soybeans (with those for rice, groundnut and maize pending). Agrigenetics' patent on high oleic acid sunflowers effectively stopped all such work in this area outside the company. The prevalence of 'driftnet patenting' is at odds with the view held by many that patents encourage innovation. It raises the possibility that the seed industry for any one crop may ultimately become dominated, or at least hostage to, one commercial enterprise.

Patent enforcement is an important tool in building corporate empires and eliminating competition. Monsanto, which has staked much on its quest to become the 'Microsoft of engineered foods', has acquired companies, and stakes in others, reflecting its belief that patents will be a key source of competitive advantage in coming years. Pioneer's moves to patent its in-bred lines on grounds that this would prevent other companies carrying out research on them illustrates that even the 'biological patent' which hybrid corn varieties are endowed with is being superseded by strengthened IPR legislation.

As patents proliferate, it will become increasingly difficult for any enterprise to conduct research in the full knowledge that it is legitimate, especially since burden-of-proof legislation makes it incumbent upon those accused of patent violation to prove that they are innocent. The possibility arises in which a company carrying out research happens, by accident, to be working with a variety in which a patented gene sequence exists. Such a company would be unwittingly breaking the law, and would be expected to provide proof of its innocence.

There is a suggestion that IPR-related concerns are driving strategic alliances in the industry. Since, increasingly, more than one form of IPR will be involved in developing a given variety, such alliances reduce the likelihood of any one IPR-holder blocking development of the product concerned. IPR-based restructuring can be expected to produce 'many casualties, some survivors, and a few successes'. The growing significance of IPR, appears to be leading to greater concentration in breeding effort, which is unlikely to promote diversity in agriculture.

Loss of Farmers' Privilege

UPOV 1991 appears to deny that farmers might also be breeders. Industry estimates suggest farmer-saved seed accounts for between one and two thirds of all seed planted in the world, though the proportion for developing countries is believed to be of the order 85 per cent. Indeed, in developing countries, the exchange of seed from farmer to farmer is probably the main avenue for diffusion of new seed varieties. Erstwhile Director General of GATT, Peter Sutherland, has suggested that such informal practices are 'generally not of interest to the owners of protected varieties', yet already in the US, court actions have been taken against farmers involved in such activity, and it will not have been lost on IPR owners that interfarm sales constitute 62 per cent of all seed purchases in India where Sutherland made his speech.

In the US, companies such as Monsanto require growers of their Roundup Ready Soybeans to be licensed to grow the material, though other companies, including AgrEvo and DuPont, do not require licensing since, in the words of one commentator, 'they seem to feel that the additional chemical usage that's tied in with [DuPont's] STS beans is enough'. Monsanto states that 'if necessary, the terms of the [Roundup Ready] contract will be enforced under the P.V.P.A., US Patent Law and general contract law'. One study estimates that within a few years, 40 per cent of all US farmers will be contract growers, or renters of germplasm from the same companies to whom they sell their product.

Farmers will also need to be alive to the possibility, especially when growing outbreeding crops (those which cross-pollinate) of falling foul of patent legislation. It is not clear, given reverse burden of proof, how the law would interpret a situation where a farmer grew a variety which through cross-pollination contained a patented sequence. Potentially, the onus will be placed upon farmers to ensure no such cross-pollination takes place. Furthermore, the possibilities for farmers to experiment with varieties covered by UPOV 1991 seem limited.

As with the case of 'essentially derived' varieties, it seems likely that case law will determine what is and is not allowed, but in the meantime, farmer experimentation with new varieties, which can create new races of out-breeding crops, may be a risky enterprise. This may have implications for diversity. More recently, the patent awarded to Delta and Pine Land Company on so-called 'terminator technology' (which prevents seeds from germinating in the next generation) provides a technological means through which to prevent farmer seed-saving.

Wide use of Agronomic Genes

It has been stated that the introduction of novel genes will increase genetic diversity in agriculture. However, if the same gene is licensed to several companies for use in a large proportion of varieties in use, and in different crops, the gene becomes a component of uniformity. The possibility arises of the occurrence of a southern corn leaf blight on a more global scale. In China, 15 million hectares are planted to hybrid rice, each plant possessing, as with US corn varieties in 1970, a common gene for cytoplasmic male sterility. Genes from the bacterium, *Bacillus thuringiensis*, which produce a protein that is lethal to some insects upon ingestion, are of great interest to corporations involved in biopesticides. Yet there are already concerns for insects' resistance to a number of strains of the bacterium.

Herbicide tolerant genes could quite conceivably be transferred into vast areas of crop land, potentially increasing the vulnerability of crops globally, and leading to heightened problems with herbicide resistant weeds. Indeed, the relay race mentality of the BCM mode is accepted as a matter of course: 'in 50 years, biotechnologists will almost certainly still be developing new batteries of pest- and disease-resistant genes'. The aim to engineer plants with tolerance to herbicides is a goal of companies that have integrated crop protection and seed production. In this way, purchasers of seed would be locked-in to the purchase of proprietary chemicals.

This strategy reflects the fact that the costs of developing new agrochemicals are increasing owing to costly approvals processes. The costs of engineering the seed to lengthen the effective life of a given chemical compound are less, whilst it is also possible to extend the patent life of chemicals coming 'off patent' by specifying use of the proprietary form of a generic compound. The industry claims to have shifted emphasis to compounds of lower toxicity, although much attention has focused on glyphosate, which was listed by the US National Academy of Sciences as a potential carcinogen in 1987.

Ecological Interactions

One of the major fears of environmentalists is that a gene which has been transferred to a variety to enhance its competitive performance may be transferred sexually into wild relatives, especially where maize, potatoes, rice, chickpeas and common beans are concerned. In these crops a wild-weed-crop complex is observed in which there is continual gene flow between wild and cultivated forms. More speculatively (the processes are poorly understood), horizontal gene flow mediated by micro-organisms may occur. In either case, the transfer of one gene may be sufficient for a plant to become invasive.

Lack of Research on Diversity

Quite apart from the tendencies remarked upon above which might exacerbate trends towards uniformity, the simple fact remains that little if any research associated with the use of biotechniques is being undertaken to encourage the use of diversity in the field. Indeed, biotechniques make it possible to clone plants and seeds so eliminating what residual variability there may have been in a crop bred using traditional methods.

Existing Seed Marketing Legislation

Perhaps most importantly, and what may in time become the most concrete expression of the way in which biotechniques affect the offerings of the seed industry to farmers in the fact that over time, varieties which are not the product of genetic manipulation will slip off existing National Lists of seed varieties. Since those on the list are the only ones which can enter into commerce, slowly it will become impossible to purchase seeds which are not genetically engineered.

This will force farmers and consumers alike, irrespective of concerns regarding genetic engineering, to purchase genetically engineered seeds. To the extent that genetic engineering concentrates on the integration of specific sequences within existing elite cultivars, irrespective of the number of cultivars made available to farmers, the genetic diversity within farming may decline, and certainly seems unlikely to increase.

Possibly, IPR will not increase uniformity. One scenario would see patented genes integrated into the background of existing varieties, and no change other than those created by the introduced genes. But this scenario assumes an unlikely scenario in which changes wrought by IPR and biotechniques will leave the seed industry unchanged in other respects, a scenario which current trends suggest is unlikely.

Our inability to measure diversity, and the fact that we do not know where we stand today, makes it impossible to assess change on the basis of any reliable baseline. However, it seems reasonable to suggest that biotechniques are taking agriculture in different directions to those which would be implied by the alternative approaches. If the BCM mode was environmentally damaging, and if there remain unanswered questions regarding the impact of the BML mode, why does this new mode appear to be gaining the support of most FAROs, public and private, especially when alternative paths exist?

ALTERNATIVE AGRICULTURE

In Mellon's words 'the hype surrounding biotechnology diverts our attention from those [pesticide free] solutions by focusing attention on technologically dazzling new products. By setting the proper goal, we will avoid the danger of spending millions trying to genetically engineer ten "better" pesticides, when for far less we could have taken our agriculture systems off the pesticide treadmill forever.' In particular, the alternatives (re-)emerged with some force in the 1980s on the back of environmental concern seem likely to remain alternatives in terms of the resources devoted to them. The transition from BCM to BML mode must be understood in this context.

This was not a transition that was inevitable. The BCM was under fire, and alternative approaches beyond biotechniques were available. If IPR was essential to the BML mode, they may be decidedly unhelpful for those seeking to do research outside that mode. Sederoff and Meagher opine that IPR 'are having a dramatic negative effect on the progress of non-profit research'. The following sections consider debates concerning genetic resources which have taken place outside GATT and UPOV.

AGRICULTURE AND PROSPECTS OF COLD TOLERANT GENES

Cold is an environmental factor that limits the geographical distribution and growing season of many plant species, and it often adversely affects crop quality and productivity. Most temperate plants can acquire tolerance to freezing temperature by a prior exposure to low nonfreezing temperature, a process known as cold acclimatization.

Plants of tropical and subtropical origins are sensitive to chilling temperature (0°C-100°C) and are incapable of cold acclimation. Many studies have suggested that cold regulated gene expression is critical in plants for both chilling tolerance a and cold acclimation.

Cold responsive genes encode a diverse array of proteins such as enzymes involved in respiration and metabolism of carbohydrates, lipids, phenylpropanoids and antioxidants: molecular chaperones, antifreeze proteins, and others with a presumed function in tolerance to dehydration caused by freezing. The change in the gene expression occur in plant during cold acclimatization a developmental process that results in increased tolerance. Since then, it has repeatedly been speculated that certain COR (cold regulated) genes might have role in freezing tolerance. To test this notion investigators have turned to isolating the characterizing genes that are expressed in response to low temperature.

These efforts have led to the identification of a number of genes such as the COR 15a KINI LTI 78 fad 7 genes of Arabidopsis thaliana. It is well known to farmers and scientists that low temperatures can kill plants. Low temperature stress is a major environmental factor that not only limits where crops can be grown but also reduces yields depending on the weather in a particular growing season.

In addition to exceptionally stressful years that cause significant yield reductions, less extreme stress almost certainly causes smaller losses over large areas to produce comparable yield reductions every year. Even in cases when freezing stress does not result in yield loses., it often results in crop quality reduction.

Each year, worldwide losses in crop production due to low temperature damage amount to approximately $ 2 billion. Some of the major losses include the 1995 early fall frosts in the US which caused losses of over $1 billion to corn and soyabeans. The occasional freezes in Florida have shifted the citrus belt further south, and California sustained $650M of damage in 1998 to the citrus crop due to a winter freeze.

The inherent cold hardiness of the crop determines in which agricultural areas it can be grown. Crops that are more resistant to freezing stress would allow some geographical regions to grow more profitable and productive crops with less environmental risks. However despite continued efforts, traditional breeding has had only limited success in imparting crop plants with better freezing tolerance due to very little was known about the mechanisms that regulate chilling and freezing tolerance. With the advent of molecular genetics and biotechnology, it is now possible to genetically engineer plants to be more tolerant to many environmental adversities, including low temperature.

NOVEL TECHNOLOGIES FOR BIODIVERSITY ASSESSMENT AND MONITORING

The analysis and characterization of genetic diversity is fundamental to any *ex situ* or *in situ* conservation strategy. Over the years the methods for detecting genetic diversity have expanded from Mendelian analyses of discrete morphological variants, to biometrical approaches, biochemical techniques based upon protein and isozyme profiles, to methods based on DNA sequence variation.

There is now a huge array of molecular marker techniques, and Chapter reviews the major technologies and their appropriate use in addressing key issues in germplasm conservation The application of molecular technologies to increase our understanding of biodiversity at several levels is exemplified by Chapter, which is concerned with the population level, which is concerned with molecular approaches to genetic relatedness.

Chapter reviews how the collection of diverse germplasm for *ex situ* purposes and the development of effective *in situ* collections both require an understanding of the spatial and temporal aspects of population structure and associated genetic diversity and the factors that control it.

Studies with low- or single-copy markers are particularly valuable in answering questions about genetic relatedness between species. Chapter illustrates how comparative molecular mapping provides new opportunities for exploring the genetical basis of phenotypic variation within and between plant species. The implication of comparative mapping is that genes controlling specific phenotypic traits will have corresponding map locations in different taxa. This provides a new framework for the identification of homologous genes in germplasm and new tools for the effective utilization of germplasm in crop plant improvement.

This approach is also valuable in the study of quantitative traits controlled by several genes since flanking markers for major genes provide markers for quantitative trait loci (QTL) manipulation and selection and the map-based cloning of at least some genes controlling quantitative traits.

Biotechnological Tools for Conservation

Biotechnology can make a substantial contribution to the quality of germplasm collections and their efficient management. For example, in the management of very large germplasm collections there is a clear requirement for procedures for fast, reliable taxonomic identification of material and the elimination of redundancy. Samples characterized by molecular criteria may prove superior to traditional analyses of phenotype and examples of the use of DNA markers to guide the management of germplasm are explored in Chapter by reference to one of the world's largest collections of germplasm at the International Rice Research Institute.

Not all plant species can be stored in the form of desiccated seed. Socalled 'recalcitrant' species have to be kept in moist, warm conditions and even then have relatively low longevity. In addition, there are many cultivated plants which must be maintained clonally, such as banana, which produce few seeds, or potato, which does not come true from seed.

To these traditional problem species we must now add the products of biotechnology, new categories of germplasm including clones of elite genotypes, and transformed cell lines, for which storage of seed is inappropriate. Biotechnology has impact on storage technologies in two quite different ways. The first of these, concerns the various types of *in vitro* culture which allow the propagation of plant material in an aseptic environment, thus ensuring the production of disease-free stocks and simplifying quarantine procedures for the international exchange of germplasm. The development of sensitive molecular probes to detect low levels of pathogens.

A more radical approach to safe storage and distribution of germplasm, and possibly the only way to significantly reduce the rate of loss in biodiverse tropical forests, within the bounds of available resources, is to develop alternatives to the costly (and inherently vulnerable) methods of *ex situ* storage of seed or tissue cultures, by conserving germplasm in the form of DNA or DNA-rich materials.

The latter include those dried specimens conserved in the vast resources of the world's herbaria.

The DNA Bank-Net, an international network of DNA repositories, has now been established with both storage and distribution functions. Although technologies for rapid removal, analysis, amplification and secure storage of small amounts of DNA are now well advanced, research is still needed on methods to amplify the entire genomic DNA of a species, which means that storage of DNA or DNA-rich materials should still be considered to be something of an insurance policy rather than a replacement for conventional modes of germplasm storage.

Biotechnology-assisted Crop Improvement

Effective use of genetic resources in plant breeding programmes requires methods to determine whether useful genetic variation exists, and costeffective methods to introduce useful genes into the material. Biotechnology has made a contribution to both these aspects. The use of genetic maps based on molecular markers linked to genes of interest facilitates the more rapid selection of both simple and complex traits thus accelerating their incorporation into breeding materials. Where traits are controlled by one or a few genes, direct isolation of genes and their incorporation into elite cultivars by increasingly efficient transgenic technologies provide an additional approach to the more traditional methods of plant breeding. This prospect also challenges one of the classical concepts in plant genetic resources work, namely the 'gene pool' concept of Harlan and de Wet (1971), which separated cultivated species and their wild relatives into different categories based upon their 'crossability'. The value of this concept in guiding the work of the plant breeder must now be re-evaluated in the light of molecular genetic technologies which have not only cast a completely new light on the relatedness of crop plant genomes, but which have effectively widened the potential gene pool of a species to include all other life forms.

What challenges does this imply for the traditional plant genetic resources programmes? Why is the emphasis on conserving species that are related to crops when more exotic and potentially valuable genes may be present in liverworts, or green algae, or other less studied taxa?

BIOPROSPECTING FOR NOVEL GENES

The value of genetic resources as a source of novel compounds for various industries has been alluded to above. For centuries biodiversity has provided sources of fuels, medicines, foodstuffs, pesticides, etc. The high returns available from introducing a new drug, for example, plus the realization that biodiversity is rapidly being lost (estimated rates of extinction range from 30 to 300 species per day) has encouraged new attitudes to the exploration and exploitation of biodiversity through 'bioprospecting'. How does biotechnology impact on this traditional area? A fundamental operation in any bioprospecting programme is the screening of thousands of extracts for compounds of interest. Gene technology now allows the speeding up of screening programmes for new compounds through the development of more sophisticated *in vitro* assays. For example, the genes encoding receptor proteins for certain classes of drug, or enzymes that may serve as targets for novel pesticides, may be cloned and expressed on a large scale in high-throughput *in vitro* assays into which thousands of plant extracts may be applied. Bioprospecting is also concerned with discovering novel genes and natural variants of genes or gene products that may either be used directly or to guide sequence improvement by molecular methods. Chapter gives examples of amino acid sequence improvement in peptide hormones as a result of screening homologous genes from the wild, and this area of biotechnology is likely to develop in parallel with new molecular methods of generating random variation through employing the mutational potential of the polymerase chain reaction (PCR) – i.e. 'forced molecular evolution' – or recombinatorial libraries of peptides.

Whilst gene prospecting has provided a new perspective on biodiversity, the area is not without political controversy because these advances tend to favour agriculture and other industries of developed countries, and may actually displace traditional commodity production in resource-rich but underdeveloped countries. Chapter also outlines the development of strategies which integrate conservation and exploitation with economic returns and sustainable development in those countries from which the bioresources were first obtained.

THE EXCHANGE OF INFORMATION AND GERMPLASM

The value of all plant genetic resources lies in their utilization in crop improvement programmes or in other bioindustries. The vast majority of conserved germplasm is available for distribution to the global scientific community but this international distribution poses considerable risks of accidental introduction of non-indigenous plant pathogens and pests.

The detection and elimination of such disease-causing agents is therefore essential.

Often, such agents will be present at very low levels and may actually be symptomless. In this context then, new technologies reviewed in Chapter have been of great value in the development of reliable and sensitive nucleic acid- or antibody-based diagnostic probes, and these will be of great importance in optimizing the health and value of germplasm collections.

Finally, if the biotechnology revolution has had great impact in developing novel molecular and cellular technologies, no less a revolution has affected radically the way in which the world organizes and disseminates information.

Powerful computers linked by the Internet have facilitated the development of sophisticated and comprehensive informatics systems which enable information on plant genetics and molecular biology to be cross-related to systematic, ecological and other data through international networks. Chapter explains how this information is organized and distributed against a background of continual development.

MOLECULAR MARKERS AND THE MANAGEMENT OF GENETIC RESOURCES IN SEED GENEBANKS

Plant germplasm has been accumulated over many decades, and is now stored in genebanks in countries around the world. The United States National Plant Germplasm System reputedly holds more than 380 000 different accessions of some 8700 species of plant, within which, for example, there are more than 110 000 accessions of wild and cultivated cereals held in the Small Grains Collection in Idaho.

The Consultative Group on International Agricultural Research (CGIAR) centres conserve about 500 000 germplasm samples of more than 30 crop species and wild relatives, mostly as *ex situ* seed collections, but also as field genebanks and *in vitro* collections. The International Rice Genebank at the International Rice Research Institute (IRRI) has a collection of more than 80 000 samples of rice alone from more than 110 countries comprising landrace varieties, breeding lines and commercial varieties of *Oryza sativa*, landrace varieties of *O. glaberrima*, and all 20 wild species of the genus *Oryza*.

Since 1973 over 740 000 packets of rice have been distributed throughout the world for use in basic and applied research, and this germplasm has contributed to improvements in many characteristics of new rice varieties. Pressure for germplasm distribution will increase over the next 30 years as plant scientists strive to meet the demands for a 70 poe cent increase in rice production by the year 2025.

The management of such collections is difficult due simply to their vast size, and there is a clear requirement for the development of procedures which utilize fast and reliable methods for the identification of material, the measurement of diversity, or for the determination of redundancy, in order to facilitate the organization and prioritization of germplasm. Even more important may be the need to rapidly and efficiently identify the most appropriate rice germplasm for research and crop improvement.

In recent years there has been an explosion of new DNA-based marker methods such as restriction fragment length polymorphism (RFLP) analysis, and those utilizing the polymerase chain reaction (PCR) such as RAPD (random amplification of polymorphic DNA). RAPD technology has been used successfully for measuring diversity in plants, and the patterns of variation observed have been shown to closely resemble those obtained using more classical characters.

A range of other PCR-based techniques suitable for the measurement of diversity have been developed over the last few years. Some of these target repeat regions of the genome to produce markers. Single primers complementary to minisatellite repeats provide multilocus markers. More widespread is the use of pairs of primers complementary to sequences flanking microsatellite repeats. This strategy produces single locus markers which, because of the hypervariability of microsatellite repeat lengths, is an efficient method for the detection of polymorphisms.

Yang *et al.* (1994) used PCR technology to identify microsatellite polymorphism across landraces and cultivars of rice, whilst Wu and Tanksley (1993) have reported the identification of microsatellite alleles that are specific to either indica or japonica rices. Recently, an additional PCR-based marker technique – amplified fragment length polymorphism (AFLP) analysis – has been developed, which results in the selective amplification of restriction fragments from within a total digest of genomic DNA to yield typically 50–100 dominant marker bands per polyacrylamide gel track. This method seems certain to find wide application in the study of plant diversity.

Germplasm collections in many parts of the world, like that at IRRI, are faced with challenges related to the various activities which must take place within the genebank, and these are exacerbated by the size of some of the collections and the number of accessions which are conserved.

It is now clear that some of the constraints could be alleviated by the application of the molecular marker technology which is becoming more and more readily available.

Examples in which molecular markers may be suitably employed to assist genebank management, organization and the way that material is accessed include:

- The accurate identification of germplasm.
- The routine maintenance of germplasm, which is a continuous processinvolving seed viability testing, rejuvenation and replenishment of stocks, which will be streamlined by the identification of duplicates and the development of core collections.
- The selection of germplasm for safety storage at other genebanks.
- The choice of germplasm for use by breeders and other researchers involved in making crosses, and mapping, identifying and isolating genes of interest.

Molecular markers are, of course, being used very successfully for the assessment of genetic diversity amongst genetic resources of an increasing number of crop species and wild relatives, and this may take place either before germplasm is accepted into a genebank for storage, or after storage has been initiated. This molecular assessment of 'biodiversity' is dealt with extensively in other chapters.

OPERATIONS WITHIN A SEED GENEBANK

Confining the discussion only to the primary concerns of a genebank manager, we can identify four principal areas of activity. First is the acquisition of seeds, which involves receiving material as a result of exploration and collection activities or exchange, its preparation for storage, documentation and quarantine procedures. Secondly, seeds are conserved in the form of active/working or base collections, or duplicate safety collections. Thirdly, there are seed management activities including multiplication, regeneration, viability testing, characterization and seed distribution. Finally various activities may take place related to the utilization of germplasm such as detailed evaluation or enhancement.

Taking a more detailed example, the International Rice Genebank at IRRI has as a primary aim the conservation and continued availability of genetic resources for rice improvement worldwide. The germplasm is freely available on request, and is used to continually restore valuable material which has been lost in the country of origin. Base and active collections are maintained, and provision is also made for duplicate 'black box' storage of germplasm at the National Seed Storage Laboratory, Fort Collins, USA. Incoming germplasm is examined by the Seed Health Unit at IRRI, and viability testing is carried out. As and when necessary, germplasm is rejuvenated and multiplied to produce high-quality seed for longterm conservation and for distribution in response to requests. Wild species receive particular attention and are always grown in pots in a quarantine screenhouse, and perennial species are often maintained as living plants for prolonged periods if seeds are difficult to produce. Seed samples – 10 g each for cultivated rices, but only 10 seeds for the wild rices – are routinely sent out on request, with more than 740 000 since 1973. When no specific samples are requested, judgements often have to be made as to which material is most appropriate and will most effectively satisfy the demands of the recipient.

No genebank should be a 'museum collection'. The material conserved must be characterized not only to distinguish species, taxonomic groupings and varieties, but also to facilitate preliminary selection of germplasm by end-users. To this end, morphological and agronomic characters are scored in small field plots at IRRI using standard descriptors which will allow the rational choice of material for exchange and distribution, while the maintenance of passport data permits selection of germplasm on an ecogeographical basis. In addition to such routine characterization, IRRI scientists have also screened thousands of accessions for resistance to pests and diseases, and tolerance for different abiotic stresses, an evaluation process which requires more or less elaborate testing of germplasm in the laboratory or field trials.

REGULATION OF AGRICULTURAL BIOTECHNOLOGY

For traditionally bred plants, regulators rely on plant breeders to conduct appropriate safety testing and to be the first line of defence against genetic alterations that might prove dangerous. The same should be true for agricultural products developed using biotechnology. There is no reason for regulators to treat biotech plants any differently than traditionally bred plants, particularly given the fact that biotechnology provides greater control over gene manipulation. Despite fears about gene manipulation, traditional crossbreeding has altered plant genes and improved the human diet for the past 30 years. In fact, people worldwide safely consume these plants every day. Examples include wheat, corn, potatoes, tomatoes, and countless other staples of the American diet. In 1984, the biotechnology industry began experimenting with "gene-spliced plants" — the more advanced approach to gene manipulation that we now call "biotechnology."

DEPARTMENT OF AGRICULTURE

USDA regulates the release of all genetically engineered agricultural plants under statutes giving USDA's Animal and Plant Health Inspection Service (APHIS) the authority to regulate plants that may be or may become weeds or other nuisances — what the statutes call "plant pests." Although the rules apply in a general sense to novel or exotic varieties of both gene-spliced and conventional plants, APHIS typically only requires field testing of conventional plants that are new to a particular U.S. ecosystem (transplanted from another continent, for example).

However, all genetically engineered agricultural plants face a higher regulatory bar — making it more difficult to expand the use of biotechnology. Genetically engineered crops must be field tested under APHIS regulations prior to commercialization. For example, a new variety of corn produced with conventional hybridization requires no government-mandated field testing, but all new varieties of genetically engineered corn do, even though there is no logical reason for the regulatory disparity. For most genetically engineered plants, APHIS requires the company producing the plants to submit notice detailing the gene or genes that have been inserted, where the plant will be tested, and other relevant characteristics of the plant prior to receiving permission to conduct the field trials. Once the company completes field testing, APHIS reviews the results and makes a determination on whether or not the product should be "deregulated" and can be released into the market.

REGULATORY SCHEME

At the time, the White House Office of Science and Technology Policy began crafting a framework wherein existing federal agencies would regulate genetically engineered organisms on the basis of their characteristics, not the method of production — thus wisely deferring to the scientific community's judgement that regulation ought to address the products, not the process of biotechnology.

Engineered organisms would not require extra scrutiny simply because genetic engineering produced them (the process). Instead, they would be subject to heightened scrutiny only if the individual organisms expressed characteristics that posed some conceptually heightened risk (the product). The federal government divided regulatory jurisdiction among agencies already involved in agricultural, food, and environmental regulation. These include the United States Department of Agriculture (USDA), the Environmental Protection Agency (EPA), and the Food and Drug Administration (FDA). While each of these agencies considers the characteristics of individual products in their regulation, only FDA followed the general scientific thinking that genetically engineered and non-genetically engineered products should be regulated similarly. Both USDA and EPA automatically subject all genetically engineered plants as a class to premarket approval requirements not ordinarily applied to conventionally bred plants.

ENVIRONMENTAL PROTECTION AGENCY

EPA regulates plants that are modified to produce substances that act like pesticides: that is, substances used by a plant to protect itself from pests, such as insects, viruses, and fungi. EPA's proposed rule for these "plant-incorporated protectants" is not yet finalized. In the interim, however, EPA has regulated such plants by applying its proposed guidelines, which are functionally similar to rules for the registration of synthetic chemical pesticides. Again, biotech products face higher regulatory hurdles, even though plant-incorporated protectants developed through conventional breeding are exempt from these requirements.

HEALTH AND HUMAN SERVICES: FOOD AND DRUG ADMINISTRATION

FDA is responsible for ensuring that food items, including foods derived from genetically modified plants, are safe to eat. Following the general regulatory framework that emphasizes product regulation rather than process regulation, FDA rightly does not treat foods derived from genetically engineered plants to be inherently unsafe. Food producers are not required to seek premarket approval from FDA unless there is a substantive reason to believe that the novel trait (or traits) in the food pose a safety question. The initial determination of safety is left to the producer, but FDA has encouraged producers to consult with agency scientists prior to marketing a food produced with biotechnology to ensure that the appropriate determination is made. Recently, FDA published a proposed rule that would require producers to notify the agency at least 120 days before marketing, which may be a signal that FDA could abandon its more reasonable approach in the future.

In addition, FDA does not require labelling of foods derived from biotechnology unless the genetic insertions so alter the food that the common name no longer describes it adequately. Examples of this phenomenon would include such alterations of the food product that would raise a safety issue or change the product's nutritional composition or its storage or preparation characteristics.

INTERNATIONAL TRADE

While U.S. consumers do not appear to be strongly opposed to biotech foods (they, in fact, seem rather indifferent), a strong anti-biotechnology movement has arisen in several European countries over the last few years. The European Union (EU) has established strong restrictions on the commercial planting of genetically engineered plants, and European food processors and retailers are reluctant to import harvested agricultural products derived from biotechnology — thus jeopardizing the marketability of U.S. commodity grain exports. And the EU is now negotiating for strong restrictions on agricultural biotech products in important international agreements governing food safety and environmental protection. Very strong restrictions were included in the Biosafety Protocol, finalized in January 2000.

Perhaps, more important, though, are the Codex Alimentarius Commission standards for food safety. Without convincing scientific evidence that genetically engineered crop plants pose a heightened environmental or human health risk, restrictions on agricultural biotech imports could be challenged under the General Agreement on Tariffs and Trade and the World Trade Organization.

Thus the EU has resorted to justifying its decisions on the basis of a risk-management philosophy known as the "precautionary principle," and the EU advocates inclusion of the precautionary principle in international environmental and food safety agreements (such as the Codex Alimentarius), as well as within GATT itself. Its inclusion would give nations greater latitude in restricting imports of U.S. agricultural products.

No single definition exists for the precautionary principle, but its general meaning is that when an activity raises threats of harm to human health or the environment, regulatory measures should be taken to prevent or restrict the activity even if the risks have not been demonstrated scientifically. Although the EU asserts that the precautionary principle is an unbiased risk-management philosophy, critics have noted that its lack of definition and evidentiary standards makes it all too easy to abuse for the purpose of masking trade protectionism, and that its very approach to risk management is inherently flawed and may, in fact, increase net risk.. Nevertheless, the precautionary principle has already been included in several international environmental treaties, including the Convention on Biological Diversity and the Biosafety Protocol.

LABELLING ISSUES

Some activists, however, argue that the government should mandate the labelling of all genetically engineered foods. They assert that consumers have a "right to know" how their foods have been altered, and that a mandatory label would best allow consumers to choose between genetically engineered and conventional foods. Biotechnology advocates have argued against mandatory labelling because such requirements raise food costs— something that mostly harms lower-income Americans and people on fixed budgets. Perhaps more important, while biotech products are not substantially different from other products, special labels would likely make consumers think these products were more dangerous. Hence, rather than serving educational or "right to know" purposes, such labels promise to simply confuse consumers.

A government-mandated label on all genetically engineered foods also would raise important First Amendment free speech issues. In 1996, the U.S. Court of Appeals, in *International Dairy Foods Association et al. v. Amestoy*, ruled unconstitutional a Vermont statute requiring the labelling of dairy products derived from cows treated with a genetically engineered growth hormone, noting that food labelling cannot be mandated simply because some people would like to have the information. "Absent ... some indication that this information bears on a reasonable concern for human health or safety or some other sufficiently substantial governmental concern, the manufacturers cannot be compelled to disclose it."

In other words, to be constitutional, labelling mandates must be based in science and confined to requiring disclosure of information that is relevant to health or nutrition. Furthermore, consumers need not rely on mandatory labelling of biotech foods to truly have a choice. Real-world examples show that market forces are fully capable of supplying information about process attributes (including kosher and organic production standards) that consumers truly demand.

The same can be said about nonbiotech foods, and the FDA recently published proposed guidelines to assist producers in voluntarily labelling both genetically engineered and nongenetically engineered foods. Additionally, the USDA's newly published rule for organic certification necessarily excludes biotech products from organic food production.

Consequently, consumers wishing to purchase nonbiotech foods need only look for certified organic products.

Even as underdeveloped nations clamor for biotechnology applications, and as countries like China continue to experiment with and use agricultural biotechnology, opponents of agricultural biotechnology in the West, particularly Europe, attack it as an unnatural process that will destroy the world, not better it. They argue that biotechnology should be heavily regulated, if not banned. Already, however, genetically engineered plants are subject to strict regulatory oversight that is equal to or greater than that advocated by the vast majority of scientific specialists.

Additional regulation will slow down research and development of genetically engineered crops, keep beneficial products off the market, and raise the cost of products that do make it to consumers. Furthermore, the inclusion of similar restrictions — or inclusion of the precautionary principle— in international agreements will greatly impact the international trade of agricultural goods and delay their introduction into the marketplace. Each of these problems could prevent this technology's benefits from being introduced to industrialized nations and, more important, the developing world.

TAXONOMIC IDENTIFICATION AS PART OF GERMPLASM CHARACTERIZATION

The accurate identification of material held in any genebank is arguably the most essential part of the germplasm characterization process, for without such information breeders will have no means of selecting material for crosses and entry into breeding programmes. Taxonomic identification is an essential first step to determine whether any germplasm is part of the primary, secondary or tertiary gene pool of the crop concerned. While such identification may be undertaken using traditional taxonomic characters, this is not always possible or indeed accurate. A very useful summary of examples of the way molecular markers have contributed to our understanding of crop gene pools is given by Gepts.

In Asian rice (*Oryza sativa*), six crossability groups have been recognized comprising the bulk of the primary gene pool. It is of particular concern to breeders that unambiguous identification of indica and japonica rices can be achieved, as these are currently the focus of plant breeders' attention for crossing and for the development of the 'new plant type', despite the fact that there is difficulty in hybridization and recombination between these two types. Breeders attempting to utilize the considerable variation represented by these groups face increasing difficulty in distinguishing material from these groups, and have regularly used isozymes to help make the necessary identifications. For example, the lines Azucena and PR 304 have been classified as indica using morphological characters, whereas they behave as japonica types in crossing studies.

When analysed using RAPD they are clearly revealed as japonica rices. Such discrepancies were apparent in another experiment by Virk *et al.* (1995a). Forty-four rice accessions which had been previously classified as indica or japonica on morphological grounds were found by cluster analysis of RAPD data to divide into two major groups. All 31 accessions of one group had been classified as indica; however, eight of the other group had been designated as either japonica or javanica, while the other five had been classified as indica. Clearly the RAPD classifications do not always correlate exactly with classifications based on morphology, but they do accord well with the classifications of rice based upon crossability and isozyme data.

Six *O. sativa* isozyme groups can now be identified using RAPD; these groups reflect precisely those defined by crossability. The same can also be achieved by other molecular marker strategies including RFLP, and microsatellites. The ease with which this discrimination can be made using RAPD markers is illustrated in Fig..

Of equal importance to genebank management is the ease with which germplasm of closely related species can be identified. In *Oryza*, and particularly within the *Oryza sativa* complex of AA genome species, there is often uncertainty with regard to the allocation of germplasm to several of the species. *O. rufipogon* and *O. nivara* from Asia, and even *O. glumaepatula*, which does seem to have a fairly distinct geographical distribution in South America, have all posed problems of identification using morphological characters. Much more precise identification can be achieved using RAPD markers. RAPD markers have even revealed the true identity of material entering the genebank with the designation *O. meridionalis* that had been misidentified when compared to holotype material.

IDENTIFYING DUPLICATES

The race against genetic erosion of crop gene pools has yielded many thousands of accessions safely stored in genebanks. However, genebanks have a finite capacity, and it is apparent that they often conserve more than one sample of the same genotype. In other words, plant genetic resources collections contain duplicate materials, yet the scale of the problem in seed genebanks cannot be determined with certainty. For many vegetatively propagated species, this situation is much more easily addressed.

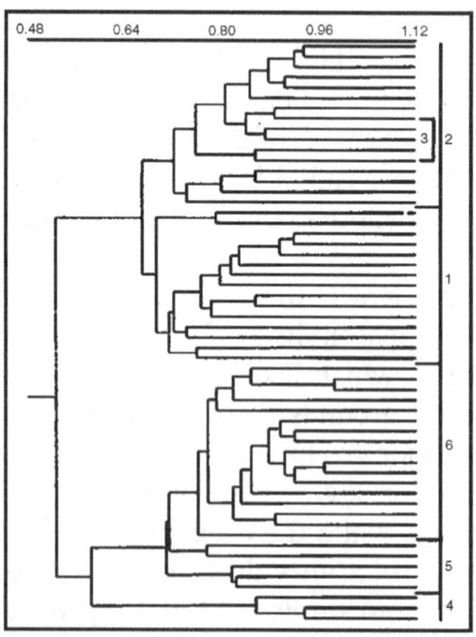

Fig. Clustering of Oryza Sativa Accessions According to Crossability Group,Based Upon RAPD Data.1-6:Rice Crossability I–VI

From a purely management point of view, there are distinct advantages in identifying duplicate accessions, and thereby focusing most effort on unique genetic materials for conservation. Until now the identification of duplicate accessions has had to rely on comparison of morphological characters, some of which are subject to environmental variation, together with passport data including (amongst others) variety name and origin.

Identification of duplicates of vegetatively propagated species, such as potato, is more straightforward than for seed-propagated crops such as rice.

Some Accessions (e.g.5,22,23,28and 36)were Subsequently Found to have Been Misiclentified. At the International Potato Centre (CIP) duplicate accessions have been routinely identified for some time by comparison of tuber proteins separated in polyacrylamide gels, and complemented by field observations of morphological characters. Duplicate clones have been eliminated from the collection of Andean potato varieties, reducing its size to more manageable proportions. In more recent work, germplasm samples of rice from IRRI including known and suspected duplicates, as well as closely related germplasm, have been subjected to molecular analysis by Virk *et al.* (1995b). Their results demonstrate that an accurate discrimination of these categories of germplasm samples, including the identification of true and suspected duplicates, can be achieved.

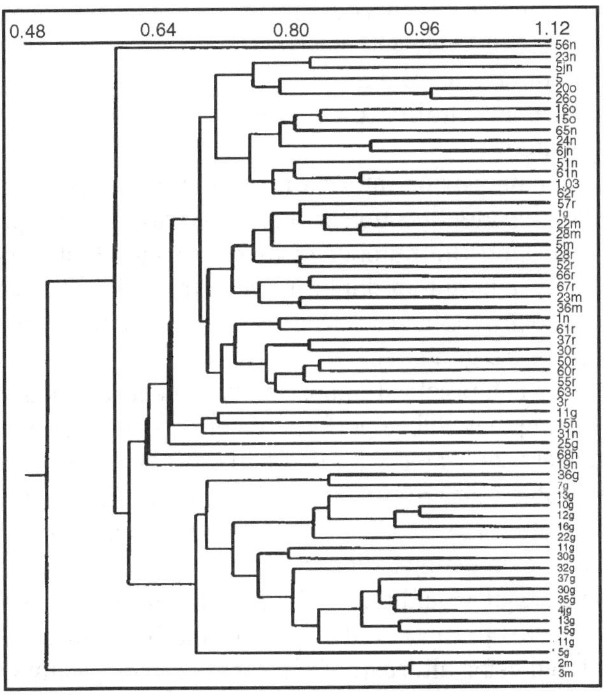

Fig. Identification of Wild Rice Species Using Cluster Analysis (Simple Matching Coefficient and UPGMA Clustering) of RAPD Data.(n = Oryza Nivara; r = 0, Rufipogon: g = 0.Glumaepatula: m = 0. Meridionalis).

Two procedures have been proposed for identifying such duplicates. These differ in the method used in the initial stages, and the relative merits of either method would have to be balanced by those persons charged with conserving a collection. Local situations will vary considerably with regard to the relative costs of field work compared to molecular biology, and the expertise available to carry these out.

Procedure:

- Select potential duplicate accessions from the collection following examination of available 'passport' data.
- Undertake initial morphological characterization of the suspected duplicates.
- Undertake a full molecular scrutiny of those germplasm samples that cannot be separated using these data.
- Designate as duplicates germplasm samples that cannot then be discriminated.

Procedure:

- Select potential duplicates from the collection following examination of available 'passport' data as in Procedure 1.
- Carry out a pre-screen of pairs of suspected duplicates using a small number of molecular markers instead of morphological evaluation.
- Undertake a full molecular analysis of those germplasm samples which then cannot be separated.
- Designate as duplicates germplasm samples that cannot be discriminated.

Such results may provide information useful in the design of procedures that permit the routine identification of duplicates within a germplasm collection. Discussions concerning the number of marker bands that it is necessary to score before the designation of duplicates can take place are complex. It will never be possible to prove that two accessions are genetically identical without sequencing their entire genomes.

Given that this is a practical impossibility, a decision must be made about the amount of testing that will be performed before two accessions are accepted as (or 'designated' as) duplicates. This decision must be influenced by the number of potential duplicates that are to be tested. However, the results of Virk *et al.* (1995b) indicate that in rice, for one very similar pair of accessions, we can be 99 poe cent confident of detecting a difference between them if we examine a total of 86 RAPD markers. It would clearly be possible to use other types of DNA-based markers for this purpose, although it would be important to ensure that the variation defined using alternative markers was biologically valid in terms of taxonomy and genetics. Moreover, the number of bands to be scored may differ depending upon the sequence types represented by the markers.

The reliable identification of duplicate accessions will provide management options for the germplasm curator. Whether it will lead to the reduction in size of germplasm collections is debatable.

In the case of CGIAR centres, their germplasm collections are held in trust, and many accessions are actually intentional duplicate materials of existing national germplasm collections. Clearly the CGIAR centres have an obligation to continue to conserve those germplasm accessions already accepted for safety storage.

With the acquisition of new germplasm accessions, however, the situation is potentially different. The study of Virk *et al.* (1995b) suggests a novel procedure which would allow the level of certainty of identifying duplicate samples to be set before those samples became part of a germplasm collection, and before they were assigned unique accession numbers.

This option is one which could have a significant impact on germplasm management, provided the PCR-based marker technology can be easily and economically utilized by germplasm curators.

CORE COLLECTIONS

A major issue in genetic resources for some time has been the size of germplasm collections in relation to their effective management and use. The large size of collections in genebanks may severely constrain many genebank practices. One answer to these problems is to develop a core collection, originally envisaged by Frankel (1984) as being a subgroup of accessions of any germplasm collection which would incorporate, with minimum redundancy, the genetic diversity of a crop species and its relatives.

To all intents and purposes the core would form the 'active collection' for germplasm evaluation and distribution, with the remainder of the germplasm being kept as a 'reserve' collection. More specifically there are perhaps two rational and practical motives to develop core collections. The first is to facilitate germplasm management, and the second to increase the use of germplasm by breeders. These two objectives may not complement each other exactly, and indeed the germplasm curator may well be biased towards the former. However, both objectives could be achieved by the use of molecular markers.

The core collection approach has already been taken for barley, cassava, sorghum, wheat, coffee and *Phaseolus*. In rice, where the problem of collection size is as great as any, various steps have been taken towards the core approach which aim at fulfilling both of the objectives referred to above. A principal objective of the International Rice Genebank is to establish a core which can help in the safe duplication of accessions representing the broad diversity of the genus *Oryza*, in several locations around the world.

The more straightforward the development of this core, the better from the point of view of the germplasm curator. In addition, current knowledge of rice diversity based upon geographic, morphological, agronomic, biochemical and molecular characteristics has resulted in the development of a small core of about 270 accessions of *O. sativa* which represents the known diversity of rice. Similarly, Vaughan (1991a,b) has designated a core collection of wild rices to enable researchers to evaluate this germplasm efficiently. Continued collecting and biodiversity studies mean that the composition of these core collections must be updated periodically.

Molecular and biochemical markers can be used to determine the degree of differentiation that actually exists between wild species themselves and between them and *O. sativa*. Questions about precisely where allelic richness can be found and whether wild species really are sources of distinct alleles can be addressed in order to determine whether new accessions should be added to the core, or whether existing accessions in the core are largely redundant because they contribute little which is genetically unique. The use of the core collection in combination with marker data at IRRI has also enabled rapid identification of germplasm possessing some much sought-after traits. For instance, studies of allozyme data enabled accessions of the small rayada group of rices to be pinpointed for resistance to leaf scald (a seed-borne disease caused by *Rhynchosporium oryzeae*).

How then is the choice of core material to be made? Brown (1989a) has identified stratified sampling as being more efficient in establishing a core than purely random sampling. This relatively simple procedure involves dividing the collection into nonoverlapping groups, and then taking samples from each group. The way in which the groups are established will probably vary by crop species, but will depend upon taxonomy, passport data and ecogeographical information. In the face of uneven distribution of diversity and differentiation of accessions, this method will ensure that the allelic richness of a core will be maximized. One constraint to this approach, however, is the dearth of accurate passport data that unfortunately typifies the situation in many germplasm collections worldwide.

Schoen and Brown (1995) have gone further than this by undertaking a set of simulations. Utilizing allozyme marker data to demonstrate how core collections might be established, they identified how allelic richness could be achieved in cores developed by six different strategies, two of which (H and M) were described for the first time.

All six strategies invoked stratified sampling from designated geographical groups, but the H and M strategies differed in that they utilized genetic marker data to guide sampling from within groups. When compared for allele retention the six strategies gave differing results, with the two strategies guided by marker data (M and H) consistently performing best, and the simplest approach, involving only random sampling, the worst.

To what extent molecular markers will be employed in the future via strategies of the H and M categories is arguable. Schoen and Brown clearly indicate that the way forward must be through stratified sampling of an entire collection using passport and other data to develop the groups to be sampled. With the increased use of different molecular marker techniques the limitations of allozyme information will be overcome, so that it will be increasingly possible to improve the selection process of material for the core from within the stratified groups of accessions. Selecting more accessions from groups of high marker gene diversity (H strategy), or targeting particular accessions that are both high in allelic richness and well differentiated (M strategy) can offer the possibility of further improvement over what can be achieved by simple stratified sampling.

In terms of the actual size of any core collection, statistical theory rather than any practical use of marker data has been used principally to determine what needs to be done. Brown has argued that the core should consist of about 10 poe cent of the whole collection, up to a maximum of about 3000 accessions, for each species. He estimates that, at this level of sampling, the core will generally contain over 70 poe cent of the alleles present in the whole collection. This seems to be a rule of thumb which many germplasm curators are adopting.

However, an alternative approach to this problem is provided by the work of Lawrence *et al.*, who considered the size of sample required to capture at least one copy of each allele at each of a number of independently inherited loci at a given probability. Their calculations indicate that, provided the sample size chosen gives a very high probability of conserving the alleles of a single locus, this size is also sufficient to give a high probability of conserving at least one copy of each allele at all other loci. Calculations based upon assumptions that the average genome of a species contains 40 000 structural loci, and that 40 poe cent of these loci are polymorphic, indicate that, even if the species is predominantly inbreeding, a sample size of only 172 will give a very high probability (>0.999999828) of conserving all of the alleles at all the polymorphic loci, even if the frequency of one allele at each locus is only 0.05.

In practice, therefore, it should be relatively easy to conserve all or very nearly all of the alleles of a population in a random sample of 172 plants! Whether core collections are made up of samples taken from 172 plants or 3000, it would seem to be clear that the sampling strategy chosen to select the material which will make up those numbers is still of overriding importance and can be substantially assisted by using stratification and molecular markers.

FACILITATING THE USE OF GERMPLASM

PCR-based molecular markers are increasingly being used to assess genetic diversity in germplasm collections. Sometimes, morphological data including those for quantitative traits of economic importance are also available from the genebank. There is an increasing desire to utilize this wealth of useful information, somehow to use molecular markers to assist identification of useful characteristics amongst conserved germplasm, and therefore to narrow the gap between genebank managers and plant breeders. One of the successes of the CGIAR conservation effort over three decades has been the close linkage between conservation and exploitation of germplasm. This has led to many outstanding examples of varietal release to alleviate hunger in developing countries.

Molecular markers are increasingly being used in marker-assisted selection programmes. Both theoretical and experimental studies have shown that marker-assisted selection can be highly effective for producing improved genotypes. However, the success of such selection programmes is largely taken to depend on genetic linkage between markers and the relevant gene loci.

Whilst such studies are invariably based on materials derived from planned crosses, could similar principles be applied to genetic resources held in germplasm collections? The work of Virk *et al.* (1996) has gone some way to achieving this by using multiple regression analysis to predict the performance of germplasm accessions of rice given the molecular marker genotypes of those accessions. From a large number of markers it was possible to pinpoint a handful which are significantly associated with a particular trait of interest.

Subsequently they were able to make accurate predictions of field performance in a range of agronomic traits such as plant height, culm number, and days to flowering in a particular environment. It is known that associations can exist because of linkage disequilibrium as well as linkage.

It is also clear that for marker-assisted procedures to work for prediction or selection of complex inherited traits, it is of benefit that a high level of linkage disequilibrium must exist, particularly if the genome is not well saturated with markers. This appears to be the case in the study of Virk *et al.* (1996), and if true, it appears that associations between alleles at quantitative trait loci (QTLs) and at marker loci has been conserved throughout the period of diversification of rice germplasm in South and Southeast Asia.

One obvious benefit of obtaining information about molecular markers and quantitative traits would be the more efficient selection of putative parents for producing populations to map QTLs for a particular trait. Also, the procedure could be used as an initial screening method for the identification of QTLs.

The established method for this is the selection of two parents that differ markedly in a particular quantitative character, and then the determination of associations between markers and that character in F2 or backcross progeny. The apparent advantages of using diverse germplasm instead are: (i) that this could allow the detection of QTLs that vary across a wide spectrum of biodiversity rather than just between two parental lines; and (ii) that QTLs for any quantitative trait can be studied in the same investigation.

Regardless of the underlying causes of the associations which have been detected, the use of molecular markers, which are more or less randomly distributed across the genome, coupled with multiple regression analysis could substantially change and improve the way in which crop biodiversity is used in the future. The combination of techniques should allow the prediction of what a plant will look like in terms of quantitative agronomic traits prior to elaborate field trials. If a diverse test array of germplasm is scored for important traits requiring specialized assessment conditions (such as stress tolerances) then marker data could provide an efficient means of predicting the value of additional germplasm for such characteristics. Such results may demonstrate the value of *ex situ* plant germplasm collections not just as repositories of useful genes, but also as sources of information about phenotypic characters.

EQUITY OF ACCESS TO GENETIC MODIFICATION BIOTECHNOLOGIES

Private investors have taken the lead for most investment in modern biotechnology, and in so doing have also had to manage the associated economic risks. In many situations such investment risks are protected by patents. An obvious concern is, therefore, that access to GM crops and trees will be controlled almost entirely by private corporations. Agreements for the use of techniques or material could be prohibitively expensive, preventing the use of the technology where it would be of value. In the past, governments were the main investors in breeding and molecular biology research. Their role may have to expand once again in this regard, in order to provide a flow of material and information that can be developed and shared by both private and public institutions (Santos and Lewontin, 1997), as opposed to being controlled largely by private companies. Nevertheless, financial constraints, rather than available technical knowledge, are probably the largest challenge ahead for applying modern biotechnology in less developed countries, although it has been shown, e.g. by the Agricultural Biotechnology Support Project (www.iia.msu.edu/absp/), that arrangements through donations are possible.

Even if financial considerations can be addressed in the research phases of biotechnology, moving the technology from the development stage to operational reality provides a new set of technical challenges with additional expenses (Polonenko, 1999). It is likely that only a few economically important species will warrant such additional investments, and it is on these economic situations and realities that the debate should focus.

Public acceptance of genetic modification

As has been alluded to previously, in terms of biotechnology research, one of the most important economic considerations separating forestry from agriculture is that trees require substantially more time for proper evaluation under both laboratory and field conditions. Rapid generation turnover in crop species has allowed genetic modification technology to develop quickly.

Although this has been desirable from an economic point of view, the ecological ramifications have been far more difficult to study, as complex experiments must be carried out over several years. Public and government acceptance of GM plants is now probably as dependent on biological risk assessment and risk issue management (Leiss, 1999) as it is on technical developments or economics alone.

For GM trees, there is clearly a need to provide scientifically sound, neutral and intelligible information to the public, so that when and if GM trees become commercially available, informed decisions can be made about their release into the environment. Foresters are generally very aware of the need for public involvement in many other areas of forest management (e.g. chemical applications in forests, game management, felling patterns and plans, allowable annual harvesting levels), so decisions concerning GM trees may simply be an additional issue to be managed in the process.

CONCLUSIONS AND FUTURE CHALLENGES

The main issues that should remain relevant to the use of any modern biotechnology in forestry, now and in the future. The following are some summary points and conclusions:

- The development time, from the laboratory to the field, for all new types of tissue culture or GM trees will be substantial, usually in the order of a decade or more. While this may provide forest scientists and managers with time to evaluate many of the issues being faced in agriculture, the economic realities of relatively long generations will continue to be a major challenge to investors in biotechnology in forest trees. It appears that genetic modification will therefore become a reality only for particularly novel and valuable traits in short-rotation species in intensively managed plantations. Herbicide resistance is likely to be the first trait for which genetic modification will become economically viable, but even this may be several years away from large-scale use.
- All modern biotechnologies require large research and development investments. The allocation of funds, through either private or public agencies, needs to achieve a balance between building scientific capabilities and knowledge and supporting more applied, well proven forestry technologies (Burdon, 1994). In this regard, the investment and use of any biotechnology needs to be assessed on a case-by-case basis, considering the specific technology in the specific ecological, political and economic environment.
- The public must be confident that regulatory frameworks are in place so that commercial development of biotechnology does not take precedence over biosafety issues. However, the development of useful

biotechnology in forestry should not be put at risk with extraordinary regulatory compliance costs that are driven by unwarranted concerns.

- Government participation in the development of biotechnology appears necessary to keep some flow of material and knowledge in the public domain. Without this, the costs of using privately owned genes could seriously limit further interest and development of genetic modification technology in trees. Concerns about monopolies arise with any new products, but governments can step in to deal with substantial inequities or corporate actions that are not in the interest of society as a whole.
- Even if GM trees are not used on a wide scale in the future, research and development in GM technology will provide a great deal of information about gene action and regulation, which could be of long-term value to conventional breeding programmes.
- Other forest management decisions with potentially more serious ecological consequences, including large-scale species introductions or inappropriate use of provenances or improved trees from even conventional breeding, need to be evaluated by foresters, managers and regulatory agencies in the same way as the products of modern biotechnology are.
- Geneticists have always attempted to create genetic options for the future. If we look carefully at where these current and future biotechnologies are heading in forestry, it is obvious that the research and development will largely be used in support of advancing clonal forestry. Therefore, it is in the context of appropriate clonal forestry that most of the issues need to be examined and evaluated.

From a genetic perspective, concerns that biotechnology is "unnatural" ignore the dynamic changes in the genetic code that occur within and across species genomes through modification of transposable genes or elements by virus vectors and through mutation. This article has attempted to point out that humans have already had a huge effect on the genetic structure of many plants and animals, using many different technologies which will continue to develop and change over time. However, while those in the forefront of any technology will promote its potential benefits, in the end it will be the economic and regulatory systems of governing bodies at the national and global levels that must evaluate the technology's relevance and appropriateness.

MODERN AGRICULTURAL BIOTECHNOLOGY

In 1973, Cohen and Boyer transferred a gene from one organism into another. In 1982, the first biotech plant, an antibiotic resistant tobacco, was developed. In January 1983, at a meeting of genetic researchers in Miami, three different teams reported success in using Agrobacterium tumefaciens, a bacterium, to carry new genes into plant cells, heralding the dawn of modern agricultural biotechnology. Agrobacterium tumefaciens, described as a "natural genetic engineer", splices its own genes into host plant cells. This pathogenic bacterium was now converted into a pack mule, to carry new, foreign genes into plant cells, minus the disease and this became the most common means of producing Genetically Engineered Organisms (GEOs).

Field tests for GE crops resistant to pests and pathogens were first conducted in the US, in 1985. A co-coordinated framework for the regulation of GEOs was established and the first GE tobacco was released, in 1986 . The US Department of Agriculture published guidelines for field trials of GE crops in 1991.On approval from the Food and Drug Administration, Flavour Saver, the first GE tomato, with a longer shelf life, was on the US markets in 1994. During 1995-96, GE soybean, corn and cotton were approved for commercialization, in the US.

A number of GE crops, developed for pest, pathogen and herbicide resistance, are now commercially cultivated in several countries. Rice with pro-vitamin A, higher iron content, or human milk proteins and potatoes with high protein content are in various stages of development. Tobacco plants producing functional human haemoglobin and bacteria that produce human insulin have been developed, as well as plants with vaccines against rabies and other viral diseases. Food grain crops that withstand drought and salinity are high priority research, and so are those for high yield. A GE tobacco plant detoxifies soils contaminated by explosive residues, providing solution to a frustrating environmental problem in countries ravaged by armed conflict.

Now more than 70 biotech agricultural crops that have been approved for use in North America, including varieties of soybeans, cotton, canola, corn, potatoes, squash, tomatoes and papaya. About six million farmers in some 17 countries now cultivate GE crop on about 125 million acres, a 30 fold increase over 1996. By end of the year 2002, six GE crops planted in the US (soybeans, corn, cotton, papaya, squash and canola) produced an additional four billion pounds of food and fibre on the same acreage, improved farm income by US $ 1.5 billion and reduced pesticide use by 46 million pounds. In 2003 in the US, 80 per cent of soybean acres will be planted with biotech varieties.

A number of activist groups vehemently attack GE technology and its products on grounds of safety to humans and the environment, and costs of technology transfer and its reach to the needy. Products of agricultural biotechnology bear the brunt of this ill-informed, unscientific and prejudiced onslaught much more than GE products related to health or industry.

In 2001, the European Community released results of a 15-year study, costing US $ 64 million, and involving more than 400 research teams and 81 projects. This report concluded that GE products pose no more risk to human health or the environment than conventional crops. So far, extensive and intensive research on the probable risks of GE technology has not brought out any adverse effects and none of the fears expressed by anti-tech activists were proved even marginally. Nevertheless, caution and examining issues case by case, is the watch word of technologists, who are aware of their responsibilities.

For a number of products, technology transfer is free of costs for developing countries, as for example Golden Rice, the rice with pro-vitamin A. It is the responsibility of the Governments of the respective countries to bear the costs of developing local varieties and to reach the products to the needy at an affordable cost. In India, three GE varieties of pest resistant cotton were approved for commercialization, a year ago. Approval for other GE varieties of cotton and GE mustard was deferred twice by the Genetic Engineering Approval Committee (GEAC), the highest authority on the issue.

In India, the level of public awareness of the realizable benefits and probable risks of GE products is abysmally low. The functioning of the GEAC leaves much to be desired. Taking advantage of this hazy situation, mixing up economic, social and political issues with science, and even using such vagaries of nature as the severe drought, several groups of vested interest have created mistrust, confusion and scare. Trashing a very promising technology this way does not augur well for the future of Indian agriculture. This results only in denying the benefits of technology to the farmers and consumers.

The next 20 years of plant biotechnology is expected to improve quality of life, through improved foods, pharmaceuticals, new industrial materials and a better environment. These benefits should reach the people of the developing world, who need them most. For India to benefit from these developments, it is urgent and essential that the public is educated on the realizable benefits and probable risks related to GE products. We also need to reorganize the mechanism of regulating GE products by providing for an expedient, transparent and responsible authority. The media have a very important role to play in this regard. Only an informed and reassured public can make viable choices. Every effort should be made to provide such an opportunity to the consumer.

4

Gene Technology in Plant Science

GENES INTO PLANTS

To genetically modify a plant, the thousands of bases of DNA comprising an individual gene are transferred into an individual plant cell where the new gene becomes a permanent part of the cell's genome. This process makes the resulting plant "transgenic." Transfer of DNA into plant cells is done using various "transformation" techniques that are the result of discoveries in basic science.

One method to transfer DNA into plants takes advantage of a system found in nature. The bacterium that causes "crown gall tumors" injects its DNA into a plant genome, forcing the plant to create a suitable environment for the bacterium to live. After discovering this process, scientists were able to "disarm" the bacterium, put new genes into it, and use the bacterium to harmlessly insert the desired genes into the plant genome.

In the "biolistic" or "gene gun" method, microscopic gold beads are coated with the gene of interest and shot into the plant cell with a burst of helium. Once inside the cell, the gene comes off the bead and integrates into the cell's genome.

It was also discovered that plant cells could be "electroporated" or mixed with a gene and "shocked" with a pulse of electricity, causing holes to form in the cell through which the DNA could flow. The cell is subsequently able to repair the holes and the gene becomes a part of the plant genome.

When using these methods, new genes are successfully introduced into only a small percentage of the cells, so scientists must be able to "pick out" or "select" the transformed cells before proceeding. This is often done by concurrently introducing an additional gene into the cell that will make it resistant to an antibiotic.

A cell that survives antibiotic treatment will most likely have received the gene of interest as well; that cell is subsequently used to propagate the new plant. There is a concern that the gene giving antibiotic resistance could naturally be transferred to bacteria once the transgenic plant is in the wild, making bacteria resistant to antibiotics that are used to fight human infection. Scientists are currently devising ways to select for transformed cells that will alleviate this issue.

Thus, the present invention has been accomplished. The present invention can be applied to any plants regardless of their species. However, effects of the present invention can be particularly obtained in plants in which gene introduction is considered to be difficult, such as trees and the like. In the present invention, the shoot to be used as the tissue for gene introduction means an elongated normal bud or adventitious bud. Also, in order to further improve the transformation efficiency, it is preferable to use, as the tissue for gene introduction, a tissue or part having high redifferentiation potency such as a young tissue having a low content of polyphenols which become the cause of browning of tissues.

From this point of view, a hypocotyl obtained by germinating a seed and an elongated apical bud or lateral bud of a plant cultured in a culture vessel are selected in the present invention as suitable tissues for gene introduction.

However, the shoot must have at least one end having a growing point and a base part having no growing point. For example, when a hypocotyl is used as the tissue for gene introduction of the present invention, the shoot can be obtained by cutting off a root and a part having a base of root from a hypocotyl after germination of a seed while leaving the apical bud and apical bud primordium as they are. In this case, the one end having a growing point means a shoot end having an apical bud and the like, and the base part having no growing point means a part having the end face formed by cutting off a root and the like.

Also, when an elongated apical bud or lateral bud of a plant is used as the tissue for gene introduction of the present invention, it can be obtained by simply cutting out from the plant. In this case, the one end having a growing point means a shoot end having an apical bud and the like similar to the case of the hypocotyl, and the base part having no growing point means a part having the face cut out from the plant. That is, the desired gene is preferably introduced into the end face of the base part of the shoot generated when a shoot is prepared as the tissue for gene introduction.

In the present invention, an adventitious bud is differentiated by introducing a desired gene into the base part having no growing point, while keeping the growing point of at least one end. As the desired gene, various genes such as a gene which can provide an industrially excellent character and a gene which cannot always provide an industrially excellent character but is necessary in studying gene expression mechanism can be selected and used.

A desired gene can be introduced into the base part of the shoot indirectly via Gemini virus, Brome mosaic virus, Agrobacterium tumefaciens (hereinafter referred to as "A. tumefaciens"), Agrobacterium rhizogenes and the like viruses and bacteria, by inserting the desired gene into an appropriate vector, or directly by the particle gun method and the like. For example, in a gene introduction method using Agrobacterium, an appropriate vector into which a desired gene has been inserted is introduced into Agrobacterium in advance, and the desired gene is introduced into a tissue for gene introduction by infection with the Agrobacterium. The infection of the Agrobacterium is carried out, for example, by soaking the tissue for gene introduction in a solution in which the Agrobacterium is suspended.

Also, since the infection of the Agrobacterium occurs in a wound of the plant tissue, when a shoot in which a wound is formed on its base part is soaked in an Agrobacterium suspension, the wound is infected with the Agrobacterium and the desired gene is introduced into the base part of the shoot.

When a hypocotyl is used as the tissue for gene introduction by cutting off roots and the like or elongated apical bud or the like is used by cutting out it from a plant, the infection occurs at the end face (cut surface) of the base part of the shoot formed by the cutting. That is, in these cases, the desired gene is introduced into the base part of the shoot by merely soaking the tissue for gene introduction simply in the Agrobacterium suspension.

Furthermore, a plant tissue is surely infected with the Agrobacterium when a tissue for gene introduction is soaked in the cell suspension and then co-cultured with the Agrobacterium for several days by introducing the tissue on a solid medium. As the coculturing medium, a well known basal medium such as MS or WPM, or a modified composition thereof to suit for the tissue for gene introduction to be infected with the Agrobacterium, can be used by supplementing it with a carbon source and a medium solidifying agent and, if necessary, with plant hormones such as auxins, cytokinins and the like appropriately.

In this case, generally, 10 to 30 g/l sucrose is used as the carbon source; 5 to 10 g/l agar or 1 to 4 g/l gelan gum is used as the medium solidifying agent; 0.01 to 5.0 mg/l zeatin, benzyladenine or the like is used as the cytokinins; and 0.01 to 2.0 mg/l naphthaleneacetic acid (NAA), indolebutyric acid (IBA), indoleacetic acid or the like is used as the auxins. Also, infectivity of the Agrobacterium is increased in some cases by adding 10 to 200 mg/l acetosyringon to the above coculturing medium.

On the other hand, when a desired gene is introduced by a particle gun, the above coculturing medium can be used as the medium for gene introduction treatment. In this case, a tissue for gene introduction is placed on the medium with its position into which the desired gene is to be introduced, namely the base part of the shoot, upside, and the gene introduction is carried out by manipulating the particle gun in the usual way. In general, when a desired gene is introduced into a tissue for gene introduction, a selectable marker gene is introduced together with the desired gene, and expression of the selectable marker gene is used as an index of the introduction of the desired gene. In the method of the present invention, the transformation efficiency can be further improved by using a cytokinin-related gene as the selectable marker gene.

Herein, the cytokinin-related gene is a gene which acts in a direction of increasing the influence of cytokinin in the introduced plant cells and thereby increases adventitious bud differentiation ability of the cells.

Examples of the gene include the ipt gene as an A. tumefaciens-derived cytokinin synthesis gene, the Escherichia coli-derived.beta.-glucuronidase gene as a gene which activates inactive cytokinin, the Arabidopsis thaliana-derived CK11 gene which is considered to be a cytokinin receptor gene and the like. Particularly, the ipt gene used in Examples of the present specification is a gene which is most well known and whose function has been revealed.

Also, the desired gene, the cytokinin-related gene and other nucleotide sequences and genes which are optionally introduced may be introduced by inserting them into the same vector or be introduced by inserting them into different vectors with no problems, so long as they are incorporated into the same cells of the tissue for gene introduction.

However, when the genes and the like are incorporated into the same vector, it is necessary to arrange them such that the presence of one side of genes and the like does not inhibit expression of the other side of genes and the like.

An adventitious bud can be differentiated from a tissue after the gene introduction by culturing the tissue using an appropriate medium. A composition of the medium suitable for the adventitious bud differentiation varies depending on each plant, but in the case of the genus Eucalyptus, the MS medium in which the concentration ratio of ammonia nitrogen and nitrate nitrogen is changed to 1:3 (hereinafter simply referred to as "modified MS medium") can be used as the medium for adventitious bud differentiation (the shoot regeneration medium) after diluting it to 1 to 4 folds and supplementing it with 10 to 30 g/l sucrose, 1 to 4 g/l gelan gum or 5 to 10 g/l agar, and 0.2 to 5.0 mg/l zeatin and 0.01 to 1.0 mg/l NAA as plant hormones.

However, when a cytokinin-related gene is used as the selectable marker gene, the plant hormones may not be added (may be hormone-free) or auxin alone may be added. Also, when a gene is introduced by the above Agrobacterium method, antibiotics such as carbenicillin, ticarcillin, cefotaxime and the like are added to the medium in an amount of 10 to 10,000 mg/l to inhibit the Agrobacterium growth. It is preferable that the temperature is from 15 to 30° C. and the light intensity is from 0 to 200 μ mol/m^2/s. Since the growing point of a shoot preserved at the time of the gene introduction is not particularly required in the subsequent steps, it can be cut off at an appropriate stage.

The tissue cultured using the shoot regeneration medium differentiation differentiates the adventitious bud generally several weeks after commencement of the culturing. In this case, a callus may grow slightly prior to the adventitious bud differentiation. The desired gene is introduced into the thus differentiated adventitious bud at higher frequency than the differentiated adventitious bud differentiated by introducing the gene into a segment based on the conventional method.

However, when the gene introduction is carried out using a cytokinin-related gene as a selectable marker gene, the adventitious bud into which the desired gene has been introduced may sometimes show morphological abnormality such as multiple bud or the like due to the induction of morphological abnormality by the gene.

Even in that case, however, an adventitious bud having normal morphology can finally be obtained by removing influence of the cytokinin-related gene when the cytokinin-related gene is used in combination with a DNA factor having leaving ability, for example, in JP-A-9-154580.

A plantlet into which the desired gene has been introduced can be regenerated by cutting out the thus obtained adventitious bud and transplanting it on a rooting medium containing, for example, 0 to 1.0 mg/ml auxins, for rooting.

The present invention is based on the knowledge that a desired gene is efficiently introduced into cells of a base part of a shoot having no growing point when the base part is subjected to a gene introduction treatment while keeping a growing point on at least one end of the shoot, and that the base part of the shoot into which the desired gene has been introduced in this manner has higher ability to redifferentiate an adventitious bud into which the desired gene has been introduced, in comparison with the case of introducing the desired gene into the base of the shoot from which growing point has been removed.

The reason for this is not necessarily clear. However, it is considered that the growing point which is present in at least one end of the same shoot is contributing to this in some forms. Since plant tissues and cells are always damaged at a certain degree in carrying out gene introduction and the recovering strength from this damage is reinforced by the presence of a growing point, it seems that active growth ability is also maintained in the gene-introduced cells, and thereby acts advantageously on the introduction of the desired gene, and/or redifferentiation of an adventitious bud into which the desired gene has been introduced.

In addition, the base part of the shoot having no growing point is generally considered to be a part suitable for rooting. Accordingly, it is considered that when a gene is introduced into the part using a cytokinin-related gene as the selectable marker gene, difference in the ability differentiating an adventitious bud becomes sharply large between the gene-introduced cells and not-introduced cells and, as a result, the cells into which the desired gene has been introduced selectively differentiate adventitious buds. Thus, the use of such a gene as the selectable marker gene more advantageously results in the differentiation of the adventitious bud into which the desired gene has been introduced.

According to the present invention, introduction efficiency of a desired gene can be improved by the method for introducing a gene into a plant. Furthermore, in the present invention, an adventitious bud into which a desired gene has been introduced are differentiated efficiently from a tissue introduced with the desired gene. Thus, according to the present invention, introduction efficiency of the desired gene into an adventitious bud is particularly improved.

The present invention can be applied to many plants, and the effect of the present invention has a great meaning particularly for a plant in which gene introduction has been considered to be difficult. That is, according to the present invention, gene introduction into industrially important tree species can be carried out and the present invention opens a way for preparing a transformant which can be practically valued. The present invention is explained below based on Examples in details; however, the present invention is not limited thereto.

GENE MAPPING

An important, though by no means an essential, step in genetic analysis is to produce genetic maps of the marker loci. Such maps are often referred to as 'framework maps' because they provide a framework within which important genes can be located, as well as providing a means of comparing chromosome organization in other closely or distantly related species.

There are two stages to mapping. Firstly, to arrange the markers in a linear sequence separated by an appropriate map distance, *i.e.* to construct a linkage map. The second is to relate the linkage maps to particular recognizable chromosomes. The latter is often the most difficult and generally not essential for breeding or conservation work. We will, therefore, concentrate on the former.

Chromosomes contain a single linear molecule of DNA and hence the markers on that chromosome occur at particular positions along that molecule. Typically, each chromosome contains 107 to 108 base pairs (bp), *i.e.* 104 to 105 kilo-base pairs (kbp) of DNA, while a typical structural gene, coding for a polypeptide chain, would be between 1 and 2 kbp long. Assuming only 10per cent of the genome is coding DNA, then a chromosome probably contains something of the order of 1000 to 10 000 genes. Fortunately, the fact that the chromosome is a linear molecule means that the genes need to be mapped in only one dimension.

Currently, the only useful method of gene mapping, at least as far as breeders and conservationists are concerned, relies on recombination between homologous chromosomes resulting from genetic exchange, which can be seen as chiasmata at diplotene through first metaphase of meiosis. A single chiasma on a particular chromosome results in half the gametes from such a meiosis being recombinant for that chromosome: half the gametes will contain a copy of that chromosome that is part maternal and part paternal in origin, while the remaining gametes will be entirely parental. Very few plant species have a sufficiently low number of clearly recognizable chromosomes that the number of chiasmata on a particular chromosome can be compared in different nuclei.

However, in many species the total number of chiasmata in each diplotene nucleus can be counted, at least in pollen meiosis, and the average number of chiasmata per nucleus calculated. The number and position of chiasmata on a particular pair of homologous chromosomes will vary from nucleus to nucleus; it is as though there are a large number of potential sites of exchange, but that in any given meiosis only a few sites are actually involved.

The longer the chromosome the more potential sites there might be. There will invariably be one chiasma and there may be two, three, four or more; a typical set of results is shown in Table 8.1a. A chromosome that has one chiasma on average is said to be 50 centimorgans (cM) long – a map length unit named after the American geneticist Thomas Hunt Morgan.

This relates to the fact mentioned above that one chiasma results in 50per cent recombinant chromosomes. By extension, a chromosome with an average of 2.5 chiasmata is said to be 125 cM (*i.e.* 2.5 350 cM) long. It appears that, as a rough rule of thumb, each chromosome has on average two chiasmata and so is 100 cM long. It follows, therefore, that providing the haploid chromosome number is known (n), the total map length will be approximately 2 3 n 3 50 cM. The average number of chiasmata per nucleus has been calculated in many species and this rule generally holds.

Table. Chiasma Fraquency and Mapping.Chiasma Frequency in Chromosomes of Secale Cereale.

Number of chiasmata per chromosome	Frequency	Percentage unrecombined Chromosomes
1	0.267	50.0
2	0.716	25.0
³3	0.017	12.5
Mean	–	22.0

This rough rule is important because it gives the geneticist an idea of the total genetic map length that should be expected as more genes or markers are mapped. It provides a guide to the extent to which the currently mapped markers cover the full genetic map. Because the map is based in chiasma units, it is not the same as a physical map. The distribution of chiasmata is not uniform over a given chromosome in a species and will vary with the chromosome and the species.

Although a knowledge of the number of chromosomes and, ideally, the mean chiasma frequency per nucleus for the species, indicates the total map length, the actual map has to be constructed from examining the frequencies of progeny in crosses; *i.e.* maps are constructed from information based on the consequences of chiasmata, namely recombination of genetic markers. Recombination of genetic markers is identified by the presence of gametes that contain recombined markers, and such gametes can be recognized from the phenotypes of progeny of a cross.

The frequency of such gametes is an estimate of the recombination frequency between the two markers concerned. The closer two markers are to each other on the chromosome, the less likely a chiasma will occur between them and the less likely they are to recombine. Providing the two markers are sufficiently close that either none or one chiasma occurs between them but never more, then the recombination frequency as a percentage is equal to the map distance in centimorgans as described above. The problem arises when the markers are sufficiently far apart for two or more chiasmata to occur between them in some meioses. It can be shown that not only does a single chiasma between a pair of markers result in 50per cent recombination between them, but so also do two, three or more chiasmata, on average.

It is necessary to say 'on average', because there are a variety of possible consequences with two, three or more chiasmata in any given meiosis but, in practice, when one is looking at the progeny of a cross or self, each progeny will be the result of gametes from different meioses.This fact implies that although map distance increases linearly with the number of chiasmata, the same is not true with the frequency of recombination, which can never be more than 50per cent for a pair of markers even though they might be 100 cM apart at opposite ends of a chromosome.

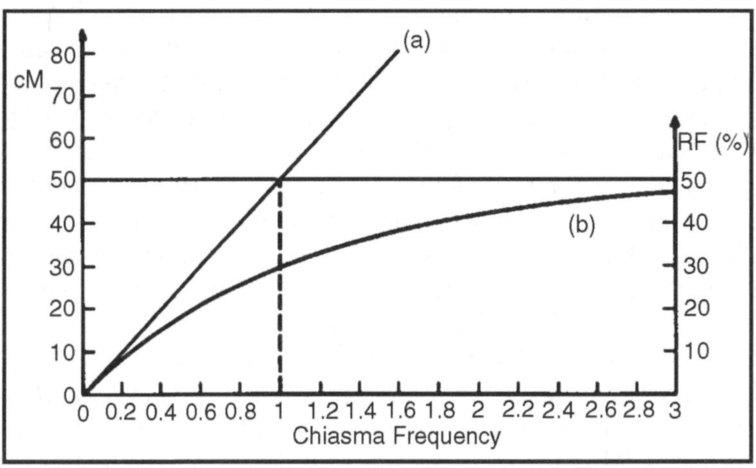

Fig. The Relationship of Chiasma Frequency with (a) Genetic Map Length and(b)Recombination Frequency.

These relationships are illustrated in Fig.. In order to overcome this problem, mapping functions have been devised to correct for multiple chiasmata. The most common of these are the Haldane (1919) and Kosambi (1943) mapping functions. Haldane's assumes that the probability of none, one, two or more chiasmata in a given interval follows a Poisson distribution, *i.e.* that the chiasmata are independent, random events. Kosambi's method allows for a certain degree of nonindependence in chiasma occurrences. It has long been known by cytogeneticists that chiasma interference occurs over quite large regions of a chromosome.

Interference means that if a chiasma occurs at a particular point on the chromosome, the next one will never occur closer than some fixed distance, the interference distance, from it. Beyond this distance, there is a short distance in which the interference disappears and the next chiasma can then occur randomly outside this range.

Observational data in various species of plants are confirming this and suggest that the interference distance is generally between 15 and 20 cM, *i.e.* 15 to 20per cent of a typical chromosome on either side of the chiasma. This has important and useful consequences for genetic and breeding work as will be shown later.

Recombination frequencies between pairs of markers can be scored in a wide variety of different crosses. The simplest to use are generations derived from an F1 because only two alleles are segregating, and their distribution in the chromosomes of the parents of the F1 can be determined. The generations that can be used are F2, backcrosses (Bc), recombinant inbred lines (RILs) or doubled haploid (DH) lines, with F2s being the most informative.

General formulae for calculating recombination frequencies are given by Allard (1956). In outbreeding species where F1s are not available, a given individual can be considered as an F1 and its selfed progeny as being an F2, even though the inbred parents do not exist. If it cannot be selfed but controlled crossing to another individual is possible, then again those genes segregating in the cross can be mapped although the situation is more complex.

There could be from two to four alleles segregating at each locus and, with respect to any one locus, the cross could represent an F2, Bc1 or Bc2. Moreover, the distribution of alleles in the two parents cannot be known and has to be inferred from the progeny. This complexity is compounded by the large number of marker loci that may be segregating in any given cross.

Fortunately, a range of software packages are available to estimate these recombination frequencies, identify linkage groups, assign the markers to the most likely order and space them in map units (cM) on these linkage groups; examples are MAPMAKER and JOINMAP. Ideally the number of linkage groups should be equal to the chromosome number in the gametes but, unless the markers available provide a good coverage of the genome, it is likely that those markers on a given chromosome may appear as two or more separate linkage groups simply because the subsets are not sufficiently close for them to be recognized as being together on one chromosome.

A typical marker framework map is shown in Fig. for the chromosomes of *Brassica oleracea*. Clearly these maps provide a very detailed coverage of the chromosome. Many of the markers are very close, *i.e.* less than 10 cM, and over this distance it matters little which of the two mapping functions is used. With the more widely spaced markers, *i.e.* recombination frequency >15per cent, Haldane's function will exaggerate their distance apart because it will allow for more double crossovers than actually occur.

As more markers are mapped, the total length should converge on the value predicted by the chiasma frequency, *i.e.* approximately 2 3 *n* 3 50 cM. It is important to remember that genetic map distances have standard errors which are dependent on the size and type of population used to construct the map, the mapping population. The map obtained is that which best fits, given the data and the assumptions underlying it, but there may well be other maps which also fit the data well, though are slightly less likely.

Any particular map always develops a greater aura of respectability once it is published, frequently without stating its reliability, and subsequent yet different maps are treated with suspicion. A recombination frequency of true value *p* has a standard error of ?[*p*(1 2 *p*)/*N*], where *N* is the size of the gamete population. For example, if *p* is 0.1 (*i.e.* RF = 10 per cent) and *N* = 100, the standard error is 0.03 or 3 per cent. In other words the 95 per cent confidence interval of the estimate lies approximately between 4 per cent and 16 per cent.

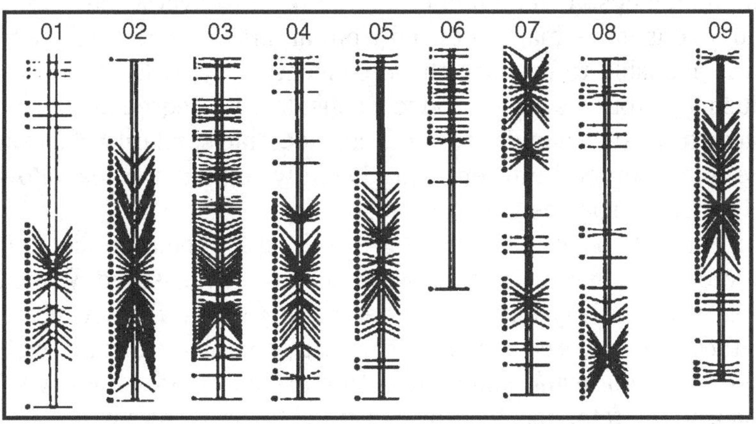

Fig. Typical Molecular Marker Framework Map. The Nine Groups of Brassica Oleracea Based on RFLP, Isozyme and Morphological Markers

There are other reasons, apart from statistical sampling, why maps could be wrong. The most obvious is the accuracy with which the data are scored and recorded. Autoradiographs and banding patterns on gels can easily be misread and research workers are loath to discard data even though its interpretation is ambiguous. All data should be checked independently by two or more people and any ambiguities removed. Wrong scores will bias the data and often exaggerate map lengths as they suggest double recombination; wherever double recombination appears to have occurred in two short, adjacent intervals, the data should be checked. Changes in methylation rather than scoring errors could cause a restriction site to appear or disappear so giving the false impression of double crossing- over, or it could simply be an error.

Where maps appear to have major inconsistencies in different crosses, different chromosomal structural arrangements may be responsible, such as translocations or inversions. There is considerable evidence that chiasma frequencies, and hence recombination frequencies, may be quite different in male and female meioses even on the same plant although in other studies this may not be so.

Thus although the order of markers on a linkage group may remain the same, the relative distances between them may be quite different if the map derives from male versus female meioses. For example, in *Brassica* it appears that the genetic map obtained from a backcross where the F1 is the female parent is 60per cent longer than when the F1 is the male parent. It is also clear that environmental factors during meiosis, particularly temperature, can affect the number and distribution of chiasmata, and this could be very critical if the crosses are set up at different times of the year.

Two or more different populations will almost certainly be segregating for different combinations of polymorphic markers, although some at least should be in common. These common ones can be used to overlay the two or more maps and a consensus map produced from an amalgamation of these. Again software is available to produce these consensus maps. Despite the various error-causing factors discussed above a considerable degree of consensus can be produced from such populations and, considerable similarity of chromosome sequence is conserved between even distantly related species, allowing the potential for cross-species genetic transfer.

It was stated above that the total map length should be dictated by the chiasma frequency, in so far as this can be accurately measured. When the first few genes were initially put onto genetic maps in the first 40 years of this century, they only covered a small part of the genome – except in wellstudied species such as *Drosophila* and maize. As more and more genes were placed on the map, so the total map lengths increased towards the asymptote dictated by the chiasma frequency. However, during the first decade following the use of molecular markers, the map lengths of some species appeared to exceed that expected, and it was even argued by some that the chiasma theory of recombination might be wrong.

No one would wish to argue that chiasma frequency data are free from error but there would have had to have been quite excessive underscoring to result in the disparity which was occasionally found. It would appear, however, that the excessive lengths were due in part to errors in scoring and to the use of small mapping populations, because subsequent map sizes have shown a progressive decrease towards the predicted asymptote. Scoring errors bias estimates of RF generally upwards, while RFs have to be large to be detectable in small populations. Our own experience with mapping in cereals and brassicas has also shown a progressive decrease in estimated map length as more markers can be found to fill some of the large gaps in the map and as errors are removed from the mapping data.

THE GENE REVOLUTION: A CHANGING PARADIGM FOR AGRICULTURAL R ANDD

The Green Revolution brought high-yielding semi-dwarf wheat and rice varieties, developed with conventional breeding methods, to millions of small-scale farmers, initially in Asia and Latin America, but later in Africa as well. The gains achieved during the early decades of the Green Revolution were extended in the 1980s and 1990s to other crops and to less favoured regions. In comparison with the research that drove the Green Revolution, the majority of agricultural biotechnology research and almost all of the commercialization is being carried out by private firms based in industrialized countries.

This is a dramatic departure from the Green Revolution, in which the public sector played a strong role in research and technology diffusion. This paradigm shift has important implications for the kind of research that is performed, the types of technologies that are developed and the way these technologies are disseminated. The dominance of the private sector in agricultural biotechnology raises concerns that farmers in developing countries, particularly poor farmers, may not benefit - either because appropriate innovations are not available or are too expensive.

Public-sector research was responsible for creating the high-yielding varieties of wheat and rice that launched the Green Revolution. International and national public-sector researchers bred dwarfing genes into elite wheat and rice cultivars, causing them to produce more grain and have shorter stems and enabling them to respond to higher levels of fertilizer and water. These semi-dwarf cultivars were made freely available to plant breeders from developing countries who further adapted them to meet local production conditions. Private firms were involved in the development and commercialization of locally adapted varieties in some countries, but the improved germplasm was provided by the public sector and disseminated freely as a public good.

The countries that were able to make the most of the opportunities presented by the Green Revolution were those that had, or quickly developed, strong national capacity in agricultural research. Researchers in these countries were able to make the necessary local adaptations to ensure that the improved varieties suited the needs of their farmers and consumers.

National agricultural research capacity was a critical determinant of the availability and accessibility of Green Revolution agricultural technologies, and this remains true today for new biotechnologies. National research capacity increases the ability of a country to import and adapt agricultural technologies developed elsewhere, to develop applications that address local needs (*e.g.* "orphan crops") and to regulate new technologies appropriately.

The biotechnology revolution, by contrast, is being driven largely by the private sector. Public-sector research has contributed to the basic science underpinning agricultural biotechnology, but the private sector is responsible for most applied research and almost all commercial development. Three interrelated forces are transforming the system for providing improved agricultural technologies to the world's farmers. The first is the strengthening environment for protecting intellectual property in plant innovations. The second is the rapid pace of discovery and the growing importance of molecular biology and genetic engineering. Finally, agricultural input and output trade is becoming more open in nearly all countries, enlarging the potential market for new technologies and older related technologies. These developments have created powerful new incentives for private research, and are altering the structure of the public/private agricultural research endeavour, particularly with respect to crop improvemen.

With the growing importance of the private transnational sector, developing countries are facing increasing transaction costs in access to and use of technologies. Existing public-sector international networks for sharing technologies across countries and thereby maximizing spillover benefits are becoming increasingly threatened. The urgent need today is for a system of technology flows that preserves the incentives for private-sector innovation while at the same time meeting the needs of poor farmers in the developing world.

The organization and impacts of agricultural research and technology flows in the period 1960-90, when the Green Revolution paradigm of international, public-sector research held sway. The movement towards the increased privatization of agricultural research and development and its consequences for developing country access to technologies as revealed in recent global trends in biotechnology research, development and commercialization. A number of questions regarding the potential of the Gene Revolution to benefit the poor. These questions are taken up in the subsequent chapters of the report.

The world population may reach 10 billion by the middle of this century. Over the next 20 years, world cereal demand will increase by 50 per cent, driven by rapidly growing animal feed use and meat consumption. With the exception of acid-soil areas in Africa and South America, the potential for expanding global crop area is limited. Future expansions in food output must come largely from land already in use. The productivity of this land must be sustained and improved. Most of the world's 842 million hungry people live in marginal lands and depend upon agriculture for their livelihoods. Food-insecure households in these higher-risk rural areas face frequent droughts, degraded lands, remoteness from markets and poor market institutions. For many of these people, food security will only come through increased agricultural production and income. Investments in science, infrastructure and resource conservation are needed to increase productivity and lower risks in marginal lands. Some of the problems in such environments will be too formidable to overcome. However, significant improvements should be possible. Biotechnology will play an important role in developing new germplasm with greater tolerance to abiotic and biotic stresses and with higher nutritional content. Continued genetic improvement of food crops - using conventional research tools and biotechnology - is needed to shift the yield frontier higher and to increase stability of yield.

Neolithic man - or much more likely woman - domesticated virtually all of our food and livestock species over a relatively short period, 10 000-15 000 years ago. Subsequently, several hundred generations of farmers were responsible for making enormous genetic modifications in all of our major crop and animal species. Thanks to the development of science over the past 150 years, we now have the insights into plant genetics and breeding to do purposefully what Nature did in the past by chance or design. Genetic modification of crops is not some kind of witchcraft; rather, it is the progressive harnessing of the forces of nature to the benefit of feeding the human race. Indeed, genetic engineering - plant breeding at the molecular level - is just another step in humankind's deepening scientific journey into living genomes. It is not a replacement for conventional breeding but a complementary research tool to identify desirable traits from remotely related taxonomic groups and transfer them more quickly and precisely into high-yielding, high-quality crop species.

The world has the technology - already available or well advanced in the research pipeline - to feed on a sustainable basis a population of 10 billion people. However, access to such technology is not assured. The range of potential barriers includes issues related to intellectual property rights, technology acceptance by civil society and governments, and financial and educational barriers that keep poor farmers marginalized and unable to adopt new technology.

THE GREEN REVOLUTION: RESEARCH, DEVELOPMENT, ACCESS AND IMPACT

The Green Revolution was responsible for an extraordinary period of growth in food crop productivity in the developing world over the last 40 years. A combination of high rates of investment in crop research, infrastructure and market development, and appropriate policy support fuelled this progress. These elements of the Green Revolution strategy improved productivity growth despite increasing land scarcity and high land values.

Public-sector Research and International Technology Transfer

The Green Revolution defied the conventional wisdom that agricultural technology does not travel well because it is either agroclimatically specific, as in the case of biological technology, or sensitive to relative factor prices, as with mechanical technology. The Green Revolution strategy for food-crop productivity growth was explicitly based on the premise that, given appropriate institutional mechanisms, technology spillovers across political and agroclimatic boundaries could be created. Hence the Consultative Group on International Agricultural Research (CGIAR) was established specifically to generate technology spillovers, particularly for countries that are unable to capture all the benefits of their research investments. What happens to the spillover benefits from agricultural research and development in an increasingly global integration of food supply systems?

The major breakthroughs in yield potential that kick-started the Green Revolution in the late 1960s came from conventional plant-breeding approaches that initially focused on raising yield potential for the major cereal crops. The yield potential for the major cereals has continued to rise at a steady rate after the initial dramatic shifts in the 1960s for rice and wheat. For example, yield potential in irrigated wheat has been rising at the rate of 1 per cent per year over the past three decades, an increase of around 100 kg/ha/year. Essentially, no research or elite germplasm was available for many of the crops grown by poor farmers in less favourable agro-ecological zones (such as sorghum, millet, barley, cassava and pulses) during the early decades of the Green Revolution, but since the 1980s modern varieties have been developed for these crops and their yield potential has risen.

In addition to their work on shifting the yield frontier of cereal crops, plant breeders continue to have successes in the less glamorous but no less important areas of applied research. These include development of plants with durable resistance to a wide spectrum of insects and diseases, plants that are better able to tolerate a variety of physical stresses, crops that require a significantly lower number of days of cultivation, and cereal grain with enhanced taste and nutritional qualities.

Prior to 1960, there was no formal system in place that provided plant breeders with access to germplasm available beyond their borders. Since then, the international public sector (the CGIAR system) has been the predominant source of supply of improved germplasm developed from conventional breeding approaches, especially for self-pollinating crops such as rice and wheat and for open-pollinated maize. These CGIAR-managed networks evolved in the 1970s and 1980s, when financial resources for public agricultural research were expanding and plant intellectual property laws were weak or non-existent. The exchange of germplasm is based on a system of informal exchange among plant breeders that is generally open and without charge. Breeders can contribute any of their material to the nursery and take pride in its adoption elsewhere in the world, while at the same time they are free to pick material from the trials for their own use.

The international flow of germplasm has had a large impact on the speed and the cost of crop development programmes of national agricultural research systems (NARS), thereby generating enormous efficiency gains. Traxler and Pingali argued that the existence of a free and uninhibited system of germplasm exchange that attracts the best of international materials allows countries to make strategic decisions on the extent to which they need to invest in plant-breeding capacity. Even NARS with advanced crop research programmes, such as in Brazil, China and India, rely heavily on cultivars taken from these nurseries for their prebreeding material and for finished varieties. Small countries behaving rationally choose to free-ride on the international system rather than invest in large crop-breeding infrastructure of their own.

Evenson and Gollin report that, even in the 1990s, the CGIAR content of modern varieties was high for most food crops; 35 per cent of all varietal releases were based on CGIAR crosses, and an additional 22 per cent had a CGIAR-crossed parent or other ancestor. Evenson and Gollin suggest that germplasm contributions from international centres enabled developing countries to capture the spillover benefits of investments in crop improvement made outside their borders and achieve productivity gains that would have been more costly or even impossible had they been forced to work only with the genetic resources that were available at the beginning of the period.

IMPACTS OF FOOD-CROP IMPROVEMENT TECHNOLOGY

Substantial empirical evidence exists on the production, productivity, income and human welfare impacts of modern agricultural science and the international flow of modern varieties of food crops. Evenson and Gollin provide detailed information on the extent of adoption and impact of modern variety use for all the major food crops. The adoption of modern varieties (averaged across all crops) increased rapidly during the two decades of the Green Revolution, and even more rapidly in the following decades, from 9 per cent in 1970 to 29 per cent in 1980, 46 per cent in 1990 and 63 per cent by 1998. Moreover, in many areas and in many crops, first-generation modern varieties have been replaced by second- and third-generation modern varieties.

Much of the increase in agricultural output over the past 40 years has come from an increase in yield per hectare rather than an expansion of area under cultivation. For instance, FAO data indicate that for all developing countries, wheat yields rose 208 per cent from 1960 to 2000; rice yields rose 109 per cent; maize yields rose 157 per cent; potato yields rose 78 per cent and cassava yields rose 36 per cent. Trends in total factor productivity are consistent with partial productivity measures, such as rate of yield growth.

The returns to investments in high-yielding modern germplasm have been measured in great detail by several economists over the last few decades. Several recent reports have reviewed and analysed the data from hundreds of studies conducted over the last 30 years that calculated the social rates of return to investments in agricultural research. These studies examined investments by national and international public-sector institutions in Africa, Asia, Latin America and the Organisation for Economic Co-operation and Development (OECD) countries as well as by the private sector. Although these studies were carried out using a variety of different methods, they showed considerable consistency. The average social rate of return to public investment in agricultural research reported in these studies is in the region of 40-50 per cent. Private-sector research was also found to generate similar rates of social returns.

The primary effect of agricultural research on the non-farm poor, as well as on the rural poor who are net purchasers of food, is through lower food prices. The wide-spread adoption of modern seed-fertilizer technology led to a significant shift in the food supply function, increasing output and contributing to a fall in real food prices:

- The effect of agricultural research on improving the purchasing power of the poor - both by raising their incomes and by lowering the prices of staple food products - is probably the major source of nutritional gains associated with agricultural research. Only the poor go hungry. Because a relatively high proportion of any income gains made by the poor is spent on food, the income effects of research-induced supply shifts can have major nutritional implications, particularly if those shifts result from technologies aimed at the poorest producers.

Studies by economists have provided empirical support for the proposition that growth in the agriculture sector has economy-wide effects. Hayami *et al.* illustrated at the village level that rapid growth in rice production stimulated demand and prices for land, labour and non-agricultural goods and services. For sector-level validation of the proposition that agriculture does indeed act as an engine of overall economic growth.

Once modern varieties have been adopted, the next set of technologies that makes a significant difference in reducing production costs includes machinery, land management practices (often in association with herbicide use), fertilizer use, integrated pest management and (most recently) improved water management practices. Although many Green Revolution technologies were developed and extended in package form (*e.g.* new plant varieties plus recommended fertilizer, pesticide and herbicide rates, along with water control measures), many components of these technologies were taken up in a piecemeal, often stepwise manner. The sequence of adoption is determined by factor scarcities and the potential cost savings achieved. Herdt provided a detailed assessment of the sequential adoption of crop management technologies for rice in the Philippines. Traxler and Byerlee provided similar evidence on the sequential adoption of crop management technologies for wheat in Sonora, northwestern Mexico.

Although the favourable, high-potential environments gained the most from the Green Revolution in terms of productivity growth, the less favourable environments also benefited through technology spillovers and through labour migration to more productive environments. According to David and Otsuka (1994), wage equalization across favourable and unfavourable environments was one of the primary means of redistributing the gains of technological change. Renkow found similar results for wheat grown in high- and low-potential environments in Pakistan. Byerlee and Moya, in their global assessment of the adoption of modern varieties of wheat, found that over time the adoption of modern varieties in unfavourable environments caught up with those in more favourable environments, particularly when germplasm developed for high-potential environments was further adapted to the more marginal environments. In the case of wheat, the rate of growth in yield potential in drought-prone environments was around 2.5 per cent per year during the 1980s and 1990s. Initially, the growth in yield potential for the marginal environments came from technological spillovers as varieties bred for the high-potential environments were adapted to the marginal environments. During the 1990s, however, further gains in yield potential came from breeding efforts targeted specifically at the marginal environments. Even while highlighting the yield breakthrough in wheat, the Government had also launched a massive programme to develop and spread high-yielding varieties for rice, maize, sorghum and pearl millet. These programmes were the drivers of the "Green Revolution" in India, which permitted striking advances in production and productivity without increasing cultivated area.

Because these high-yielding varieties require inputs such as fertilizer and irrigation water, social scientists criticized the Green Revolution technologies for not being resource neutral. Environmentalists attacked the Green Revolution because of potential damage to long-term productivity as a result of excessive use of pesticides and fertilizers and monocropping. Despite the success of the Green Revolution in raising millions of people out of misery, the incidence of poverty, endemic hunger, communicable diseases, infant and maternal mortality rates, low birth-weight children, stunting and illiteracy remain high.

The concerns of social scientists and ecologists and the remaining urgent problems of poverty and hunger led to my developing the concept of an "evergreen revolution" to stress the need for enhancing crop productivity in perpetuity without associated ecological or social harm. An evergreen revolution can be achieved only if we pay attention to pathways that can help to achieve revolutionary progress in enhancing productivity, quality and value-addition under conditions of diminishing per capita arable land and irrigation water availability, expanding biotic and abiotic stresses, and fast-changing consumer and market preferences. This will require mobilizing the best in both traditional wisdom and technologies and frontier science. Among the frontier technologies relevant to the next stage in our agricultural revolution, the foremost is biotechnology.

The apprehensions relating to molecular genetics and genetic engineering fall under the following broad categories: the science itself, the control of the science, access to the science, environmental concerns, and human and animal health. A disaggregated approach to the study of these issues will be important for a rigorous analysis of risks and benefits. Dealing with these issues in a composite manner for all applications of genetic engineering will result in inappropriately broad conclusions, such as the general condemnation of GMOs expressed by non-governmental organizations (NGOs) at the World Food Summit: *five years later* held in Rome in 2002.

The benefits of molecular breeding techniques such as the use of molecular markers and undertaking precision breeding for specific characters through recombinant DNA technology are immense. The work already performed in India has revealed the potential for breeding new GM varieties possessing tolerance to salinity, drought and some major pests and diseases, together with improved nutritive quality. A new era of integrated Mendelian and molecular breeding has begun.

An evergreen revolution will blend these frontier technologies with the ecological prudence of traditional communities to create technologies that are based on integrated natural resource management and that are location specific because they are developed through participatory experimentation with farm families.

This is the only way we can face the challenges of the future, particularly in the context of the growing water scarcity and the urgent need to step up productivity in semi-arid and dry farming areas. Accelerated agricultural progress is the best safety net against hunger and poverty, because in most developing countries over 70 per cent of the population depend on agriculture for their livelihood. Denying ourselves the power of the new genetics will be doing a great disservice both to resource-poor farming families and to the building of a sustainable national food and nutrition system.

LOCATING GENES OF MAJOR EFFECT

The use of extensive molecular and other markers as described above provides a general framework map. Although this has value in its own right for comparing linkage groups in different species as well as possible structural variation within a species, the main value lies in providing a set of markers to locate genes of economic or special scientific interest.

Locating individual major genes is a straightforward development of the procedures above. Let us assume that a cultivar is identified which contains an allele of interest which changes the phenotype in a clearly recognizable fashion and which appears to segregate as a single gene in crosses to other cultivars. Using a knowledge of the positions of existing marker loci, a small subset of markers can be chosen which provide a reasonably even coverage of the genome, say every 20 to 30 cM, *i.e.* approximately five per chromosome. These then have to be shown to differ between the two cultivars and any which are monomorphic replaced by polymorphic markers.

A rough position of the gene of interest can then be obtained by bulked segregant analysis. This involves taking the mapping population, *e.g.* a Bc, and dividing the plants into two groups depending on whether or not they show the phenotype of interest. Bulk samples of DNA are taken from each group, and the two samples assayed for the marker systems chosen as in the paragraph above.

Any marker which is unlinked to the gene to be located will produce similar banding patterns in both samples because it will be segregating independently of the target gene. Conversely, should one of the markers be very closely linked to the target gene, it will co-segregate and the two samples will show quite different patterns for that marker.

In practice, complete co-segregation with any one marker is not very likely but, given a reasonable spread of markers over the genome, one or two are likely to show a strong association with the two groups. Thus although bands associated with both marker alleles will be found in each sample, their relative intensities will be quite different because of the association with the target gene.

Having located the chromosome and the region on the chromosome, the actual position can be identified in a segregating population using the principal marker associated in the bulked segregant analysis together with other polymorphic markers situated about 10 cM on either side of the principal marker.

The target gene is then mapped with respect to the three markers. This approach can be used for a wide variety of populations and traits. Even if the trait has poor penetrance, the affected group will clearly show the marker banding pattern even if the nonaffected group consists of a mixture due to misclassification.

MAPPING GENES CONTROLLING QUANTITATIVE TRAITS

Quantitative traits such as yield, quality, height and flowering time, which are controlled by several genes and are greatly influenced by the environment, create particular difficulties for gene mapping. The problems arise because the genotype for the trait concerned can never be clearly identified from the phenotype; many different genotypes could produce the same or very similar phenotypes whilst the same genotype can result in different phenotypes depending on what may be elusive factors in the environment.

Whereas with a single gene difference controlling a major effect, two or more Mendelian phenotypic classes might be recognized in ratios of 3:1, etc., quantitative traits resulting from the joint segregation of many genes show a continuous, often normally distributed range of phenotypes. The total phenotypic variation in an F2, for example, VP, is made up of genetic and environmental components, VG and VE. The genes underlying such traits are variously called polygenes, effective factors or, more recently, quantitative trait loci or QTL.

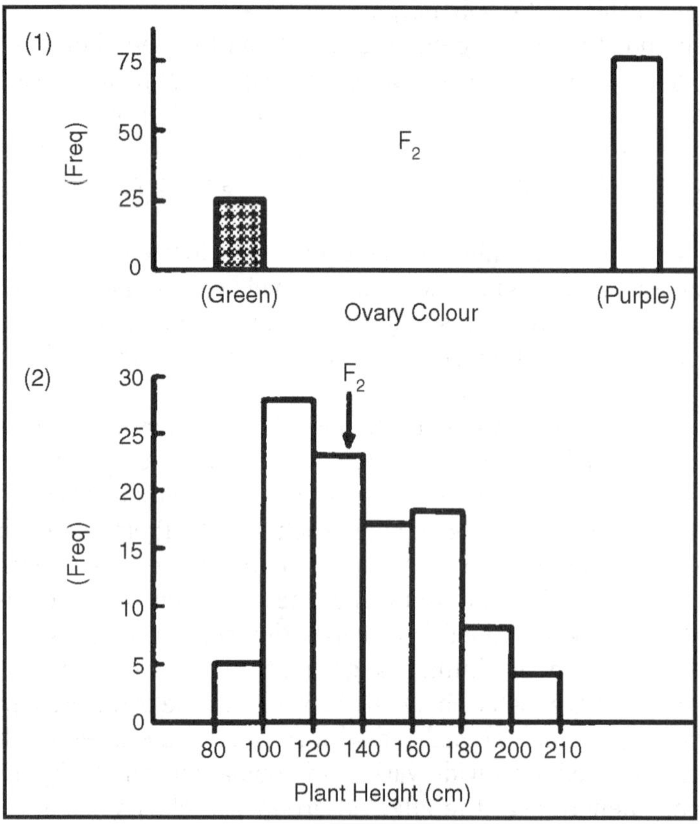

Fig. Contrast Between Distribution of a Qualitive and Quantitative trait in an F_2. Data for NIcotiana Rustica,Ovary Colour and Height.

However, it is still possible to map and measure the effects of the QTL by techniques analogous to those used for single major genes, *i.e.* by looking for the effects of co-segregation of particular marker loci with differences in the quantitative trait. Various mapping populations can be used as before, F_2, Bc, RIL, DH or open-pollinated full sib (FS) or half sib (HS) families, and the individuals or families are scored both for their phenotype for the quantitative trait(s) and for their genotype at the marker loci. The type of population that is used will depend on the traits studied and the breeding system of the plant species but, other things being equal, an F_2 population will normally provide the most information for a given size. Unfortunately, most traits of economic importance such as yield, quality and disease resistance are not usefully scored in an F_2 population because one is interested in how the material performs in something similar to the high-density monoculture that it will encounter in agricultural practice. Measuring yield in an F_2 population with spaced plants is straightforward and statistically very powerful but it may well not provide any insight into how yield is controlled in agricultural practice.

It is therefore necessary to use a plot structure for most traits in an attempt to simulate commercial conditions. For this reason most plant breeders will prefer to work with genotypes that can be extensively and easily replicated such as RILs, DHs or F_3s. It is also easier to explain the principles underlying QTL mapping using a population of DH lines, and so for both reasons the methodology will be illustrated with DH lines, although the same principles apply to all populations. Because several trait loci are concerned, probably on different chromosomes, bulk segregant analysis is of limited applicability, at least initially.

THEORY OF QTL MAPPING IN DOUBLED HAPLOID LINES

Doubled haploid (DH) lines can be produced parthenogenetically from an F1 by microspore or ovule culture and, in some species, such as wheat and barley, by wide outcrossing. Each original DH plant is derived from a single gamete, which subsequently becomes homozygous and disomic either naturally or by treatment with the drug colchicine, which inhibits spindle formation, and hence chromosome disjunction, at metaphase of mitosis.

Selfed seeds from such a plant produce a DH line of identical individuals so the line becomes effectively immortal. If we consider an F1 that is heterozygous for a marker locus M1/M2 and for a linked QTL locus Q1/Q2, where the 1/2 indicates the allele with an increasing or decreasing effect on the trait, this will produce the gametes and hence the DH lines shown in Table.

If the DH lines are scored for some trait for which the Q+/Q+ homozygotes increase the trait above the overall mean, *m*, by *a* units while the Q2/Q2 homozygotes decrease the mean by *a*, then the means of the four gametic types are as shown in Table, where *R* is the recombination frequency between the QTL and the marker. From this it follows that the mean trait score of all those DH lines which are M1/M1 (M1M1) is *m* 1 *a*(1 2 2*R*), while the mean of all those DH lines which are M2/M2 (M2M2) is *m* 2 *a*(1 2 2*R*). In other words, half the difference between these two means (M1M1 2 M2M2) is *a*(1 2 2*R*). Clearly, if the marker is so close to the QTL that it never recombines, *R* will be 0 and the marker difference will represent the QTL effect, *a*. Conversely if the two loci are unlinked, *i.e.* $R = 0.5$, then the difference will be zero. This effect is, of course, the basis of bulked segregant analysis. In general, the magnitude of the marker effect will decline linearly with the distance of the marker from the QTL in terms of recombination frequency, and so with several linked markers, their individual trait effects will be as shown in Fig.. Similar Arguments Apply to Other Populations Such as F_2s and RILs.

Table. QTL model for market analysis in doubles haploid (DH) lines from and F_1.(a)The F_1 constitution; M=Market and Q = QLT genotype.(b)DH genotypes,frequencies and genetic values. (c) Expertations of marker means and differences.R=recombination frequency between M and Q; m and a are the mean and additive genetic effects of the QLT;d is the additive genetic deviation at the market locus

(a)

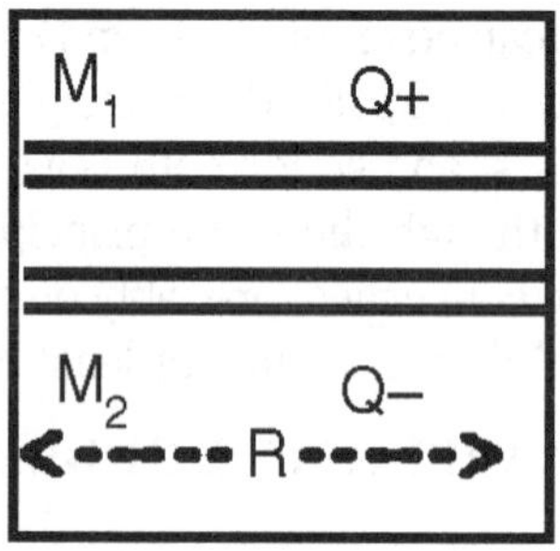

(b)

	F_1 gametes			
	$M_1 Q^+$	$M_1 Q^-$	$M_2 Q^+$	$M_2 Q^-$
Frequency	a(1–R)	a(R)	a(R)	a(1–R)
DHgenotype	$M_1M_1 Q^+Q^+$	$M_1M_1 Q^-Q^-$	$M_2M_2 Q^+Q^+$	$M_2M_2 Q^-Q^-$
Geneticvalue	m+a	m–a	m+a	m–a

(c)

$$\begin{aligned} M_1M_1 &= \{a(1-R)(m+a)+a(R)(m-a)\}a \\ &= m+(1-R)a \\ M_2M_2 &= \{a(R)(m+a)+a(R)(1-R)(m-a)\}a \\ &= m-(1-R)a \\ \delta &= (\overline{M_1M_1}-\overline{M_2M_2})/2 \\ &= (1-2R)a \end{aligned}$$

If the trait means of the markers and their map are known very accurately, and only one QTL existed on the chromosome, then as Fig. shows, it would be relatively easy to locate the QTL. Figure shows the distribution of flowering time among a number of doubled haploid lines of *Brassica oleracea* containing alternative alleles at a marker located very close to a QTL.

In practice neither the map positions of the markers nor their means are known with very great accuracy because the number of DH lines and replicated plots are generally few and the heritability of quantitative traits is generally low. Moreover, linked QTLs create difficulties because their individual effects can combine either to create the impression of an intermediate, ghost QTL, or to conceal their effects altogether.

Various analytical procedures have been developed to maximize the efficiency and accuracy of locating individual QTLs and to separate linked QTLs. They all rely on the relationships described in the previous paragraph and use either maximum likelihood or weighted least squares regression procedures to estimate the QTL locations and effects that best fit the observed trait scores for each genotype.

The main difficulty is that there are two unknown parameters with respect to each marker score; the QTL effect, *a*, and map position. This means that it is necessary, in all methods, to use an iterative approach which involves trying all possible QTL locations along each chromosome and identifying the position or positions of the QTL which best fit the observed data as indicated by the size of the likelihood or the residual regression variance. As a result, a very large number of statistical tests are performed with the concomitant risk of false positive results.

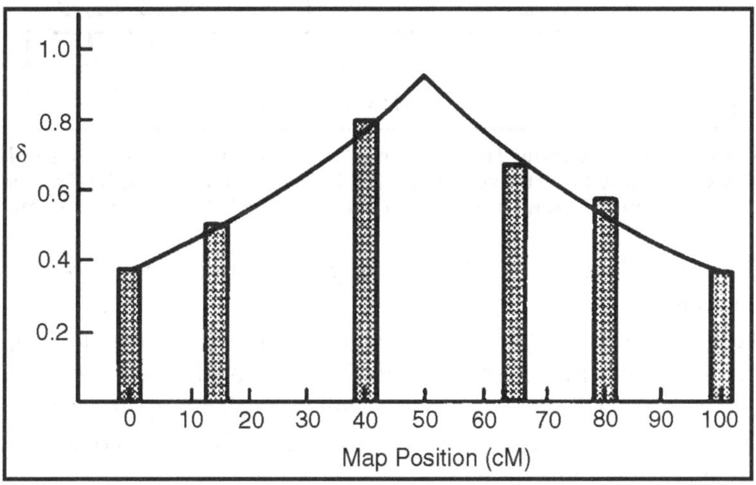

Fig. Quantitative Trait Effects (d) Associated with Marker Loci.The Bars Indicate the Observed Effects,While the Curve Illustrates the Expected Relationship if a QTL of d = 1.0 Existed at 50 cM

All methods provide estimates of QTL locations and effects with similar precision in terms of confidence intervals, and this precision drops rapidly as the heritability of individual QTLs declines, *i.e.* the estimates are accurate only when there are just two or three unlinked QTLs with large effects. For example, Hyne *et al.* (1995) showed that the 95per cent confidence interval for the location of a QTL contributing 10per cent to the phenotypic variance of an F2 could be as large as 35 cM.

However, conservationists and breeders who are concerned with introducing QTL into commercial cultivars from other cultivars or more distantly related species, would normally only be interested in those cases where a few QTLs of major effect are to be manipulated. There is no evidence that dense maps are required for initial

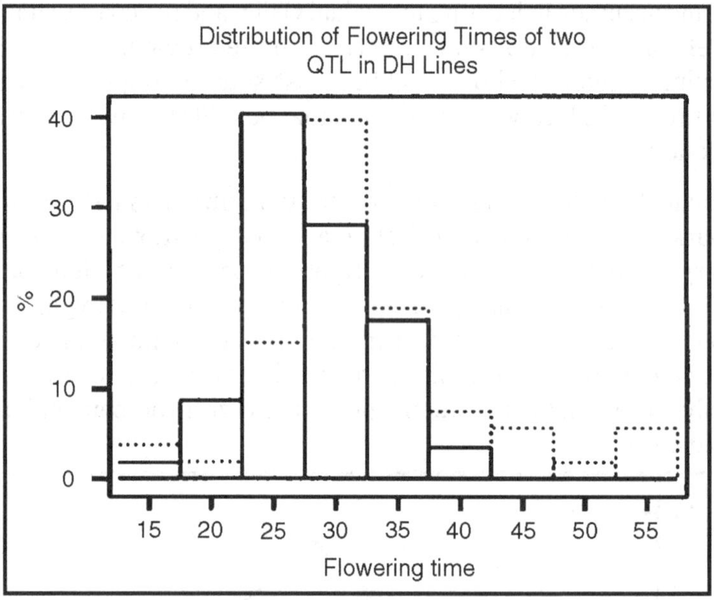

Fig. Distribution of Flowering Times Among Doubled Haploid (DH)Lines in Brassica Oleracea Containing alrernative Alleles of a Marker Linked to a QLT.The two sets Significantly Different Means Indicating a QLT with an Effect of Apporoximately ± 2 Days

QTL location. It is better to have large populations and a few (5–10) markers per chromosome.

There is evidence that many of the larger QTLs located so far map closely to previously known major genes. This is often used as evidence that there are, in fact, very few QTLs, but this may well be misleading for several reasons. The QTL map locations are sufficiently imprecise that they could well appear spuriously close to candidate loci.

Only QTLs of large effect will stand much chance of being located and, by definition, they are likely to be major genes. It is likely that many cases where QTLs have very large effects could be due to the chance association of alleles of like effect at several QTLs along a chromosome, which cannot be separated as individual effects.

Much effort has been devoted to increasing the precision of QTL location, in particular in improving the power of the test to reduce the failure to detect genuine QTL. A long-established problem with breeding for quantitative traits in plants is the genotype 3 environment interaction. Using QTL location techniques, it is now possible to explore these effects at the level of individual QTLs although it is often difficult to know whether genotype 3 environment or poor repeatability is responsible.

ALIEN GENE TRANSFER

Once useful genes have been located on a framework map, it is then possible to use the molecular markers to facilitate the transfer of the useful genes between species and strains in an efficient manner. As stated earlier, recombination is not a frequent event along a chromosome and, furthermore, two recombination events rarely occur within a distance of 15 to 20 cM of each other.

It therefore follows that if a useful target gene (a major gene or a QTL) is known to be located within a region of chromosome flanked by two markers that are no more than 15 cM apart, then these markers can be used to follow and control the progression of the target gene through successive stages of a breeding programme.

Clearly, the alleles at the marker loci have to differ in the donor and recipient cultivars, but any chromosome that contains A1 and B1 at the marker loci will also contain T1, the target allele to be transferred.

Should single recombination occur close to T, then the chromosome will contain A1 or B1 but not both. Providing the markers are less than 15 cM apart double recombination resulting in A1T2B1, and hence loss of T1, can be safely ignored. If the location of T is not accurately known, then three or more marker loci may be needed to be sure of safely bracketing the region containing T without fear of double recombination.

There are conflicting requirements in marker-assisted gene transfer: the need to keep the bracketed region large enough to be sure of holding the gene to be transferred whilst not having it so large that too many other linked but undesirable alleles are transferred with it. Let us now consider the basic procedure of gene transfer.

Traditionally, breeders have introduced a useful allele into their cultivars by backcrossing the original F1 to the commercial cultivar whilst selecting at each generation for the desired allele.

As was shown earlier, this can result in a very large region of chromosome around the target gene surviving even after ten generations of backcrossing – a phenomenon known as linkage drag. With markers, however, it is possible to reduce linkage drag considerably whilst simultaneously reducing the number of backcrossing generations to two or three instead of the normal six to eight. Moreover, the amount of plant material raised at each generation can be reduced because much of the molecular genotyping can be achieved at the juvenile stage.

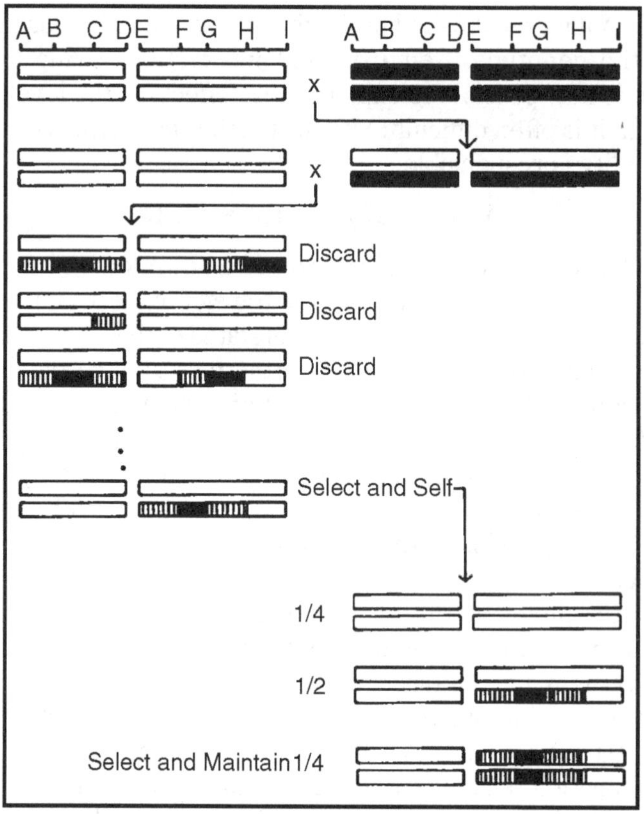

Fig. The use of Molecular Markers to Facilitate Gene Introgression.

The aim is to hold the target allele, T1, heterozygous throughout the backcrossing process, by selecting for the flanking markers, whilst simultaneously selecting for the genotype of the recurrent, commercial cultivar at all other loci. The latter can readily be achieved by having as few as four or five well-spaced markers on all chromosomes and selecting for recurrent parent alleles at each generation.

Depending on the chromosome number, some individuals in the first backcross generation will be homozygous for several whole chromosomes whilst still being heterozygous for the target gene. Those which have most of the recurrent parent chromosomes are selected and backcrossed again, and if necessary the process is repeated for a third generation. For each generation, fewer markers need to be screened as those found to be homozygous in the previous backcross no longer have to be checked.

Unless the position of the target gene is very accurately defined, it is prudent to have several markers covering its likely position, again with the proviso that they need to be separated by no more than 15 cM. At the final stage of the introgression process, a backcross individual which is homozygous for most of the recurrent parent alleles is selfed or, if this is not possible, intercrossed with a similar genotype. From among the progeny, individuals are chosen which have different combinations of markers bracketing the target region as shown in figure.

These can be selfed again and homozygotes for the different sequences identified and multiplied for reassessment for the target trait. Some will fail to show the target trait because of recombination, but of those that do, the lines with the shortest sequence of donor markers are selected for further trials. The approach can obviously be adapted to the simultaneous introgression of several genes: no new principles are involved but the screening process becomes more elaborate.

It is always possible at a later stage to reduce the length of the introgressed region around T by looking at other, more closely linked, markers. Similarly, it is wise to check the origins of the other chromosomes to avoid having introgressed unwanted regions through missing double recombinants in the central regions of individual chromosomes or single recombinants towards their ends.

If such errors are found they can easily be corrected by another round of backcrossing to the recurrent parent. When attempting to reduce the length of donor chromosome surrounding the target gene, it is important to realise that this should be done in two successive generations. If the sites of the required recombination events are less than 15 cM apart, then the simultaneous double event will not occur.

Even if they are far enough apart for the double event to occur, the probability of such an event may be too low to be practicable. If two recombination events each have a probability of 0.05 (*i.e.* 5per cent), their combined probability is at best 0.0025.

Therefore, it would be necessary to genotype approximately 1200 individuals to be 95per cent sure of having at least one double recombinant. If it was tackled in two stages then it would require only the genotyping of 86 (= 28 in the first round plus 58 in the second; less being needed in the first round because the recombination could occur on either side of T), a very considerable saving in effort of 93per cent, or 1114 plants!

Marker-aided selection has been successfully tried for quantitative traits in several crops while the efficiency of the procedure has been examined by Lande and Thompson (1990).

MAP-BASED GENE CLONING

Providing a target gene can be located to within a pair of marker loci, it should be possible to use standard cloning techniques to walk along the chromosome between the markers and to find the target gene. The feasibility of this depends on the distance between the markers and the ease with which the target gene can be recognized from among the clones. Depending on the species, a genetic length of 1 cM can, on average, consist of as few as 280 kbp of DNA in rice, or 2220 kbp in maize. It is not always easy to locate a target gene with sufficient accuracy to be able to say that it is between two markers as close as 1 cM, and hence the actual lengths of DNA between them can often be much greater. With QTLs one can seldom achieve anything approaching this accuracy by existing methods of conventional mapping. Furthermore, these are average distances per centimorgan; some chromosomal regions could be much longer or, indeed, less.

Increases in mapping reliability can be obtained by using what are called near isogenic lines (NILs), which are lines obtained by selfing or backcrossing and are known to differ from some standard genotype by just a short defined section of chromosome. Providing this section is delineated by accurately mapped flanking markers, and the NILs which do and do not contain this region can clearly be shown to differ for the target gene, then the target gene has to be in that region. The genetic and physical proximity of the markers can be established by conventional mapping and by gene cloning.

Map-based gene cloning involves starting from one of the markers and walking to the next by means of a series of partially overlapping cosmid clones. Cosmid clones are necessary because they can be up to 50 kb in length and hence require fewer cloning steps in order to cover the DNA between the markers (280 to 2200 kb per cM in the examples cited above). Identifying the target gene is more difficult. Clearly, if the gene is between the two flanking markers it has to be on one of the cloned regions and could possibly be on two overlapping clones.

It could be identified by transforming plants with each clone separately and identifying which transformant expresses the effect. If the gene concerned exhibits classical dominance it would be necessary to transform the homozygous recessive with the dominant allele or the transformant would not be recognized. Alternatively, the gene has to be introduced into a species that does not normally express the gene.

Once the effective cloned fragment is identified, the actual location of the gene within the fragment could be identified by successively transforming sub-fractions of the clone or, following sequencing, by looking for potential open reading frames. In those cases where the target gene produces a known product, cDNA clones derived from tissue likely to be rich in the appropriate mRNA of the target gene can be used to identify likely sequences in the previous overlapping clones.

Once the gene has been identified by one of these methods, its structure and activity can be studied in detail. Disease resistance genes are obvious candidates for such studies as their action is very specific and their location on the genome of many species is well documented. However, unless suitable candidate loci.

NEW CHALLENGE OF GENE TECHNOLOGY

With few exceptions classical bioprospecting for drugs has not proved economically beneficial for developing nations, but nor has it directly damaged these economies. In contrast, bioprospecting for genes may soon pose a real threat to the economic survival of these biodiversity-rich countries. The farmers of the North are currently suffering under low prices and overproduction of certain traditional crops and livestock. As a result they are looking for new markets and products.

Modern biotechnology promises to aid Northern farmers in this endeavour, eliminating or displacing traditional export commodities from developing countries and transferring production or substitutes from the farm fields of the South to those of the North. Quite possibly the transfer may even skip the Northern farms and jump straight into bioreactors. In Africa alone, US$ 10 billion in exports are vulnerable to industry-induced changes in raw material prices and requirements. Most developments in plant biotechnology have been achieved with crops cultivated mainly in industrialized countries. This denies the farmers of developing countries the chance to benefit from the new agricultural opportunities of gene technology.

As mentioned above, Calgene's high-laurate canola oil may displace coconut and palm kernel oil, posing a threat to the economic survival of millions of farm families in the South. In the Golfito region of Costa Rica the government has started a programme to grow oil palms on banana fields deserted by the US fruit multinationals. All these efforts, which are partially financed with Northern developmental aid, may vanish into thin air as a result of Calgene's new rape seed.

Thaumatin, a sweet-tasting basic protein from the tropical plant *Thaumatococcus*, has been traditionally used in West Africa as a sweetener. With the collection of *Thaumatococcus* fruits for the British food industry, the population in this region earned a large part of its income.

However, the thaumatin gene has now been cloned and the sweetening protein can be produced by large-scale fermentation of brewer's yeast at low cost. The same applies to natural compounds like indigo, which can now be produced by the fermentation of *Escherichia coli* engineered with genes from a toluol-degrading subspecies of the soil bacterium *Pseudomonas putida*.

Products like vanilla, pyrethrum and rubber may follow this same path. Along the same lines, the extension of patent laws to the developing nations through the GATT could mean that the biotech industry obtains a monopoly on genetically engineered livestock and crops, which farmers in developing countries must cultivate under constraint in order to remain competitive. For example, US-Patent No. 5 159 135 of Agracetus covers all genetically engineered cotton. This patent is a warning of potential future problems, and has already caused an outcry in developing countries like India. If the biotech industry continues developing without adequate controls, the consumers and farmers of developing countries may even end up paying royalties on biotech products that were originally developed from their very own resources and knowledge.

As a step in the right direction, the US Patent and Trademark Office reversed its decision to grant the Agracetus patent at the end of 1994, but primarily as a result of pressure from the US biotech industry which argued that patents like this will inhibit research and development. Such issues bring important questions to light: How can developing nations be motivated to conserve their biodiversity under these threatening circumstances?

Are the regulations and tools of modern bioprospecting, as described above, sufficient to face this challenge? The Biodiversity Convention attempts to address this new threat by requiring that access to biodiversity's genetic potential be combined with the biotechnology transfer to the South in order for those countries to develop their own methods of sustainable biodiversity utilization. Nevertheless, the Convention suffers from three major drawbacks.

- The treaty was not ratified by the USA, the leading country in biotechnological research, development and application.
- It specifically excludes (under US pressure) *ex situ* genebank material collected before the enactment of the treaty. As a result, huge stocks of germplasm collected by the North, mostly in tropical and subtropical countries, are not restricted by the Convention. The recent transfer by the Consultative Group on International Agricultural Research of its 12 genebanks to the auspices of the UN must be the first step in keeping the access of developing countries open to their own resources.
- It still remains nearly impossible to control the illegal transfer of genetic material into the North. Genes can be cloned from minute amounts of DNA or RNA and isolated from biological material that easily fits into an airmail envelope. Genes do not have tags designating their country of origin, and once they are cloned, they are no longer controlled by their source country. This is quite different from the isolation of natural compounds from plants, where larger amounts of plant material must be collected and, for the process of isolation and structure elucidation, must be re-collected. In this last case, controlling the flow of biological material is possible simply because industry will eventually require sample resupply at a given point, and for this the industry needs reliable partners in countries of origin.

These issues must be approached in an active and more aggressive manner than traditionally used. During collaborations with traditional pharmaceutical and biotech companies like Merck and Co., Bristol-Myers Squibb and the British Technology Group, INBio used such an approach, proving to industry that fair partnerships are mandatory and conducive to success. More importantly, INBio demonstrated that reliable applied research is possible in a developing country, and that technology transfer to acquire necessary know-how and equipment works to the advantage of the industrial partner as well.

The same applies to the biotech industry. In this case, the transfer of gene technology is not critical, in contrast to natural compound chemistry, because it does not require million-dollar investments in laboratory infrastructure. Rather, the industry relies on the know-how already existing in many source countries. Gene technology also represents a promising development opportunity for countries that do not have large research budgets at their disposal. Moreover, gene technology is a very straightforward approach, relying on natural history observations (e.g. whether certain plants show natural resistance to pathogens), and not involving the automated random screening of thousands of samples. Biodiversity inventories, which are already in place in countries like Costa Rica, are a reasonable and advantageous prerequisite for successful 'gene prospecting'.

Costa Rica's aggressive strategy to foster collaborations with the international industry and academic institutions in drug research, gene technology and agriculture actively seeks to develop and patent natural compounds, proteins and genes in Costa Rica based on a foundation of national research.

Collaborations of this nature will help launch Costa Rica onto a scientific and technological plane that offers services and goods that are both competitive and compatible with those of industrial nations.

Simultaneously, INBio works within this strategy to increase knowledge about Costa Rican biodiversity in general, access revenues for further conservation efforts, and to assign biodiversity a higher value than it has had in the past.

COMPARATIVE MAPPING IN OTHER PLANT GROUPS

Comparative mapping in the *Solanaceae* showed that when RFLP maps of potato (*Solanum tuberosum*) and tomato (*Lycopersicon esculentum*) were compared, the two genomes showed a high degree of conserved synteny. Five whole-arm inversions were identified that distinguished the two genomes.

Comparative mapping has also been reported extensively in legumes, where several studies show conserved linkage groups between species such as mungbean (*Vigna radiata*) and cowpea (*V. unguiculata*) and pea (*Pisum sativum*) and lentil (*Lens culinaris*). However, the soybean genome appears to show considerable internal duplication and rearrangement relative to other legume genomes.

Comparison of *Brassica* species shows that there are a greater number of chromosome rearrangements than commonly found in cereal species, but there are still large segments with conserved marker orders. Importantly, *Brassica* species and the model plant species *Arabidopsis thaliana* are both in the *Cruciferae* family and are sufficiently closely related for *Arabidopsis* clones to be used directly for mapping and gene isolation in *Brassica*.

Comparison of the genomes of *B. oleracea* and *Arabidopsis* reveals extensive rearrangement, including at least 17 translocations and nine inversions, but homoeologous segments can still be recognized. Thus, *Brassica* researchers can readily exploit the vast mapping, sequencing and genetic resources of *Arabidopsis*.

GENOME RELATIONSHIPS AND WIDE-HYBRIDIZATION

Domestication of crops is likely to have captured only part of the allele diversity present in the wild progenitor gene pools, leaving many desirable traits absent from the crop. In some cases, such as barley 3 *Hordeum spontaneum* or maize 3 teosinte, crops can be crossed easily with wild relatives and there are few barriers to the incorporation of novel alleles since the genomes are essentially homologous.

In contrast, recently evolved polyploids such as hexaploid wheat will have had few opportunities to enlarge the gene pool. This is reflected by RFLP studies of wheat, barley and maize which show that wheat is less polymorphic.

The frequency with which allopolyploid species occur in nature shows that the processes of pollen germination, correct pollen tube growth, fertilization of the egg cell, and embryo development are often conserved in related species. This can be exploited to synthesize new combinations of genomes. Such interspecific or intergeneric hybrids are often referred to as wide-hybrids or wide-crosses. Wide-hybrids can be used to enlarge crop gene pools for any character but are most valuable where suitable genetic variation does not exist within a species, or where screening for a desired character would be difficult, time consuming or expensive. Disease resistance or abiotic stress tolerances are the characters most frequently targeted for transfer.

Barriers to Wide-hybridization

The ease of producing wide-hybrids, and the ease with which genetic material may be transferred by backcrossing to a crop parent, varies greatly from cross to cross. Successful gene transfer may be restricted by prefertilization, postfertilization or chromosome instability barriers. These are not necessarily directly related to the degree of genetic divergence between two species and the genetical control of hybridization barriers may be simple.

Prefertilization barriers usually involve failure of pollen germination or inhibition of pollen tube growth, which may be at the surface of the stigma, in the stigma, or in the transmitting tissue of the ovary wall. The best studied incompatibility reactions affecting cereal wide-hybrids are due to the *Kr* genes of wheat where dominant alleles inhibit the growth of pollen tubes of other species at the base of the style. These genes are best studied in crosses with rye and *Hordeum bulbosum*. No method is known for overcoming the effect of these genes, so wide-hybrids involving bread or durum wheat usually use genotypes with recessive, noninhibiting alleles. In grasses there are, of course, well-characterized genes for self-incompatibility (the *S* and *Z* loci) but it is not known whether there is homology between these and the *Kr* loci or whether a similar mechanism of inhibition operates.

Postfertilization barriers leading to seed abortion are common in widehybrids. In cereals, this is frequently due to defective development and subsequent collapse of the endosperm. Embryos, however, are often viable, and hybrid plants can be recovered if the embryos are excised from developing seeds and cultured on suitable nutrient agar. In the *Triticeae*, using these techniques, a huge catalogue of wide-hybrids have been produced.

For example, hexaploid, tetraploid and diploid wheats and their wild relatives can be intercrossed but embryo rescue is usually unavoidable, especially if different ploidy levels are crossed. Similarly, ovule or embryo culture has been used to recover hybrid plants from *Brassica* crosses. In some crosses, for example between bread and durum wheats, hybrid necrosis kills the developing seedling. This is due to the epistatic interaction of independent genes in the respective species. Crosses within the separate gene pools are unaffected, probably because selection has eliminated necrotic combinations of alleles. Incompatibilities between nuclear genomes and the cytoplasm donated by the female parent may also contribute to lethality or sterility in certain crosses.

In vitro fertilization using isolated gametes, and the subsequent recovery of plants, is a means by which prefertilization and some postfertilization barriers might be overcome. Currently, *in vitro* fertilization has only been used successfully in maize, but such systems have potential for wide-crosses.

A further barrier to the production of wide-hybrids is the phenomenon of chromosome elimination. In such cases fertilization results in the formation of a hybrid zygote, but the complete chromosome complement of one of the parents is lost during subsequent mitotic divisions, leading to the production of a haploid embryo. The best characterized system is the cross between barley and its wild relative *Hordeum bulbosum*. However, the phenomenon occurs in a wide range of other cross combinations in the *Triticeae* including wheat 3 *H. bulbosum*, wheat 3 barley and wheat 3 maize, sorghum or pearl millet.

The mechanism of chromosome elimination is not understood but involves the inability of the chromosomes of one genome to successfully attach to the mitotic spindle. In some crosses, the occurrence of chromosome elimination is affected by the parental genomes since some barley 3 *H. bulbosum* crosses produce a high frequency of barley haploids while others give a high frequency of relatively stable hybrids. The genetical control of this regulation is probably fairly simple. Although chromosome elimination is detrimental to gene transfer, efficient haploid production methods benefit breeding and genetic analysis programmes and are extensively used by workers in these areas.

Utilization of Hybrid Plants

Once hybrid plants have been produced there are usually problems of sterility due to the inability of the hybrid to form viable gametes, particularly male gametes. Usually, this is due to a lack of chromosome pairing and chiasma formation at meiosis between the respective genomes and the subsequent formation of unbalanced gametes from random combinations of univalents.

This can occur even in cases where close homoeology between the genomes can be shown by comparative chromosome and genetic analysis. Clearly the mechanisms controlling the pairing of chromosomes can be so precise as to prevent homoeologous pairing between different genomes. Such mechanisms may be genetically simple, as in wheat where pairing between the A, B and D genomes of bread wheat is restricted by a few pairing control genes of which the most powerful is *Ph1* on chromosome 5B.

Fertility can be partially or fully restored in many hybrids by doubling the chromosome number with colchicine or an equivalent spindle inhibitor as this provides identical pairing partners for each chromosome. In such cases new allopolyploid ‘species’ can be established.

The major motivation for creating wide-hybrids is the introgression of useful genes into cultivated species. Generally, there are three levels at which the transfer of genetic material can be considered. Firstly, a whole genome can be introduced into an existing crop species to create a new allopolyploid species.

For example, the introduction of a diploid rye genome into durum or bread wheat to create hexaploid or octaploid triticales, respectively, has been highly successful. New genome combinations such as *Tritordeum* (the amphiploid between wheat 3 *Hordeum chilense*) are also being investigated as crops for specific environments.

Secondly, whole chromosomes can be introduced to create substitution or addition lines. Thirdly, translocations can be induced between homoeologous or nonhomoeologous chromosomes so that only individual arms or small segments are introduced. The starting point for introgression is to backcross a wide-hybrid to the cultivated species.

In many cases this is possible using the hybrid as the female parent, even if the hybrid is highly sterile, since a proportion of egg cells contain a nonreduced or otherwise viable chromosome complement.

If there is homoeologous pairing between genomes, translocation products can be identified using *in situ* hybridization or molecular markers and stabilized by further backcrossing. Backcrosses can also be tested for the presence of the desired trait or traits in the 'alien' chromosome segment. Combinations of cytological and marker techniques can then be used to characterize suitable lines. In some hybrids pairing between homoeologous genomes does not occur, or occurs too infrequently to be useful. In wheat, this can be overcome by using the *ph1* mutant or plants lacking chromosome 5B (carrying the *Ph1* pairing control gene). This allows higher levels of homoeologous chromosome pairing. Alternatively radiation treatments can be used to induce translocations. In such circumstances it is usually more effective to develop single chromosome addition or substitution lines prior to selecting recombinants, particularly as such stocks can confirm that the desired trait is carried by a single chromosome.

Plant breeders are generally interested in maintaining the characteristics of the cultivated species while adding one or more useful genes from the alien source. Generally the aim is to reduce the size of the alien chromosome segment as much as possible since donor species may have many alleles that are undesirable in the crop.

This is greatly assisted by the use of molecular markers which define the alien segment. *In situ* hybridization using total genomic DNA or species-specific repeats is also valuable because it reveals the number and size of the alien segments. *In situ* methods are also increasingly able to locate low-copy sequences on plant chromosomes, thereby linking the molecular and physical mapping.

Comparative mapping information is important for the manipulation of alien chromosome segments because of the importance of producing plants with genetically balanced genomes.

For example, substitution of rye chromosome 1 for a wheat group 1 chromosome is successful because these chromosomes show conserved synteny throughout their length.

In contrast, rye chromosome 7 is translocated with respect to the wheat chromosomes of groups 2, 4 and 5 and substitution of rye chromosome 7 for a wheat group 7 chromosome produces a plant with reduced doses of group 7 segments and extra doses of group 2, group 4 and group 5 segments.

Comparative mapping therefore provides a framework for the assembly of new chromosome segments in genetically balanced combinations. This principle also applies to other crop groups such as *Brassica*.

PLANT TRANSFORMATION

Genetic engineering of plants is much easier than that of animals. There are several reasons for this: (1) there is a natural transformation system for plants (the bacterium *Agrobacterium tumefaciens*), (2) plant tissue can redifferentiate (a transformed piece of leaf may be regenerated to a whole plant), and (3) plant transformation and regeneration are relatively easy for a variety of plants.

The soil bacterium *Agrobacterium tumefaciens* ("tumefaciens" meaning tumour-making) can infect wounded plant tissue, transferring a large plasmid, the Ti plasmid, to the plant cell. Part of the Ti (tumour-inducing) plasmid apparently randomly integrates into the chromosome of the plant. The integrated part of the plasmid contains genes for the synthesis of (1) food for the bacterium, and (2) plant hormones.

Genes from the Ti plasmid that are integrated in the plant chromosome are expressed at high levels in the plant. Overproduction of the plant hormones leads to continuous growth of the transformed cells, causing plant tumours. Rapid, cancerous growth of the transformed plant tissue obviously is advantageous to the bacterium: more food gets produced.

The Ti plasmid has been genetically modified ("*disarmed*") by deleting the genes involved in the production of bacterial food and of plant hormones, and inserting a gene that can be used as a selectable marker. Selectable marker genes generally are coding for proteins involved in breakdown of antibiotics, such as kanamycin. Any gene of interest can be inserted into the Ti plasmid as well. In principle, one can thus transform any plant tissue, and select transformants by screening for antibiotic resistance. However, unfortunately, there are some complications: (1) it has proven difficult to transform some monocots (grasses, etc.) by *Agrobacterium*, and (2) regeneration of plants from tissue culture or leaf discs is not always possible.

A number of genetically engineered plant varieties have been developed. Traits that have been introduced by transformation include herbicide resistance, increased virus tolerance, or decreased sensitivity to insect or pathogen attack. Traditionally, most of such genetically engineered plants were tobacco, petunia, or similar species with a relatively limited agricultural application. However, during the past decade it now has become possible to transform major staples such as corn and rice and to regenerate them to a fertile plant.

Increasingly, the transformation procedures used do not depend on *Agrobacterium tumefaciens*. Instead, DNA can be delivered into the cells by small, micrometre-sized tungsten or gold bullets coated with the DNA. The bullets are fired from a device that works similar to a shotgun.

The modernised device uses a sudden change in pressure of Helium gas to propel the particles, but the principle of "*shooting*" the DNA into the cell remains the same. This DNA-delivery device is nicknamed "*gene gun*," and has been shown to work for DNA delivery into chloroplasts as well. Over the last several years, use of the "*gene gun*" has become a very common method to transform plants, and has been shown to be applicable to virtually all species investigated.

For example, transformation of rice by this method is now a routine. This is a very important development as rice is the most important crop in the world in terms of the number of people critically dependent on it for a major part of their diet. Another method to get foreign genes into cereals is by electroporation: a jolt of electricity is used to puncture self-repairing holes in protoplasts (*i.e.*, the cell without the cell wall), and DNA can get in through these holes. However, it is often very difficult to regenerate fertile plants from protoplasts of cereals. Nonetheless, significant advances in overcoming these practical difficulties have been made over the years. Now even transgenic trees have been created: for example, the gene for a coat protein of the plum pox virus has been introduced into apricot.

The plum pox virus leads to the feared Sharka disease for which there is no cure. The resulting transgenic tree shows a markedly decreased sensitivity to this virus. The reason why continuous exposure of the tree to the viral coat protein leads to tolerance against viral infection is not yet understood, however.

FEATURES OF PLANT BIOLOGY

Plant genetics is different from that of animals in a few ways. Like mitochondria, chloroplasts have their own DNA, complicating pedigrees somewhat. Like animals, plants have somatic mutations regularly, but these mutations can contribute to the germ line with ease, since flowers develop at the ends of branches composed of somatic cells. People have known of this for centuries, and mutant branches are called "*sports*." If the fruit on the sport is economically desirable, a new cultivar may be obtained.

Some plant species are capable of self-fertilization, and some are nearly exclusively self-fertilizers. This means that a plant can be both mother and father to its offspring, a rare occurrence in animals. Scientists and hobbyists attempting to make crosses between different plants must take special measures to prevent the plants from self-fertilizing.

Plants are generally more capable of surviving, and indeed flourishing, as polyploids. Polyploidy, the presence of extra sets of chromosomes, is not usually compatible with life in animals. In plants, polyploid individuals are created frequently by a variety of processes, and once created usually cannot cross back to the parental type. Polyploid individuals, if capable of self-fertilizing, can give rise to a new genetically distinct lineage, which can be the start of a new species. This is often called "*instant speciation*." Polyploids generally have larger fruit, an economically desirable trait, and many human food crops, including wheat, maize, potatoes, peanuts, strawberries, and tobacco, are either accidentally or deliberately created polyploids.

Hybrids between plant species are easy to create by hand-pollination, and may be more successful on average than hybrids between animal species. Often, tens of thousands of offspring from a single cross are raised and tested to obtain a single individual with desired characteristics. People create hybrids for economic and aesthetic reasons, especially with orchids.

DNA

DNA is a nucleic acid that contains the genetic instructions used in the development and functioning of all known living organisms and some viruses. The main role of DNA molecules is the long-term storage of information. DNA is often compared to a set of blueprints or a recipe, or a code, since it contains the instructions needed to construct other components of cells, such as proteins and RNA molecules.

The DNA segments that carry this genetic information are called genes, but other DNA sequences have structural purposes, or are involved in regulating the use of this genetic information.

Geneticists, including plant geneticists, use this sequencing of DNA to their advantage as they splice and delete certain genes and regions of the DNA molecule to produce a different or desired genotype and thus, also producing a different phenotype.

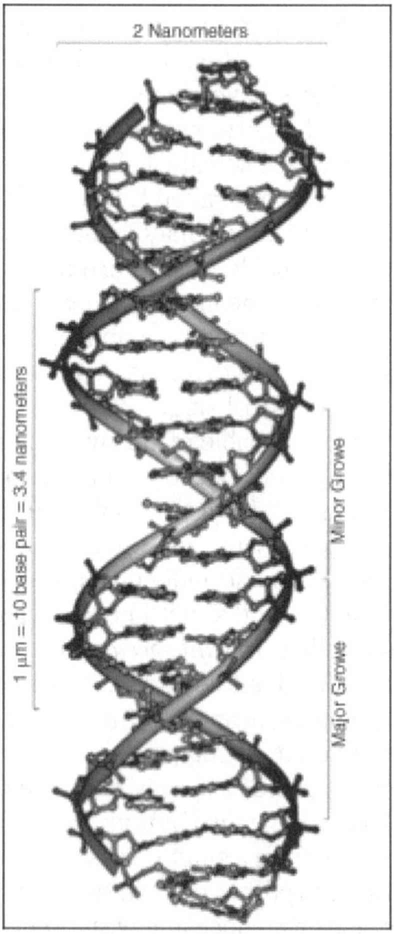

Fig. The Structure of Part of a DNA Double Helix

Gregor Mendel

Gregor Mendel was an Augustinian priest and scientist born on 20 July, 1822 in Austria-Hungary and is well known for discovering genetics. He went to the Abbey of St.

Thomas in Brno. He is often called the father of genetics for his study of the inheritance of certain traits in pea plants. Mendel showed that the inheritance of these traits follows particular laws, which were later named after him.

The significance of Mendel's work was not recognised until the turn of the 20th century. Its rediscovery prompted the foundation of the discipline of genetics allows geneticists today to accurately predict the outcome of such crosses and in determining the phenotypical effects of the crosses. He died on 6 January, 1884 from chronic nephritis.

MODERN WAYS TO GENETICALLY MODIFY PLANTS

There are two predominant procedures of transforming genes in organisms: the "*Gene gun*" method and the *Agrobacterium* method.

"Gene gun" Method

The "*Gene Gun*" method is also referred to as "*biolistics*" (ballistics using biological components). This technique is used for *in vivo* (within a living organism) transformation and has been especially useful in transforming monocot species such as corn and rice.

This approach literally shoots genes into plant cells and plant cell chloroplasts. DNA is coated onto small particles of gold or tungsten approximately 2 mm in diameter.

The particles are placed in a vacuum chamber and the plant tissue to be engineered is placed below the chamber. The particles are propelled at high velocity using a short pulse of high-pressure helium gas, and hit a fine mesh baffle placed above the tissue while the DNA coating continues into any target cell or tissue.

Agrobacterium Method

Transformation via *Agrobacterium* has been successfully practised in dicots, *i.e.*, broad-leaf plants, such as soybeans and tomatoes, for many years. Recently, it has been adapted and is now effective in monocots like grasses, including corn and rice.

In general, the *Agrobacterium* method is considered preferable to the gene gun, because of a greater frequency of single-site insertions of the foreign DNA, which allows for easier monitoring. In this method, the Ti region is removed from the T-DNA and replaced with the desired gene and a marker, which is then inserted into the organism.

This may involve direct inoculation of the tissue with a culture of transformed *Agrobacterium*, or inoculation following treatment with microprojectile bombardment, which wounds the tissue. Wounding of the target tissue causes the release of phenolic compounds by the plant, which induces invasion of the tissue by *Agrobacterium*.

Because of this, microprojectile bombardment often increases the efficiency of infection with *Agrobacterium*. The marker is used to find the organism which has successfully taken up the desired gene. Tissues of the organism are then transferred to a medium containing an antibiotic or herbicide, depending on which marker was used.

The *Agrobacterium* present is also killed by the antibiotic. Only tissues expressing the marker will survive and possess the gene of interest. Thus, subsequent steps in the process will only use these surviving plants. In order to obtain whole plants from these tissues, they are grown under controlled environmental conditions in tissue culture. This is a process of a series of media, each containing nutrients and hormones.

Once the plants are grown and produce seed, the process of evaluating the progeny begins.

This process entails selection of the seeds with the desired traits and then retesting and growing to make sure that the entire process has been completed successfully with the desired results.

GENETICALLY ENGINEERED CROPS

The use of genetically engineered crops has helped many farmers deal with pest problems that reduce their crop production. The impact of pest-resistant crops has led to a much higher yield for farmers in today's world. They can use fewer pesticides, which reduces the chemicals that they put into the ground.

Certain engineered crops have led to farmers all over the world and in the United States to increase crop yield exponentially in recent years. Farmers can use a glyphosate herbicide to kill weeds, yet the genetically engineered corn is resistant to the herbicide and is left unaffected. Thus, fields are produced that are virtually weed-free.

Genetically engineered crops can also benefit farmers when dealing with potentially harmful viruses and bacteria.

In the 1990s, a mutant strain of virus was decimating the commercial corn fields of the United States. Scientists found a virus resistant strain of maize in the highlands of Mexico and extracted the part of the maize's genome that coded for resistance against the virus and incorporated it into their existing strain of commercial corn. This allowed the commercial strain to produce progeny that were resistant to the virus. Thus, the crops were saved from decimation.

Potential Detrimental Effects of Genetically Engineered Plants

According to Vaughan A. Hilder and Donald Boulter at the Department of Biological Sciences, University of Durham, there have been serious failures in resistance to targeted pests in *Bacillus thuringiensis* (Bt) cotton; most plant-derived resistance factors produce chronic rather than acute effects; and many serious pests are simply not susceptible to known resistance factors. According to John E. Berringer, the outcome of releasing genetically modified organisms into the environment is still not known.

5

Genetic Engineering in Plant Breeding

In order for genetic resources to be efficiently utilized in plant breeding programmes, it is first necessary to determine whether useful genetic variation exists in the material and secondly to develop the most cost-effective method of introducing the potentially useful genes into commercially acceptable material. Until recently, genetic analysis has relied on conventional segregation analysis in controlled crosses for qualitative traits determined by major genes, and on standard biometrical methods for quantitative traits controlled by polygenes. Now, however, the use of DNA techniques has opened up new possibilities for genetic analysis and breeding, and in this chapter the general principles and applications of these molecular marker-based procedures will be discussed.

In its specific sense, marker-assisted or marker-aided selection (MAS) is the use of appropriate easily recognizable genetic markers to facilitate or accelerate the selection of linked genes controlling useful agronomic traits. The nature of the marker loci themselves is not important and they are used simply to indicate the presence of useful alleles of genes of commercial or practical importance. However, the term has also come to include the wider use of markers to improve breeding programmes.

GENETIC ENGINEERING TECHNOLOGY

Genetic manipulation for biotechnological plant improvement relies on the ability to modify genes as DNA sequences. This requires utilization of standard, well-documented techniques. Enzymes are the basic tool kit of the genetic engineer, and can be used for basic functions of manipulating nucleic acids in such a way as to copy (DNA polymerase), transcribe (RNA polymerase), cut at specific sites (restriction endonucleases), cut and degrade in a nonspecific nature (nucleases) and join pieces together (DNA ligase).

DNA libraries are another important resource for genetic manipulation as they represent a means of storing and screening sequences from an entire organism. The two basic types of library are the cDNA library and the genomic library. A cDNA (complementary DNA) library is based on expressed sequences. Such a library is made by extracting mRNA from a particular plant tissue of interest, then making first single-stranded DNA (via reverse transcriptase) then double-stranded DNA (via DNA polymerase). A cDNA library will then represent only those genes which are expressed in the tissue of interest (or in response to a stress of some sort), and will be devoid of introns and regulatory sequences.

A genomic library represents the entire genome (often including the organellar genomes), and will include all sequences whether they are expressed in the donor tissue or not, and even sequences which are not expressed at all. Genomic libraries will therefore include highly repeated sequences such as rRNA genes and nonexpressed simple sequence repeats (SSRs), as well as genes and their regulatory sequences and introns.

Genomic libraries may be cloned and stored in lambda bacteriophage, cosmids or larger fragments which will often include multiple genes in bacterial artificial chromosomes (BACs) and yeast artificial chromosomes (YACs). BAC library inserts may average 150 kb, whereas YAC inserts are larger on average, at up to 1 Mb of DNA.

Overall, the ability to genetically engineer a plant relies on:

- A genetic transformation system for transferring DNA into plant cells.
- A plant regeneration system to produce plants from the transgenic cells.
- A gene construct which can modify the plant phenotype in a desirable and controlled manner.

GENETIC TRANSFORMATION TECHNIQUES

A number of genetic transformation techniques have been developed since the first report of a transgenic plant. With few exceptions, these methods rely on the ability to transfer DNA to a single cell or small group of cells, such as nondifferentiated leaf tissue or meristems. Then some form of tissue culture is required to regenerate whole, fertile transgenic plants for the process to be completed. Therefore, the first requirement for plant transformation is a reliable plant regeneration system using explants/tissues which are amenable to the DNA delivery system.

This requirement was, more than any other factor, the greatest impediment to genetic engineering of members of the cereal and grain legume families – the two most important crop groups. Regeneration in most of these species is difficult, particularly for species like soybean and chickpea.

At best, regeneration is extremely genotype-dependent for species such as rice, where the *japonica* rices are easier to manipulate *in vitro* than the *indica* rices, which are more important on a worldwide basis. However, international effort in most of the important cereal species has largely overcome the problem, although there are still genotypes of a number of species which remain recalcitrant to regeneration.

The cereals and pulses (grain legumes) proved somewhat recalcitrant to genetic transformation until the late 1980s, when rice became the first cereal to become reliably transformed at high frequency, with soybean first successfully transformed via *Agrobacterium* in 1988, although repeatable transformation was only achieved later using a microprojectile approach.

There are three basic transformation systems being routinely used in most genetic engineering laboratories internationally.

These are:

- *Agrobacterium*-mediated transformation.
- DNA uptake via protoplasts.
- Microprojectile bombardment.

Agrobacterium-mediated Transformation

Agrobacterium tumefaciens, the causal agent of crown gall disease, transfers a portion of its own DNA (T-DNA) to the plant, where it is stably incorporated into the plant genome. The T-DNA is delimited by specific 25 bp repeat sequences, known as the border sequences. None of the T-DNA is actually required for transfer, and whatever DNA is bound by the borders will be transferred, hence the pathogenic DNA can be effectively replaced with genes of interest, allowing delivery of these desirable genes to cells from which whole plants can be regenerated. The ability of *Agrobacterium* to transfer genes to plants is known as virulence, and is largely controlled by the *vir* genes on the Ti plasmid, and to a lesser extent, the chromosomal virulence (*chv*) genes. The *vir* genes are induced by the plant's wound response, in particular the production of phenolic compounds such as acetosyringone.

The Vir proteins then facilitate the identification, copying, transfer and targeting to the host nucleus, and incorporation on a plant chromosome of the T-DNA. Generally the simplest manner by which to replace the T-DNA is to remove it, leaving a 'disarmed' Ti plasmid with the *vir* genes, then introducing a binary plasmid which contains the gene(s) of interest within the border sequences. Hence, with the induction of virulence, the Vir proteins will effect the transfer of the engineered T-DNA.

Most dicotyledonous species are hosts to *Agrobacterium* and as such can be genetically engineered in this manner. *Agrobacterium*-mediated transformation remains the method of choice for most species of the *Solanaceae* and *Brassicae*, and other top 20 most important plants such as grape, orange, cotton, sugarbeet and apple. While there are a number of reports of successful transformation of important grain legumes including soybean, chickpea and *Phaseolus* bean, there is generally great genotypedependence of such methods. It has been shown repeatedly that plant genotype 3 *Agrobacterium* strain interactions can vary from complete absence of transformation to 100 poe cent transformation within many species, including soybean, pea and sugarbeet.

GENETIC EFFICIENCY IN COLLECTING FOR EX SITU CONSERVATION

Full technical guidelines on collecting are given by Guarino *et al.* (1995). This section concentrates just on evolutionary considerations, to facilitate adaptation of general guidelines to suit individual species, objectives and constraints.

The objectives of an expedition to collect germplasm from a region may be to:

- Acquire the maximum genetic diversity of targeted taxa within the region, within the constraints of limited available resources.
- Acquire germplasm with the maximum novelty value with respect to a collection already held *ex situ;* i.e. the greatest number and diversity of genes and genotypes that have not previously been collected.
- Combat genetic erosion.
- Acquire genes or genotypes most likely to benefit a particular breeding or research objective.
- Acquire germplasm for analysis of agroecogeographic patterns of the distribution of biodiversity.

This section deals mainly with the first of these objectives. Limiting resources vary from expedition to expedition, and may be time (time to travel to a site, time to collect overall site data, time to collect each seed or plant at a site, speed of returning live plants to base); space available in the collecting vehicle; or labour and facilities to process samples at base.

Resource limitations influence optimal collecting strategy in a way that depends on population structure. The principles outlined in this section also apply to objective 2 above, except that additional information is needed on the diversity and origins of the pre-existing collection.

Where the primary objective is to combat genetic erosion, sampling strategy can and should still be designed to satisfy objective 1. However, painstaking planning to maximize the diversity collected may be counterproductive where the rate of erosion is so high that diversity is lost whilst planning is in progress.

In these circumstances, speed of undertaking a collecting expedition is of overriding importance. Objective 4, a breeder-driven collection to support a particular breeding objective, requires a totally different sampling strategy, to locate particular genes rather than maximize diversity of genes. Nevertheless, knowledge of evolutionary patterns can aid identification of sites most likely to contain the desired genes or genotypes.

Objective 5, collection for agroecogeographic analysis, requires yet another strategy, namely an appropriately randomized sampling procedure. This fact is not sufficiently recognized, as many published analyses are based on collections made for conservation or breeding purposes. Yet any sampling strategy that aims to maximize diversity or to target specific genes can generate incorrect and misleading estimates of components of variance. For example, suppose two collections are undertaken in two different regions, both with a sampling strategy to maximize the diversity sampled within each region. A comparison of both collections will then incorrectly suggest that there is less difference between them and more variation within each region than is really the case.

A specific example of an erroneous conclusion may be the widely accepted latitudinal cline in genetic diversity of *Trifolium repens* across Europe, according to which southern populations are believed to be highly diverse and northern ones uniform. This is likely to be spurious, and at least partly a consequence of collections being specifically targeted at well-managed pastures in the north but at highly diverse habitats in the south. A northern collection targeted at diverse habitats contained as much diversity as southern populations.

Appropriate randomization for agroecogeographic analyses need not mean full randomization. Collecting for maximum diversity can be compatible and even beneficial to agroecogeographic analysis. Seeking to collect maximum genetic diversity by targeting maximum environmental diversity improves the sensitivity of analysis of the relationship between genetic diversity and environmental diversity. This of course requires collection of all relevant environmental data so that they can be incorporated as independent variables in statistical analysis.

Spatial Scale of Biodiversity: Overall Distribution of Sites

The general scaling properties of biodiversity have two immediate implications. First, it is important to cover as large an area as possible. Second, adjacent sites should not be further apart than the genetic patch size, since increasing the geographical distance beyond this does not increase the expected genetic distance between two populations. For this purpose, genetic patch size should be measured using neutral genes, to provide a general baseline sampling strategy that is not influenced by any particular pattern of environmental diversity. The baseline can then be refined following the principles outlined in Section. At the lower end of the scale, the genetic neighbourhood area defines the minimum possible scale for taking distinct population samples, at least of the seed population. At smaller scales mating is random and Hardy–Weinberg equilibrium is expected, with no possibility of division into genetically distinct subpopulations.

This applies only to seeds: the population of adult plants may show genetic subdivision at smaller scales if the environment is heterogeneous at smaller scales, imposing smallerscale heterogeneity of pressures within the genetic neighbourhood. Thus there may be merit in finer-scale sampling of adult populations than seed populations.

However, the genetic neighbourhood area of most wild plant species is remarkably small, far smaller than the unit regarded as one population by the collector. For the insect-pollinated self-incompatible perennial *Trifolium repens* the reproductive genetic neighbourhood area is 2 m^2; for the wind-pollinated self-incompatible perennial *Lolium perenne* it is 8.4 m^2 (ignoring contribution of seed dispersal: calculated from pollen dispersal data).

In practice, therefore, each sample of a wild population in an *ex situ* collection almost invariably comprises genotypes from what were originally numerous distinct genetic populations. In some species, such as *Trifolium repens*, there is good evidence for genetic differentiation between such genetic populations, each showing distinct local adaptation to microenvironmental variation within the collecting site. To date it has never been considered justifiable to maintain these genetic populations as separate accessions for conservation purposes – although it is of course necessary for detailed investigations of evolutionary responses to environmental heterogeneity.

Combining many genetic populations into a single accession has consequences for its maintenance *ex situ*. These consequences will be considered in Section.

The genetic neighbourhood area of crop plants is closely related to the type of farming. In primitive farming communities it is generally much smaller than for modern agriculture. Farmers in such communities usually maintain and select their own seed, with limited 'dispersal' (by seed exchange) between isolated communities or even between farmers within communities. It is essential, when collecting, to determine what are the local customs in relation to seed selection and exchange, especially (i) whether a formal centralized system exists for exchanging seed, or whether exchange is informal and centralized through the market, or informal and localized to individual farmer–farmer interactions; and (ii) how much farmers rely on their own farm-saved seed, and if so whether they consciously make their own selections. Only with such local knowledge can the collector judge the probable scale of distribution of diversity.

Adaptation to Environment: Targeting by Habitat

Targeting the maximum diversity of habitats for collection will maximize the diversity of genes contributing to adaptation to the selection pressures imposed in the environments sampled. It will also maximize diversity of genes closely linked to the adaptive genes and of pleiotropic characters. It will have no effect on the diversity of genes that are neutral for the particular environmental diversity sampled – this includes not only genes that appear neutral with respect to all known selection pressures, but also genes that are non-neutral for different types of environmental diversity.

In this context, for agricultural and horticultural species, enviro-nmental diversity and the corresponding diversity of selection pressures necessarily includes diversity of selections made by farmers and gardeners. It includes, for example, local variation in subjective preferences for flavour and appearance, as well as local variation in perception of agronomic quality.

Effective environmental targeting in this manner depends on the collector having good knowledge of environmental diversity in the region, and of the distribution of the target taxa in relation to environmental diversity. Much of the planning phase of a collection should be devoted to identifying contrasting environments, using as many sources of information as possible, preferably in map form: not only conventional geographical maps, but also maps of surface geology, soil, temperature, rainfall, vegetation and land use. Much additional information is not available in map form, and may not be readily available prior to the expedition, being in the knowledge domain of local extension scientists and farmers. Relevant local knowledge covers not only natural variation between fields but also diversity in farmer-selection pressures resulting from variation in crop usage and variation in preferred crop characteristics.

Phylogeny: Targeting Centres of Diversity

It is now widely accepted that evolution does not progress at a uniform rate, but involves periods of relative stability interspersed with periods of rapid change. Exactly how and how much the rate of evolution changes is still the subject of debate. Nevertheless, for most species and genera it is possible to identify centres of diversity, associated with a phase of rapid diversification at some stage in their evolutionary history.

Centres of diversity are most strongly developed for crop species, leading to the famous pioneering work of Vavilov (1951). These centres are associated with early agricultural developments. They are attributed to disruptive selection caused by the simultaneous action of natural selection for fitness and artificial selection for agronomic value, combined with diverse artificial selections applied by different farmers in different environments, and with introgression between conspecific crop and wild relative. By definition, a collecting expedition will obtain the greatest diversity if it is located within the centre of diversity of the target taxa. The content of *ex situ* collections should therefore contain a bias in favour of populations from the centre of diversity.

Phylogenetic Constraints: Targeting Limits of Species Distribution

All species have a limited distribution. Ultimately, this may be attributed to phylogenetic constraints on the extent to which a species may adapt genetically to different environments. Of course, absence of a species from a location does not imply it cannot adapt to that environment – very often absence is attributable rather to dispersal barriers preventing colonization.

However, even in these cases, the barriers to dispersal are environments to which the species cannot adapt. Here we are concerned with the actual boundaries to the distribution of a species: the transition from an environment in which the species can and does persist, to one in which it cannot persist. The boundaries to the distribution of a species logically define the limits to which the species can adapt genetically. Populations near the boundary therefore are likely to contain genes not present in the centre of the species range. These genes are likely to be particularly valuable in breeding to broaden the range of adaptation of a crop, particularly for improving tolerance to stresses such as cold, heat, drought, etc.

For these purposes, it is important to consider not just the geographical distribution of the species, but all dimensions of its ecological distribution: its range for pH tolerance, temperature, nutrients, light, moisture, defoliation, competition, and so on. The upper and lower limits to the species distribution on each environmental gradient will each be associated with different genes for extreme adaptation. It is therefore important to target as many as possible of these limits.

Some of the ecological limits will be encountered geographically within the centre of diversity; in these cases targeting the centre of diversity can be combined with targeting extremes of adaptation. However, other ecological distribution limits and geographical distribution limits are geographically distant from the centre of diversity. The centre of diversity will not contain the genes for adaptation to these extremes. It is then necessary to target these extremes separately from the centre of diversity.

Connectivity: Targeting Isolated Populations

Although the genetic neighbourhood area of most plants is small, pollen and seed dispersal are usually highly leptokurtic and small numbers can be dispersed large distances. For example, although *Lolium perenne* has a pollen neighbourhood of 8.4 m2 (a standard deviation for dispersal of 2.9 m), pollen has been detected at distances over 1000 m. Other species, such as *Cacao*, show even greater leptokurtosis of dispersal, with occasionally seeds being dispersed thousands of kilometres over the sea.

Whilst genetic differentiation between populations is possible at any scale above the genetic neighbourhood area, the potential for differentiation increases with the degree of isolation. A completely isolated population cannot exchange genes with other populations and therefore there is no possibility for genetic divergence by drift or selection to be counteracted by the repeated sharing of genes.

This is most obvious at the species level, illustrated by the high level of endemism in isolated islands – indeed, this was a major factor leading Darwin (1859) to his theory of evolution by natural selection. The same applies to genetic variation within species, both wild and cultivated.

It is therefore important to target populations that are most isolated. For crop species, isolated farmers or farming communities should be sought. For wild species, areas should be sought where their natural habitats are highly fragmented, especially if the intervening habitats form effective barriers to gene flow.

Drift: Targeting Small Populations

The rate of genetic drift is inversely proportional to the number of individuals in a population. It affects all genes, but especially those that are not subject to strong selection pressures. Thus, two small populations are likely to differ through random drift in all polymorphic parts of their genome. This is in marked contrast to variation associated with adaptation to particular environments, which generally involves relatively few genes. Targeting small populations is therefore an efficient means of acquiring maximal nonspecific genetic variation between populations. Against this, small populations are more inbred and contain less within-population variation.

Scales of Environmental Heterogeneity: Stratified Sampling

The environment is a multidimensional entity. Genetic adaptation to environment is correspondingly multidimensional. Different environmental variables show different patterns of variation in space and time. Therefore genetic variation for adaptation to different environmental variables also shows different patterns. For example, Sackville Hamilton (1980) collected *Trifolium repens* from an area of high diversity of soils and grassland management but uniform climate. Relative to the global diversity present in the entire gene pool of *T. repens*, genetic diversity between populations was high for vegetative and morphological characteristics important for adaptation to soils and management but low for time of flowering.

More generally, it may be, for example, that populations from adjacent fields differ mainly in genes affecting response to management; ones from nearby fields differ mainly in genes for response to aspect; ones further apart differ mainly in genes for response to soils; ones from different altitudes differ mainly for response to temperature; ones from different villages for local human preferences; ones from different latitudes for response to daylength; and so on.

Given this situation, a stratified sampling strategy will not just maximize the genetic diversity collected; it will maximize the diversity of different types of genetic diversity collected. A particular advantage of stratified sampling is that it does not depend on prior knowledge of the different scales of heterogeneity of different environmental attributes.

Although such knowledge helps, nevertheless the fact that different environmental variables show different scales of heterogeneity is itself sufficient to make a stratified sampling procedure more efficient in obtaining qualitatively different types of genetic diversity. For some purposes, the stratification of sampling procedure should be extended to sampling individuals within sites, at least for natural populations.

Certainly this will maximize within-accession diversity sampled from such populations. For example: sampling several individuals from a single genetic population will sample diversity in genes that are truly polymorphic at the genetic population level; sampling from different quadrats within a field will acquire diversity in genes responsible for microscale adaptation to patchiness of the vegetation, soil characteristics, and microflora, microtopography, etc.; and samples from the boundaries of the field are more likely to contain immigrant genes from nearby, differently adapted populations.

Stratification of sampling procedure within a population is rarely appropriate for crop populations and market populations. Even for natural populations it may not always be appropriate. In particular, by maximizing within-population variance, a stratified sampling procedure will invalidate agroecogeographical comparison of different populations.

If this is important, a random sampling procedure is more appropriate, unless the individual plants are maintained separately. For species where it is difficult, or even impossible as routine practice, to distinguish plants from each other – as with most perennial herbaceous species communities – it may be impossible to take a truly random sample. In these species, only inflorescences, or leaves, or some other part of the plant, can be sampled at random. This inevitably introduces a size bias into the sample in favour of those plants with the most inflorescences, leaves, etc.

Breeding System: Adjusting Sampling Procedures

The breeding system has a major influence on the distribution of genetic diversity. A number of mechanisms operate to fix particular variants in lines: inbreeding fixes it through homozygosity; apomixis fixes variants even in heterozygotes; some complex chromosome linkages, like those in *Oenothera*, operate to minimize recombination. All these cases reduce within-population variance, so that a correspondingly increased proportion of the total gene pool is represented by variation between populations.

In contrast, outbreeders show higher withinpopulation variance. Sampling procedures must be adjusted correspondingly, to take relatively few individuals from many populations of inbreeding, apomictic and similar species, and many individuals from each of fewer populations of outbreeding species.

Vegetative propagation (by stolons, bulbs, rhizomes, etc.) is functionally equivalent to apomixis in that it can generate numerous genetically identical plants. However, vegetative propagation is often associated with outbreeding, generating a complex two-level population structure. There is high genetic variance among the individuals originating by sexual reproduction through different zygotes, and zero genetic variance (ignoring somatic mutations) among the vegetative progeny derived solely by mitotic division from a single zygote.

In many such vegetatively reproduced species it is impossible to know at a glance whether two plants are derived from the same or from different zygotes. In these species the two-level population structure can present very considerable problems for collection for efficient conservation. The commonest approach is to ensure a large enough distance between sampled individuals so as to be reasonably confident that they are genetically distinct.

However, a single clone of even small herbaceous species can cover hundreds or thousands of square metres. The distance between adjacent samples therefore has to be undesirably large, in that it eliminates sampling the genetic diversity that is expressed at smaller scales. For many species there is no satisfactory resolution to this problem, as the only resolution may be intensive sampling followed by genetic fingerprinting to determine the genotypic composition of the population sample, which of course is unjustifiably labour intensive.

An additional problem in such species is the sampling bias referred to in Section. Populations of these species often show a highly skewed distribution of physical size of genotypes, with a few large genotypes and many small ones. When population samples are based on a random selection of inflorescences or leaves, the sample will be strongly biased in favour of the few large genotypes.

Temporal Scale of Biodiversity

Little attention has been paid to the temporal scale of biodiversity for conservation purposes. Whilst the importance of cyclic fluctuations, chaotic changes and continuous directional shifts are all well acknowledged and documented, it has rarely if ever been considered justifiable for conservation purposes to return to the same sites for repeat collections.

The only common reason for returning to a site or region is to test specific hypotheses, for example to test the extent of genetic erosion.

The problem is probably least in species with long-lived seed banks and/or adults. In these, genes for adaptation to one temporary environment may persist in the population long after that environment has been replaced by another.

Indeed, in some situations most of the genetic variation in a population may be a relict of past selection pressures, and provide a broad pool of adaptability to new unpredictable events. In these situations, a single sample, if large enough, may be sufficient to include diversity of adaptation to the environments experienced by the population over many years.

Notably, species with long-lived adults are often outbreeders that maintain high levels of diversity within populations. Acknowledging that much of this diversity may represent adaptation to temporal environmental diversity adds to the importance of conserving the within-population diversity.

STRUCTURAL ANALYSIS OF GENOME RELATIONSHIPS

Cytogenetic analysis of somatic or meiotic preparations by conventional staining, chromosome banding and *in situ* hybridization ('chromosome painting') provides information on the overall organization of plant genomes. Meiotic chromosome pairing behaviour in natural or artificial hybrids is also an important guide to genome relationships, and until recently was the only method for assaying genome similarities. However, pairing can be strongly affected by single genes such as *Ph1* in wheat, and the degree of pairing between related genomes may not directly reflect overall genome homology.

Nevertheless, chromosome pairing has been shown to be in good agreement with the conventional taxonomy of the *Triticeae*. Analysis of hybrids is important because allo- and autopolyploidy are common in plant evolution.

Allopolyploids include crop species such as oilseed rape (canola), oat, wheat, groundnut, upland and sea island cotton, some bananas, strawberry, arabica coffee and tobacco. The cultivated species derived from polyploid hybrids are generally thought to have combined agronomically important traits from their diploid ancestors.

GENOME SIZE AND COMPOSITION

The amount of nuclear DNA varies greatly between plant species. For example, the haploid genome of hexaploid wheat contains about 1.7 3 1010 bp, each of the three genomes being roughly similar in size to that of barley. The genome of rice, in contrast, contains only about 4 3 108 bp.

Variation in DNA amount, other than variation due to ploidy level, is probably due almost entirely to variation in the copy number of repeated DNA sequences. Large genome species contain higher proportions of short tandem repeats, often present as blocks of heterochromatin, and a higher proportion of dispersed repeats.

A major component of the latter in wheat and barley are retroelement-like sequences such as Wis-1, Wis-2 and Bis-1. Long-range analysis of repeat structure in large cereal genomes suggests that chromosome evolution has involved duplication of blocks of repeats.

Nuclear DNA amount can also vary within plant species. This variation, and variation between species, can often be correlated with ecogeographical variables suggesting that DNA amount is of adaptive significance. In maize, for example, DNA amount declines with increasing altitude or increasing latitude.

This variation can be as much as 30 poe cent of the genome size and involves variation in the amount of C-band heterochromatin as well as variation in dispersed repeats. Proposed mechanisms producing changes in repetitive sequences include unequal crossing over, transposition, retrotransposition, gene conversion and replication slippage. They are rapid in evolutionary terms and consequently the copy number of specific sequences can vary greatly in different species or lineages. New sequence variants also arise frequently, probably because there is little or no selection constraining the DNA sequence of many classes of repeats.

As a result, there are families of repeated sequences which are specific to individual grass species, or genomes, or which differ greatly in copy number between closely related species, and in general the divergence of repetitive sequences can be used to indicate the degree of relatedness of whole genomes in diploid species and their polyploid derivatives. Repeat sequence plasticity can be exploited for Southern blot analysis and for chromosome painting using dispersed repeats or total genomic DNA. For example, wheat chromosomes can be clearly differentiated from those of rye or barley in hybrids and the three genomes can be differentiated in hexaploid wheat.

Fluorescent *in situ* hybridization (FISH) is emerging as the method of choice for such work because several probes, each labelled to give a different colour when illuminated by specific wavelengths, can be hybridized to the same preparation. *In situ* hybridization is extremely valuable for identifying hybrids and introgression lines in crop species and for studies of plant evolution in general.

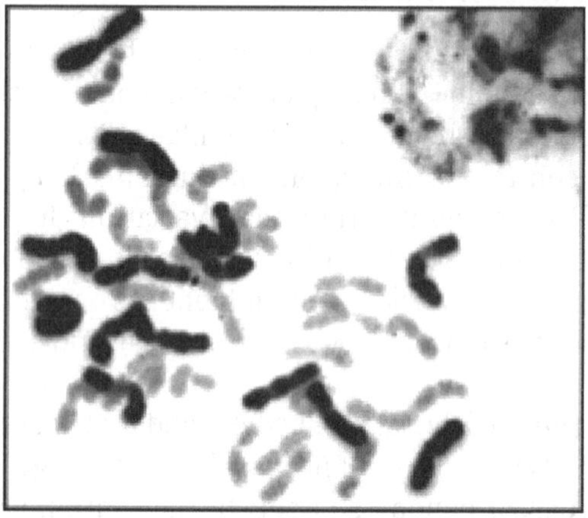

Fig. Genomic in Situ Hybridization Using Digoxigenin-labelled rye DNA on root-tip Metaphase Chromosomes of Hexaploid Triticale. The 14 rye Chromosomes are Brightly Stained.

DNA FINGERPRINTING AND GENETIC MAP CONSTRUCTION

The rapid evolution of repeated sequences results in high levels of polymorphism that can be detected by methods based on Southern hybridization or polymerase chain reaction (PCR). Repeated sequences are, therefore, potentially useful as fingerprinting markers. Polymorphism in repeated sequences can also be used for genetic map construction. For example, it has recently been possible to map very rapidly almost 500 interspersed repetitive sequences in mouse. Larger sequences, such as retrotransposons, can also be used for fingerprinting or genetic mapping. Specific PCR primers enable some repeats to be used as single locus markers.

The most useful of these are the simple sequence repeats (SSRs or microsatellites), which are usually di- or tri-nucleotide repeats, since these are often flanked by unique sequence DNA. This allows the use of flanking primers for PCR detection of length polymorphism within the SSR itself. SSRs are highly polymorphic and are therefore useful for studies of genetic diversity as well as for mapping.

Although repeated sequences can be used for genetic mapping, the genetic maps of most crop species and model organisms are based on polymorphisms of low- or single-copy sequences. Most reasonably detailed maps are based on analysis of restriction fragment length polymorphism (RFLP), which is detected by the hybridization of cloned DNA segments to genomic DNA digested with restriction enzymes.

Most RFLP probes are either small fragments of genomic DNA (selected for low copy number) or cDNA clones which represent transcribed genes of the tissue from which they were isolated. In future, however, most maps will probably be based largely on amplified fragment length polymorphisms (AFLP), simple sequence repeats (SSRs or microsatellites) and, to a lesser extent, random amplified polymorphic DNAs (RAPD).

These PCR methods will be favoured because of their speed of assay, degree of polymorphism and their requirement for only small amounts of DNA from the plants of interest. This is illustrated by the analysis of the haploid megasporophytes of conifers. These yield sufficient DNA for the construction of genetic maps using RAPD markers.

Genetic maps are usually constructed to locate genes of biological or economic interest in relation to molecular markers. For single genes of large effect this is straightforward since a chromosome location can easily be established by segregational analysis. However, many traits of interest to plant breeders and researchers are controlled by several to many genes acting in combination. The availability of high-resolution genetic maps and increasingly sophisticated statistical methods has greatly increased the ability to identify the number and chromosome locations of such quantitative trait loci (QTL).

There are two main applications for genetic maps. The first is the use of DNA markers linked to the gene or genes of interest to select individuals in breeding programmes without the need to assess the phenotype at each stage of development. Such markerassisted selection (MAS) is already used in breeding programmes. It is particularly useful for the manipulation of quantitative traits, which are the most difficult to evaluate phenotypically in conventional breeding programmes. The second application of genetic maps is the isolation of genes by a map-based cloning approach such as chromosome walking or chromosome landing. This approach has been used, for example, in the isolation of genes conferring disease resistance from tomato.

COMPARATIVE MAPS AND CONSERVED SYNTENY

The cloned DNA fragments used to detect RFLP often hybridize well to DNA of related species, giving a series of common markers that enable the respective genetic maps to be aligned. This is termed comparative mapping. cDNA clones are particularly useful for inter-species work, probably because the accumulation of mutations in expressed gene sequences is limited by selection.

In contrast to the plasticity of genome structure revealed by studies of repeated sequences, including nucleolus organizer regions, the order of single-copy sequences on genetic maps of different species often shows a high degree of conservation. Conservation of marker order is termed conserved synteny, since two genes or markers are syntenic if they lie on the same piece of DNA, usually a chromosome or chromosome fragment, irrespective of whether they are genetically linked.

COMPARATIVE MAPPING IN GRASSES

Comparative mapping shows that the three genomes of hexaploid wheat have very similar marker orders with the exception of a few major chromosome rearrangements. Additional translocations are found when comparison is made with the rye or barley genome, but marker order is still highly conserved. Comparative maps of maize and sorghum are well established and comparative maps of *Oryza* species are being developed.

Remarkably, comparison of the genetic maps of wheat and rice, rice and maize and wheat and maize also reveals large linkage segments that can be aligned. Furthermore, equivalent linkage blocks can be recognized in rice, wheat, maize, sorghum, sugar cane and millet. Thus a model of cereal genomes can be constructed using segments of the rice genetic map as the building blocks. This aligns the maps of the various species, revealing the genetic relationships between their chromosomes.

This analysis also confirms the tetraploid nature of the maize genome since the duplicated chromosome segments can be organized into two separate genomes. An additional important result from comparative mapping is that RFLP detected by cDNA clones from single rice YAC (yeast artificial chromosome) clones either cosegregate in barley or are tightly linked in the same map order as in rice.

Thus, fine-scale organization of gene order is also conserved. Mapping studies also suggest that there are duplicated regions in diploid species, suggesting that duplication and subsequent divergence of chromosome segments has been an important feature of plant evolution even in the absence of polyploidy. Examples include regions containing dwarfing genes in rye, segments of rice chromosomes 1 and 5, duplications in sorghum and soybean (*Glycine max*) that may reflect ancient polyploidy, and duplications in the diploids *Brassica oleracea* and *B. nigra*.

USES OF MARKER-AIDED SELECTION

In an ideal world, selection should be aimed directly at those genes which control the trait to be improved, but this assumes that genotypes at these target genes can be recognized easily, unambiguously and at an appropriate time without impeding the breeding programme. There are several reasons for using marker genes to improve selection efficiency.

Difficulties in Identifying Trait Genotype

The genotypes at a single-trait locus may be difficult to recognize because the desired alleles may have poor penetrance, be recessive, or interact with other genes or the environment. Environmental variation itself may impede accurate genotyping by causing the phenotype of different genotypes to overlap; this is particularly a problem with polygenic traits.

Some phenotypes such as resistance to a particular pest, pathogen or abiotic factor, like drought or salinity, may only appear under particular conditions which are difficult to define or control.

These difficulties may be more severe during the early stages of a breeding programme when plant numbers are too few to allow adequate replication or the breeder does not want to place his valuable yet scarce breeding material at hazard.

Earlier Selection

Another major reason why it may be valuable to use marker or indicator genes is to reduce the time from sowing to selection. Many traits of economic importance are only apparent in the mature plant and so may not occur for months or even years after sowing. This is particularly a problem with biennial species and long-lived crops such as trees. Indicator genes, on the other hand, may be detectable in seedlings within days of sowing, if not in the seeds themselves, so avoiding the waste of valuable resources involved in raising and harvesting plants, most of which may prove to be of no value to the breeding programme.

More Intense Selection

Selection at a juvenile stage, particularly among seedlings, may also allow much larger populations to be studied and hence more intense selection to be applied. Indeed, selection may be possible in tissue culture without raising plants or running trials.

Nondestructive Scoring

Many characters are scored before maturity and the act of scoring may prevent seed being collected. Imposition of selection for pests and diseases may result in plants which are less able to produce seed. The use of certain types of marker loci may only require the removal of small quantities of leaf or other material to permit full genotyping of the plant.

Linkage Drag

Conventionally plant breeders have used backcrossing with selection to introduce a useful allele, such as one conferring disease resistance, from one strain into another. The source of the allele is often an alien or exotic species, but may be a related cultivar. The F1 is backcrossed to the recurrent parent, and backcrossing is repeated for five to ten generations, the intention being to introduce the desired donor allele whilst returning to the recurrent parent's genotype at all other loci. At every backcross generation, the breeder selects for the phenotype of the allele to be introgressed whilst simultaneously selecting for the phenotype of the recurrent parent for all other traits.

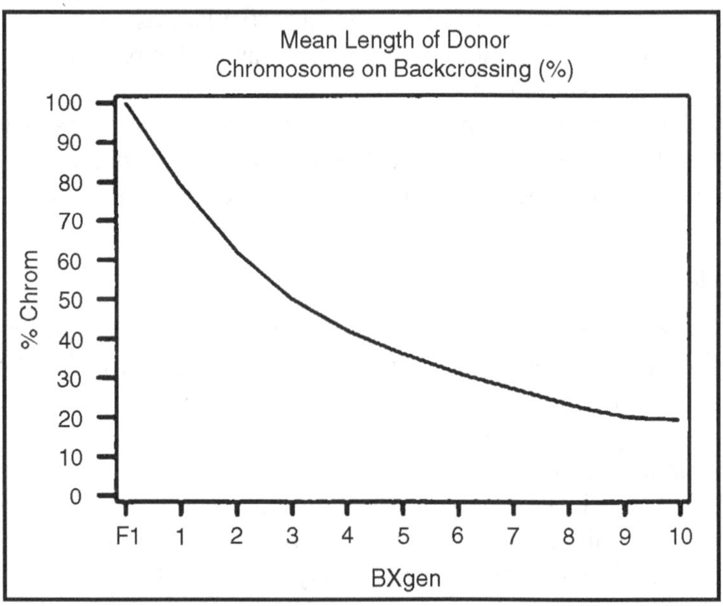

Fig. Linkage Drag.the Average Amount of Chromosome Surrounding a Selected GeneFollowing Backcrossing.Bxgen=Generation of Backcrossing from F_1

It was shown by Stam and Zeven (1981) that this procedure results in the leaving of very large amounts of donor genome associated with the selected alien allele. On average, about 20per cent of an average chromosome will consist of a donor segment even after ten backcross generations, while in any given line the actual size could be much more. Thus, as well as introducing useful alleles, the breeder will also be introducing very many undesirable alleles from the donor and so the gain in one trait may be at the expense of losses elsewhere. The use of marker loci can enable the breeder to reduce the size of this unwanted associated region as well as to accelerate the speed of return to the desired genotype of the recurrent parent.

Heterosis Potential

Heterosis, or hybrid vigour, is due mainly to the presence of dispersed dominant alleles in the parental cultivars. Breeders concerned with developing lines for hybrid production are often concerned to identify lines which differ at as many gene loci as possible. It is often assumed that this implies that the lines also have very different genotypes at marker loci and hence such loci can be used to assess the genetic distance between potential parents, but this can be a dangerous assumption unless a wide range of different types of markers are used.

Quality Control

A major problem faced by most conservationists and breeders is to maintain the integrity of the material they are working with and to keep it free from contamination. Contamination can occur in various ways. It may arise through errors due to carelessness in labelling or sorting seed or plant material, mistakes in maintaining the pedigree, or as a result of outcrossing due to stray pollen from unrelated material. Such errors are difficult to recognize when only the gross phenotype is available as a guide, but genetic markers can provide a clear genetic fingerprint which can be used at any time to confirm the origin of the material. It can also be used to confirm that the intended cross or self has occurred, or to separate the desired progeny from a particular parent which might be a mixture of selfs and crosses.

Other Uses

Marker information can be used in other situations also: distinguishing haploids, diploids and aneuploids, for example, particularly following wide interspecific crosses or tissue culture. Doubled haploid lines produced by microspore or ovule culture can be recognized in this way and possible aneuploids arising during chromosome doubling can be identified and removed. Not only can genetic markers identify contaminant material but they also allow closely related or identical material to be identified in genebank collections, so reducing unnecessary duplication.

TYPES OF GENETIC MARKERS

Up to the mid-1960s the only easily usable genetic markers were those that produced clearly visible effects on the plant's phenotype such as colour, size, shape, disease resistance, etc.. All these were the result of mutations to alleles which had, obviously, a major effect on the development of the plant. Such mutants were rare and generally not desirable in commercial material although they may be selected by breeders of decorative plants for their novelty value. As genetic markers, however, they have many disadvantages. They often affect the fitness of the individual and could well have effects on important agronomic traits, *i.e.* they exhibit pleiotropy. Because they are rare, it would be necessary to bring several such mutants together in the same material in order to use them, and they would not normally be present in commercial cultivars or, indeed, in wild material. This not only takes time, but the combined effects of many such alleles would mitigate against their value as a genetic tool.

In the 1960s through to the 1980s, naturally occurring genetic polymorphisms at the protein level came to be easily recognized. Variation in enzymes and storage proteins was detectable on an electrophoretic gel for a very high proportion of gene loci which could be studied in this way and, unlike major mutants, such naturally occurring allelic variants had much smaller effects on fitness and hence were more common and existed naturally in populations. The number of such polymorphisms was, however, insufficient to provide the coverage of the genome desired by breeders and conservationists. The development of techniques during the late 1980s up to the present revealed the vast amounts of genetic polymorphism which exists at the DNA level and has revolutionized genetic analysis, opening up a whole new range of genetic tools.

As described elsewhere in this book, this variation arises through the existence of occasional base changes in the DNA which can be recognized by restriction enzymes or by primers used in analysis involving the polymerase chain reaction (PCR). Much of the DNA, possibly as much as 90per cent in many species, is noncoding and hence reliance on morphological or biochemical variants was only using the restricted variation that plants could tolerate within the coding parts of their DNA.

Clearly, there are limits to the tolerance of such variation. Variation in the noncoding regions, whether in the intergenic regions or within introns, is probably under far less constraint by natural selection, and hence the number of polymorphic sites is potentially enormous. Because most of these sites are outside coding genes they are referred to as loci rather than genes; genes are coding regions whilst most marker loci are not.

Restriction fragment length polymorphisms (RFLPs) have been the most widely used until now. They have the advantage of being codominant and hence all genotypes in a cross can be identified. On the other hand they require quite large quantities of DNA, and hence plant material, and generally rely on radioactive probes, although fluorescent techniques are becoming more popular.

As a result, RFLPs are expensive to use. The use of radioactive probes incurs safety considerations, as well as taking longer to visualize the bands. PCRbased markers are now being used more widely.

Randomly amplified polymorphic DNAs (RAPDs), for example, require far less material because only small amounts of DNA are needed and the required sequences are then amplified. They do not involve radioactive probes and are both quick and cheap. On the other hand RAPDs are dominant although it is possible to distinguish heterozygotes from homozygotes by the amount of product. Amplified fragment length polymorphisms (AFLPs) offer even cheaper and more easily identified and extensive PCRbased polymorphism.

Because of the number of polymorphisms which exist on individual gels and the speed of production, computer software is necessary to scan the banding patterns on the gels and to analyse and assimilate the information that is generated. Other popular markers are microsatellites (short sequence repeats) and cleaved amplified polymorphisms (CAP)-based primers.

All the genetic variants discussed above will be subsumed under the broad title of marker loci. They may be of little interest in their own right, and this is particularly true of the molecular polymorphisms; their value lies in their use as markers of more useful genes.

PARTICIPATORY PLANT BREEDING

Certain trends have made the world ripe for adoption of participatory plant breeding methods.

- The shielding of crops from environmental stress in high-input systems is facing increasing economic and ecological problems. Scientists are changing their attitudes, and a new paradigm is being formulated. Instead of modifying the environment to suit the requirement of high yielding varieties, the varieties need to be modified to suit the environment.
- The claim that modern varieties can be made broadly adapted and be superior across most farming environments within an ecogeographic region is being challenged.
- If relevant diversity exists in a locale, the combined action of natural and artificial selection within a local environment may be an efficient breeding method. Experiments show that this may work also in a fertilizer-intensive system.
- In recent years farmer groups working with local seeds have been organized all over the world. They are not primarily conservers of old seeds. They want their seeds to be improved, in an evolutionary, slow and steady way, and under their own control.
- Finally, participatory methods have been developed in order to facilitate the involvement of farmers together with scientists as active

and equal partners in research to generate relevant farm technology. Such methods can be applied also to plant breeding.

Commercial seeds often diffuse into areas where the traditional seed supply system is still predominant. Farmers try them with an open mind and adopt or reject them according to their own criteria. If grown and multiplied in the villages, diversity will start to appear within them and local reselection will be possible. In that way commercial varieties eventually might become like landraces. It is also commonly observed that farmers change and exchange seeds. A traditional farming system rarely functions as an environment for the static preservation of old landraces.

Often practices range from neglect to very simple mass selection, but with a few scattered individuals who devote an exceptional amount of effort to the maintenance or improvement of seed quality. These exceptional persons, very often women, may be the source of good seeds for others in the community. Once such a community is organized for seed management and improvement, it becomes possible for it to establish links to scientific institutions.

Community resources for participatory approaches to seed management and breeding are not limited to indigenous culture and traditional practices. The educational status and experiences of modern farmers may also be turned into a resource for community action. In the Philippine group, a number of the members were high school graduates and a few had a university degree. And moreover, most of them had a couple of decades' experience with modern input-intensive farming. Seed activities opened their minds towards the traditional societies, towards themselves and towards the modern world. The traditional societies supplied them with their seeds, and through the seeds, they learned to appreciate the values and achievements of these societies. They discovered their own potential, and saw that the outside world could bring more than technology packets: it could bring knowledge and ideas to be exploited and further developed by themselves.

REASONS FOR PRIVATISATION OF PLANT BREEDING

Plant breeding can be conceptualised as an investment that develops improved varieties with the potential to generate future benefits in the form of improved crop productivity, reduced costs of production, and/or higher returns. Potential value from improved cultivars will be realised only if and when farmers adopt these cultivars in their cropping systems, AND when consumers willingly purchase the food or other crop products in a competitive market. Growers will only adopt these new varieties if they provide real financial benefits that exceed the costs of adoption, including any additional costs of acquiring the improved variety.

Similarly, consumers will only knowingly purchase food from these new varieties if by so doing they derive a net benefit in the form of enhanced attributes and/or lower prices relative to available alternatives. In common with other forms of investment, the rate of return will depend on the discounted value of the flow of future benefits net of present value of all costs neccessary to generate such benefits. Arguably the most important reason for the growing trend to privatisation of plant breeding has been significant changes in the ability of plant breeders to appropriate at least some of the benefits from improved grain varieties that otherwise would be captured by growers. Specifically, the incentive for firms to invest in plant breeding depends on the ability to exclude grain growers, and often competing plant breeders as well, from commercial exploitation of a breeder's varieties unless they pay to do so.

For some crops such as corn, the development of hybrid technology provided genetic copy protection that enabled plant breeders to capture much of the value from heterosis as well as other superior traits. For other crops, it has been the expanding scope of intellectual property rights that has enabled the capture of some of the value created by plant breeding. So while it has been the application of modern science to plant breeding that has generated much of the potential for value creation in the grain supply chain, it has been extensions to the legal framework for intellectual property rights that have made possible private capture of enough of the value created by plant breeding to provide the private sector with an incentive to invest more in plant breeding.

The most significant of these intellectual property rights are patents and Plant Breeder's Rights. In recent decades, both court decisions and legislative changes have expanded the scope and impact of these two types of intellectual property right appreciably.

Complementing these developments in the institutional framework have been scientific discoveries that led to greater potential for value creation by improvements in plant breeding methods. Apart from hybrid technology, other recent advances include new technologies that improve the efficiency of all plant breeding, including both conventional and transgenic plant breeding.

Such techniques include double haploidy, plant regeneration systems, molecular based hybrid technologies, and marker assisted selection. Use of these techniques in conventional plant breeding is already reducing the time lags from initial crosses to release of new varieties. Potentially beneficial outcomes from the application of these technologies to plant breeding include one or more of the following:

- cheaper development of improved crop varieties.
- faster/earlier development of improved crop varieties.
- development of superior improved crop varieties, that are more productive, produce better quality grain, or both.

In addition, there has been the more controversial development of transgenic technologies used to produce GMO's. Potential beneficial outcomes from transgenic technologies include:

- development of improved crop cultivars with novel agronomic/input traits that enable lower average costs of production.
- development of improved crop cultivars with novel quality-enhanced traits for which consumers are willing to pay a price premium.

On the other side of the coin, publicly funded rural research has been under pressure for at least the last two decades. In part, this been due to a growing perception that grain growers have been the primary beneficiaries of the traditional plant breeding programmes.

Historically these programmes have been funded mainly from consolidated revenue. To some extent, this concern has been addressed by the relatively recent evolution of the GRDC and similar bodies that rely heavily on collective industry funding to support much of their investments, but the fact remains that a significant part of plant breeding programmes is still publicly funded.

Governments also now demand greater accountability at the same time that they reduce funding for agricultural research and extension. As a result, many "public" institutions are under pressure to become at least partially self-funding, and are starting to charge for selected goods and services. Public research institutions also seek to patent and/or commercialise discoveries made in the course of government funded research, or pursue opportunities to license technologies to the private sector.

Public plant breeding programmes have not been immune to government pressure to generate revenue from their activities. Like private business, their capacity to capture a high proportion of the net benefits of new varieties depends on:

- a legal basis to establish ownership of the intellectual property embodied in the variety,
- the capacity to exclude potential users who are not willing to pay the nominated price,
- the costs of monitoring and enforcing compliance,
- the capacity for price discrimination.

Pricing practice by public institutions is still evolving. If they start charging significant fees at levels approaching full cost recovery, and exclude farmers unwilling to pay these fees from access to new varieties, then they cease to be public plant breeding organisations within the meaning of the term in this chapter.

Finally, agronomic practice by grain growers has become increasingly sophisticated and much more tactical. In particular, many growers now make decisions about which varieties to grow each season on the basis of the latest possible information about the climatic outlook and other seasonal indicators, such as soil moisture levels as well as weed and disease threats. Consequently they are less likely to use seed saved from the previous harvest, and more likely to purchase new seed of the desired variety from a seed merchant. This change in farming practice will increase the size of the seed market, and improve the economics of private plant breeding.

SCOPE OF PLANT BREEDING

Improving Crop Performance

The dramatic gains in agricultural productivity seen in the second half of the last century are often linked uniquely to increased mechanisation and the widespread introduction of fertilizers and agrochemicals.This 'Green Revolution' would not have been possible without the huge genetic improvements made to the crops. Plant breeding has contributed around half of the threefold increase in UK wheat yields recorded from 1947 to 1992 and yields have continued to improve since then. Plant breeding can directly improve the performance of crops in different ways.

Productive Yield

Developing crop varieties, which convert more of their biomass into productive yield, is the single biggest contributor to improved crop output. The introduction of shorter-strawed cereals is a striking example of how this has been achieved, by transforming more of the crop's productive energy into valuable grain.

Physical Characteristics

Changing a crop's physical structure can also contribute to increased yields. For example, the development of semi-leafless varieties of field peas helped to boost intrinsic yield. It also stimulated the growth of stabilising tendrils, which significantly improved the crop's standing ability so reducing crop losses at harvest.

Disease Resistance

Genetic resistance to disease enables crops to realise their yield potential– it can also mean reduced use of agrochemicals. Plant breeding has significantly improved the genetic resistance of crops against the threat of viral and fungal infection. Key examples include resistance to blight in potatoes, rhizomania in sugar beet, and barley yellow mosaic virus in cereals. The challenge of breeding resistant varieties is constant because new strains of disease develop naturally.

Timing of Maturity

Plant breeding technology has brought major improvements in the uniformity with which crops ripen ready for harvest. This not only reduces potential crop losses at harvest (as in the case of pod shatter in oilseed rape), but has also improved growers' ability to mechanise harvesting operations. In the field vegetable sector, for example, the labour intensive (and unpleasant!)

task of picking crops such as cauliflower, broccoli and Brussels sprouts has been transformed by the development of improved varieties.

Other Agronomic Factors

By improving crops' ability to cope with a range of other agronomic pressures, advances in plant breeding continue to underpin progress in agricultural productivity.

They include:

- Genetic resistance to pests, such as nematode resistance in potatoes
- Shorter crop life-cycle, to expand a crop's geographical growing areas
- Stress tolerance, such as frostresistance in field vegetables, to extend the seasonal availability of home-grown fresh produce.

Improving Crop Quality

Alongside progress in keeping Britain's farmers competitive through improved productivity, developments in plant breeding have also brought significant gains in crop quality. Gone are the days when Europe encouraged farmers to concentrate solely on maximising output, assured that their crops would find support under the Common Agricultural Policy.

Today, as production-based support and barriers to international trade continue to be reduced, Britain's growers must compete with the best in the world, on quality and cost of production. Consumer expectations of food are also more exacting. This is reflected in tighter quality specifications along the food chain. Plant breeders have responded with a continuous stream of new varieties, tailored to the needs of specific end-markets.

Cereals

Britain's historical dependence on North American imports of wheat for bread making has been reversed over the past 40 years thanks to improvements in baking technology and the development of high protein, hard milling varieties of wheat.

Until the mid-1960s, imported wheat typically accounted for up to 65 poe cent of our bread making requirements. Today as much as 90 poe cent of our bread is produced from homegrown varieties. In barley, improvements in malting quality have underpinned a fourfold increase in the volume of beer brewed per tonne, from around 2,000 liters in 1950 to nearly 8,000 liters today.

The quality of malting barley varieties grown in Britain is internationally recognized, with export markets from the near Continent to the Far East.

Oilseed Rape

Oilseed rape has become a major homegrown source of vegetable oil and animal feed within the past 30 years. The rapid expansion of this important break crop from 50,000 hectares in the mid-1970s to around 500,000 hectares today can be linked to two major breakthroughs in plant breeding. Varieties for food use were first developed in 1970. These contain reduced levels of erucic acid in the oil, making it more suitable for human consumption. This was followed in the late 1980s by 'doublelow' varieties, which also offered reduced levels of glucosinolates in the residual meal to improve quality for animal feed.

Potatoes

In the potato sector, plant breeders have responded to increasing consumer demands for quality and choice. Fresh produce counters now boast a bigger range than ever, from traditional main-crop varieties through to baby new, baking and salad potatoes. Breeding programmes have also targeted processing and catering markets, with specific varieties used to produce crisps, instant mash and chips.

Sugar Beet

Higher root yield and increased sugar content have combined to double sugar production per hectare in the past 50 years. The introduction of monogerm seed, allowing seeds to be planted individually rather than in clusters, ensures genetic improvements are fully realised in the field.

Pulses

As interest in home-grown sources of protein has increased, breeders have developed varieties of peas and beans for specialist markets–freezing and canning for human consumption, flaking for pet food, and tannin-free varieties for animal feed.

Creating a New Variety

The creation of each new variety is a complex, costly and skilled operation. It is also painstakingly slow–today's breeding programmes are already looking ten years ahead to the needs of farmers, consumers and the environment at the end of the next decade and beyond. Techniques vary between crop species, but the scientific principles of plant breeding remain unchanged from Mendel's first discovery that selected parent plants can be cross-pollinated to combine desired characteristics.

Genes–units of hereditary material that are transferred from one generation to the next, determine plant characteristics. Since each plant contains many thousands of genes, and the breeder is seeking to combine a range of traits in one plant, such as high yield, quality and resistance to disease, developing a successful variety is an extremely lengthy process–up to 12 years in the case of cereals, even longer for potatoes. Plant breeding has been compared to playing a fruit machine–not with three reels, but several hundred. The skill of the plant breeder lies in improving his chances of hitting the jackpot by combining all the desired characteristics in the same variety.

PLANT BREEDING AND GENETICS

Breeding, the art and science of changing and improving the heredity of plants, has been practised by humans since plants were first cultivated. *Selection*, the process of choosing those plants considered superior and harvesting them for seed, was one of the earliest techniques. This ability of the breeder to select plants for economically valuable characteristics is the art of breeding.

The scientific aspect comes with knowledge of genetics and other plant sciences. *Genetics* can be defined as the science that studies heredity and variation or those things that make individuals similar (*heredity*) and different (*variation*).

Genetics seeks to explain how and why variation and heredity occur. This knowledge has allowed breeders to manipulate the heredity of plants to produce higher yields, resistance to insects and disease and tolerance to drought. The changes that are made are not temporary so the benefits can be reaped over and over. First published in 1866, *Gregor Mendel* proposed that hereditary units, now called genes, controlled the expression of an individual's characteristics or traits. For instance, Mendel proposed that if you have two apple trees and one has a higher yield, the tree with the higher yield possessed genes for higher yield. He went on explain that each individual inherits an equal number of genes from each parent and the expression of a trait is determined by the combination of genes inherited from both parents. This concept was more fully developed by *Walter S. Sutton* who, in 1902, proposed the *chromosome theory of inheritance*. This stated that the hereditary factors Mendel wrote of, genes, were located on chromosomes which split during meiosis.

Chromosomes are located in the nucleus of a cell and carry and transmit genetic information of an individual from one generation to the next. In plants, the number of chromosomes in vegetative cells is the same for an entire species. This number is referred to as the *2n* or *diploid* number. In alfalfa, it is 32 and in corn, 20. In the sex cells, *gametes*, the number is one-half of the diploid and is known as the *1n* or *haploid* chromosome number. This information is invaluable to the breeder working to develop better cultivars.

So far, the only method or breeding technique mentioned has been selection. Planned *hybridisation* between carefully selected parents is now one of the dominant methods available today. The breeder must find two parents that will combine to produce a superior first-generation offspring or *F_1 hybrid plant*. The F_1 should be superior to both of its parents measured in terms of increased vegetative growth, cell size, plant height, etc.

PLANT BREEDING STRATEGIES AND PRIORITIES

Plant breeding has been recognised as the major tool for green revolution and has been instrumental in boosting crop production in both developed and developing countries. High-yielding varieties of rice and wheat were at the heart of the spectacular increase in food production in Asia and elsewhere. In CWANA region, modern varieties contributed up to 43 per cent of barley at production during the first phase of adoption. A recently published study estimating the value of the CGIAR's activities, reports a benefit-cost ratio of 9.0 for the $7.12 billion invested. The benefit-cost ratio rises to 17.3 when extrapolated through 2011 under the assumption that research benefits will continue to be realised at present rates. This confirms similar finding at INRA Morocco were every dollar invested in the development of Hessian fly bread wheat resistant cultivars, generates $9 additional benefit with profitability ratio in such investment of 37 per cent.

Plant breeding has been a central core of the research agenda at INRA and was the major driving force for funding research activities. The existence of a formal system of variety cataloguing and well organised private seed sector and adequate incentive policies for certified seed uses had a significant impact on the overall adoption of improved cultivars.

It is not sure, however, that the allocation of resources counterbalanced such performance since there is no formal resource allocation par activity and the existing financial system provides very little flexibility to the overall management of INRA. Hence, it always remains difficult to objectively assess breeding priorities in terms of predetermined resources allocation.

Nevertheless, in terms of human resources, research activities and outputs, we can easily assess that breeding of bread wheat, durum, barley for cereals, pate palm, olive, and citrus, forage species such as oats and alfalfa and food legumes such as faba bean chickpea and lentils received the highest priority in the research agenda. Other species such as fruit trees (walnut, apricot, pistachios, pomes, pear, etc.) and industrial crops such as sunflower, safflower, colza, or sugar beet, as well as vegetables occupied ambiguous position, which largely depended on the research policy of the moment, the performance of researchers, and the availability of finding resources.

CEREALS

Cereals are the most widely grown and consumed food crop in Morocco, with an average of 5.3 million hectares, which represents 60 per cent of total cultivated area. 90 per cent of cultivated area is rain-fed. Barley is the main crop with 45 per cent of the total cultivated cereal area, bread wheat represents 30 per cent, and durum 22 per cent. Cereal breeding has been the main research component of INRA breeding programmes. It is one of the strongest breeding programmes in the WANA region. A major breakthrough was achieved in the field of Hessian fly resistance. Genetic gain due to wheat improved varieties in rain-fed area is about 1.55 per cent per annum and water use efficiency improved from 1.5 to 2.5 kg/mm.

Potential yields of released of bread wheat varieties was improved from 8 to 12 quintals/ha to more than 75 quintals/ha. International centres such as CIMMYT and ICARDA played a key role in sustaining such achievements through provision of genetic resources and training of NARS scientists or technicians. Devolution of breeding activities to Morocco through decentralised selection of barley and wheat Hessian fly and rust resistance, helped foster furthermore the INRA role in the region. INRA scientists are actually posted at ICARDA and are playing a key role in the international wheat and barley breeding programmes.

Bread Wheat

Bread wheat breeding programme was initiated in 1921 at Rabat station. Early selection of genetic stocks was made among local landraces and introduced varieties from the Mediterranean region (Tunisia, Algeria) or other regions (India, Australia, USA, France, and Italy). At that time, barley (*Hordeum vulgare*) and durum (*Tricticum durum*) predominated the cropping system along with other species such as rye (*Secale cereal*), *Triticum monococcum, T. dicoccum* and *T. polonicum L.* Then, afterwards, hybridisation was initiated to further exploring the existing genetic variability. Potential yield, rusticity, and grain quality then starting from 1930, processing standards (W over150) used to be criteria of selection.

The first-released cultivars were 4 varieties from Algerian origin, then 3 from Indian (Pusa type), 2 from Australia, and 1 from France (Florence). Bread wheat cultivars such as "*Cailloux*" and "*Florence*" and "*Florence Aurora*" were among the most widely used cultivars during early century.

During the 1960s, the first semi-dwarf genotypes and early maturing varieties of Mexican origin were introduced. The success of such varieties among which Siete Cerros and Potam, was limited and they were immediately discarded because of their susceptibility and the huge damage caused by *Septoria tritici.*

In 1973, the first high-yielding cultivar "*Nasma*," which originated from a cross of Florence and *Aurore C.* was selected. This was a major breakthrough of the national breeding programme at INRA since it dominated the wheat production spectrum for more than two decades. Nasma is semi-dwarf genotype, it has a high potential yield and enhanced stability, it is resistant to lodging and to Leaf rust *(Puccinia recondita f.sp. tritici)* and has good grain quality (large seeded) that farmers appreciated most.

Since early 1980s, a wide breeding programme was initiated using CIMMYT varieties as parental lines in the hybridisation programme. This resulted in release of more productive cultivar such as "*Merchouch*" and "*Jouda,*" which in turn outweighed "*Nasma.*" The CIMMYT genetic material used to have useful common features such as, earliness and good adaptation to Moroccan diverse conditions, probably resulting from the concept of mega-environments broad-based strategy of plant germplasm resources used at this centre. Most materiel received from this centre was spring or facultative wheat not sensitive to photoperiod and resistant to rusts. It has, however, two major problems: it is susceptible to septoria (Septoria *tritici, Mycosphaerella, graminicola*) and to Hessian fly (*Mayetiola destructor (Say)*).

A specific breeding programme through hybridisation was then initiated since 1987, one for semi-arid and the other for sub-humid conditions, with major objectives of improving resistance to Hessian fly and Leaf rust for the first environment and septoria and Leaf rust for the second. The USAID/MIAC project on dry land played a key role in fostering the breeding programme on Hessian fly resistance at INRA Morocco. That was a formidable challenge for agricultural science and for improved wheat production technologies of this century. Indeed, Hessian fly does cause serious damage, which is about 30 per cent loss and reaches 100 per cent in case of late sowing. The overall economical loss is about $200 million per year. In 1997, two bread wheat cultivars "*Aguila*" and "*Arrehane*" both resistant to Hessian fly were released to farmers.

Of the 32 genes expressing different level of tolerance or antibiosis against Hessian fly larvae, identified in the USA, 10 genes (H5, H7H8, H11, H13, H14H15, H21, H22, H23, H25, and H26) were selected as conferring resistance to Hessian fly in the field in Morocco. Most of these genes are located on the A or the D genomes. The H22 gene has been successfully introgressed into adapted Moroccan bread wheat cultivars.

Alternative pests which are gaining importance are Sunn pest and Russian aphids. Accessions of *Triticum monococcum* from ICARDA combining both Hessain fly and Russian aphids resistance are being used in the breeding programme.

Since 1995, a comprehensive breeding programme was initiated on drought tolerance and heat stresses using conventional breeding and most advanced tools in biotechnology.

Durum Wheat

The objectives of the initial plant breeding programme were to improve productivity, response to high inputs, and good pasta quality. Mountainous and oasis local land races provided basic genetic resources. Since 1930, five cultivars were released, among which durum wheat "*Zemarek*," "*Sbei*," and "*BD-2777*" continued to be grown in mountainous areas until recently because of their outstanding quality. These genotypes were, however, late maturing and susceptible to lodging.

During the 1960s, a new breeding programme was initiated through intraspecific hybridisation with bread wheat and wild relatives for improving yield components, disease resistance to rust, and Hessian fly. Such effort did not lead to any significant progress because of narrow genetic base for selection and depreciation of durum quality due to interspecific crosses.

Since the 1970s, collaboration with CIMMYT then with ICARDA helped provide new genetic resources and led to increasing genetic variability of durum germplasm and led to release of two cultivars "*Cocorit*" and "*Haj-Mouline.*" These varieties were high yielding and more adapted than previous cultivars but their quality limited their adoption in mountainous areas.

Since 1980, the breeding programme took a step forward through improvement of diseases resistance to root rote caused by *Fusarium sp.* and Helminthosporium (*Pyrenophora tritici-repentis*) and rusts: Stripe rust (*Puccinia striiformis*), Leaf rust (*Puccinia recondita*), and Stem rust (*Puccinia graminis*). Selection within introduced germplasm and hybridisation led to release eleven cultivars among which "*Karim*" from Tunisian origin, "*Acsad-65*" from "*ACSAD,*" "*Tassouat,*" and "*Oum-rabia*" from the national breeding programme.

As for bread wheat, Hessian fly became the most important common feature of the wheat breeding programme. Genetic resistance to Mediterranean Hessian fly biotypes has not been found initially in durum wheat, although large numbers of durum accessions were screened. Genes for resistance were found in common wheat, some of which were transferable to durum. Genetic base was further broadened through introgression of resistance genes from selected *Triticum araraticum* and *T. carthlicum* accessions using multiple backcross methodology.

This led to release of the most significant result through registration of the first cultivars resistant to Hessian fly and tolerant to drought in 2003.

Durum wheat grain is used to prepare various products including pasta and couscous. The acceptability of a durum variety is greatly influenced by its quality characteristics. These products generally require large vitreous kernels with high protein, good yellow pigment, and strong-to medium-strong gluten. Thus, the incorporation of quality characters essential for end-use products is now days the major challenge for durum breeding programme. In response to mil industry, a new project was initiated since 2005 in order to improve the yellow pigment related to the carotenoid and protein content of the endosperm.

Barley

About 80 per cent of barley grain production is used as animal feed on farms and the remaining is used as food. Morocco is ranked first worldwide for food and feeds barley consumption with the estimated average of 54.8 kg per capita consumed annually and is the 7th largest barley producer in the word. Morocco is considered one of the major centre of diversity for cultivated barley (*Hordeum vulgare* L.). This is evidenced by the large genetic variability found between and within barley landraces and existence of numerous wild relatives including populations of the wild progenitor of cultivated barley *H. vulgare sp. Spontaneum* collected in the low Atlas region.

Barley breeding began in 1924 with the selection of lines from introduced two row barley accessions from the USA, Australia, and Europe as well as selection among sex row domestic landraces. The later programme led to selection and release of two cultivars "*077*" and "*071*" among local populations which are widely grown up today by most farmers. This clearly shows the importance of the genetic diversity in Morocco.

Several accessions have been introduced afterwards and a new breeding programme was initiated since 1970 for improving two-row barley, which led to release of several cultivars among which Brasserie Maroc, Tamellalt, and Asni. This type of genotype was not easily accepted by farmers who prefer the six-row barley though there is now evidence of their superiority.

Since 1980, new accessions were introduced from the USA in order to improved earliness and harvest index. Several accessions were also introduced from ICARDA with the main objective of improving forage and grain yields and disease resistance for leaf rust *Puccinia hordei,* powdery mildew *Erysiphe graminis f.sp. hordei*, net blotch *Pyrenophora teres f.sp*. teres, and BYDV virus.

During this period, there was a renewed interest for barley as a food component for human consumption. Its high soluble dietary fibre and â-glucan content compared with other cereals as well as the new prospect of hull-less barley genotypes that can be easily separated from the hulls after threshing could facilitate the use of barley grain for bread making and other nutritional foods. An additional programme to develop hull-less food barley varieties was initiated in 1985. Hull-less barley helped initiate a wide food processing programme for diverse dietary uses. However, the difficulty of handling the seeds and their reduced seed-germination capability limited their inscription in the national catalogue.

Since 2003, a new breeding programme was initiated with two major components, one for marginal rain-fed land (dry lands and high elevations areas), the second for favourable conditions. For the first production system, participatory breeding of local land races population, *in situ* conservation and on farmer seed production are the main features of the breeding programme. For the second, the objective is to develop pure lines through conventional breeding methods within a long term strategy aiming at the introduction of male sterility for production of synthetic varieties.

Production of varieties with increased mineral and vitamins contents is another recent area of research lunched with collaboration with ICARDA. A new breeding project on biofortification aiming to mitigate the problem of micronutrients deficiency by improving zinc and iron contents was recently initiated. But this programme remains relatively limited though its importance.

Maize

Cultivated area is about 267,000 ha 75 per cent is rain-fed, 15 per cent irrigated, and 10 per cent in mountains. Total rainfall in cultivated area is about 350 mm with 50 to 100 mm during the cropping season. Therefore, drought is a serious limiting factor to corn production. Dew may, however, play a key role since most cultivated region area is in coastal region. Initial breeding programme started at Rabat in 1947 with collection of 75 local land races. The bulk of this collection was constituted of early maturing populations with short stalks and ears, white or yellow grain, semi-flour, or dent corn.

From 1947 to 1965, an extensive breeding programme was under way with introduction of nearly 256 lines and 281 varieties from various countries mainly USA and France but also from various countries such as Portugal, Brazil, and Spain. During this period, the programme benefited also from an early FAO project aiming at introduction of US corn technologies in Europe. Breeding strategy evolved over that period as follows:

- Selection of corn varieties among local populations and introduced accessions;
- Selection of adapted hybrids among American corn hybrids;
- Creation of top- cross varieties resulting from hybridisation of locale varieties (such as "*Doukkala*" and "*Agouraï*") and simple hybrids;
- Selection of double hybrids from simple hybrids; and
- Creation of double hybrids with large genetic basis for rain-fed agriculture.

High yield potential, adaptation to rain-fed agriculture, earliness, and dent yellow seeds were used as selection criteria. TX-21 was among the most important corn top-cross variety selected released during this period and continued to be used for almost two decades.

During the 1970s, the programme put more emphasis on top-cross varieties and simple hybrid production. Selection criteria included diseases resistance to rust (*Puccinia sorghi*) and corn-stalk borer, Sesamia (*Sesamia non-agrioides*).

Since 1980, high priority was given to selection of variety populations for rain-fed agriculture. Three varieties "*Kamla*," "*Mabchoura*," and "*Doukkalia*" in 1991 were released from such programme. A small project was also initiated in the dry land programme for production of synthetic populations but was resumed. Up to now, 16 corn hybrids and five variety populations were registered in the catalogue. Yet, their utilisation remains problematic because of the overall situation of corn production. Some breeding work was also devoted to grain sorghum, but such work was neither systematic nor continued.

Triticale

Triticale varieties were initially introduced during the 1960s from CIMMYT. Adaptation trials carried out by INRA confirmed the superiority of triticales in coastal sandy acid soils of favourable areas and in mountains. During the 1980s, cultivated area was about 10,000 to 15,000 ha. Actually the production of triticales is abandoned apart from few acres of forage mixture. The number of released cereal cultivars and the difficulty of commercialisation have been two major handicaps for triticales production in Morocco.

During the first period, breeding objectives were high grain yield and adaptation to different agro-ecological conditions. Forage biomass and straw quality (Dry matter yield, haying requirement, regeneration after cutting and MAT) were afterwards included because of the increasing interest to triticales as forage crop. Screening for diseases resistance was carried out for rust, septoria, and Hessian fly.

Five cultivars of triticales were released in 1988 among which were Juanillo and Beagle. A new dual purpose triticales, "*Basma,*" for forage and grain production was released in 2002. Triticales still remains a secondary crop for cereal breeding though its high potential for chicken feed industry and forage production. An outreach regional breeding programme for Mediterranean areas was initiated with CIMMYT during the 1980s but this project was resumed afterwards.

Cereal Genetic Gain

In order to evaluate the genetic gain in cereal breeding, we requested from a panel of INRA plant breeders currently working in Morocco or in International centres such as ICARDA to evaluate genetic towards achievement of breeding goals.

Table. Major Breeding Goals and the Genetic Progress*

Major crops	Major breeding goals	Genetic progress *
Bread wheat	• Potential & yield stability	1
	• Leaf rust resistance	2
	• Septoria resistance	3
	• Hessian fly resistance	2
	• Russian aphids resistance	5
	• Multiple resistance	3
	• Drought tolerance	2
Durum wheat	• Potential &yield stability	2
	• Root rote and Helminthosporium resistance	3
	• Rusts resistance	2
	• Hessian fly resistance	1
	• Drought tolerance	3
	• Quality (yellow pigments and protein)	4
	• Multiple resistance	3
	• Specific adaptability (High lands)	3
Barley	• Potential &yield stability	2
	• Leaf rust resistance	2
	• Powdery mildew resistance	3
	• Net blotch resistance	3
	• BYDV virus resistance	4
	• Feeding values	2
	• Biofortification (Zn, Fe)	5

Note:

* Genetic progress: Estimated by breeding panel: 5 least progress to 1 major achievement

FOOD LEGUMES

Food legumes are important pulse crops in the Moroccan agriculture. Total cultivated area is about 350,000 hectares, 43 per cent of which is Faba bean, 19 per cent chickpea, 14 per cent lentil, and 9 per cent peas. They are important sources of protein and they play an important role in the cereals rotation system. This role becomes, however, limited because of the decline of cultivated areas and the overall changes in human diet.

Faba Beans

The major objective of early faba bean breeding programme in 1949 was to collect and identify genetic source of high-yielding varieties with acceptable seed size among local land races and introduced accessions from Spain or Italy. The result was the release during the 1950s of six cultivars of faba bean minor and major and collection of more than 200 accessions.

In 1979, the food legume breeding programme knew fundamental changes through collaboration with ICARDA. In 1988, the CGIAR decided to develop an outreach faba bean programme in Morocco and transferred the ICARDA faba bean breeding programme to INRA. This programme was resumed in 1991 and collaborative projects were developed afterwards with GTZ for the North Africa region, through networking projects such as REMAFEVE, REMALA, and Parasitic weeds.

Breeding for high yield and tolerance to orobanche (*Orobanche crenata* Forsk), resistance to botrytis (*botrytis faba*) and nematodes (*Ditylenchus dipsaci*), became the most challenging issues to the faba bean breeding. *Orobanche crenata*, a parasitic weed, has been the most limiting factor for faba bean production. In Morocco, infestation of *Orobanche* in cultivated food legumes crops, evolved from 12 per cent in 1981, 26 per cent in 1994, to 30 per cent in 2001 and reached 51 per cent in 2003 with estimated average yield losses in 1998 of 37.4 per cent. However, the expected progress remains very slow because of the major difficulty for combining quantitative characteristics involving resistance to Orobanche, yield per plant and large seed size. Today, in spite of some genetic progress (release of one line Sel.88 Lat 18105, faba bean minor tolerant to orabanche), the main challenge that remains is the introgression of multiple resistance to both orobanche and botrytis in large seeded type.

Chickpea

Chickpea is the second most important legume. It is cultivated as a spring crop. Early breeding programme, initiated since 1920, put more emphasis on germplaThe human species' 46 chromosomes are arranged in 23 pairs; in females all 23 pairs are homologous (correspond structurally and derivatively); in males, there are 22 homologous pairs plus a nonhomologous *X* and *Y* chromosome.

The chromosome pairs, when properly stained, differ in length and appearance so that they can be individually recognized, much as one might recognize the familiar face of a friend. During meiosis, when the gametes are formed and the chromosome number is halved, all the chromosomes in a single gamete differ because each gamete has only one member of each chromosome pair. At fertilization, two gametes unite, each containing a set of 23 chromosomes to form the fertilized cell (zygote) with 23 pairs or 46 chromosomes once again.

The number of chromosomes contained in gametes (23) is referred to as the haploid number and that of zygotes ($2 \times 23 = 46$) is the diploid number written as n and 2n, respectively. The process of meiosis affords a significant opportunity for increasing genetic variability. The most important consistent mechanism of change is crossing over, which takes place during the early phases of meiosis. Homologous chromosomes, lying next to each other, break, exchange equivalent portions of their genes, and reanneal to produce combinations that differ in arrangement from the genetic endowment provided by parents.

This phenomenon is omnipresent in the normal meiotic process; when it fails to occur in at least one parent, the resulting organism is often sterile. Crossing over is the most consistent of a number of processes that provide the raw material upon which natural selection may act.sm collection of spring chickpea of large seeded Kabuli types. Breeding work was strengthened in 1943 with germplasm collection of both kabuki and specific Desi-types maintained during winter season for ascochyta blight evaluation.

Do the plant breeder at that time was aware of the potential of winter chickpea? There were no evidence for this, nevertheless a winter hardy types were maintained at this collection and two spring chickpea of Kabuli type, "*PCH34*" and "*PCH37,*" were released from such programme. The requirement of larger seed size and competition of Mexican and Spanish types in the commerce, were a serious handicap for any further progress in the breeding programme. In 1978, INRA initiated a new breeding programme on horizontal resistance of spring chickpea to ascochyta blight (*Ascochyta rabiei (Pass.)*) with FAO collaboration. Extensive works on screening and breeding methodologies were developed within the frame work of such programme. About 26 lines with durable resistance were selected from this programme. However, the specific nature of blight pathogenic variability in Morocco limited the release of such breeding material. It is worth mentioning that such programme was afterwards extended to cover durable resistance of Faba bean to *Orobanche* and tomato to

Fusarium wilt, but this programme was not continued. The renewed interest for chickpea became, however, very clear through collaboration with ICARDA-ICRISAT programme posted at ICARDA since 1979. The introduction of the concept of winter chickpea completely changed the configuration of the breeding programme. Winter planting in November of winter hardy genotypes was suggested as an alternative option for increasing yield and stability in semi-arid environments. Resistance to Ascochyta Blight become a prerequisite in such situation because of the major blast of the disease during the winter season. Potential yields of winter chickpea was increased to more than 2.0 tonnes/ha as compared with 0.6 tonnes/ha of spring chickpea. This represents an overall increase of nearly 210 per cent with an earliness of 25 to 45 days.

Two varieties, ILC 482 and ILC 195, were released in 1987 but their seed size was too small to be accepted by farmers. Further collaboration with ICARDA led to release since 1992 of five new winter chickpea cultivars with larger seed size among which were "*Douyet*" and "*Moubarak*." The high seed cost of such cultivars has been one of the major handicap for any further uses of this genetic material by farmers.

In view of the lesson learned, farmers' field and difficulty to recognise winter sown chickpea from spring type and the high risk for planting the latter at early season, a new breeding approach was initiated in 1987 aiming at selection of dual season chickpea adapted for both spring and winter seasons.

PLANT BREEDING *VERSUS* PLANT GENETICS

The cultivation of plants is an important aspect of today's society in many ways. Not only do plants supply us with a major food resource and flow of nutrition, but they are also an important source of chemicals and other non-food products such as drugs, oils, latex, pigments, and resins. Because of the high value of plants, farmers are constantly trying to increase the yield and quality of their products by using more effective production techniques.

The physical appearance of plants is determined by genetic factors that are inherited from parental crops. The natural transfer of genes can be altered using many genetic techniques such as selection or genetic engineering. One of the important goals of plant breeding is maintaining genetic diversity. This can be achieved through selection and hybridisation. More recently, new techniques have evolved to allow us to grow crops artificially in culture from plant parts to produce whole plants. Modern science has allowed us to get away from the traditional method of plant breeding to more artificial asexual methods. This allows plants to be produced using individual, or groups of cells, which grow into full plants. Genetics has also allowed us to modify our crops in a more scientific way to select for traits that are more desired by society. We will be discussing the differences between the more traditional techniques of plant breeding and the modern advances of plant genetics.

PLANT BREEDING

Selection

Plant improvement by man began many thousands of years ago. Pre-agricultural man learned that seeds put into the ground at a certain time of the year produced similar seed-producing plants. This was the beginning of domestication of plants and led to the production of the first crops. Soon after, it was discovered that some crops of the same species grew better or tasted better than others, which became the starting point of selection breeding.

Later in the 19th century, Charles Darwin brought to light his theory of *Natural Selection*, which he described as *"survival of the fittest."* It was this statement that showed how selection occurred naturally with the environment selecting which plants would grow in specific parts of the world. Selection by farmers was an extension of this process. Farmers were able to further select for plants that not only grew well, but had better taste and looking more pleasing to the eye. Later, selection was further applied to find crops that produced the highest yield with the best possible quality.

In order for one to undergo selection, the breeder must have a large population of seeds to select from. This population should contain several different genetic types that are well adapted to the seasonal changes in the environmental conditions they are grown. These populations are called landraces and are generally found in regions where older plants have been growing for a long time. Natural selection has already occurred leaving only those crops that have been able to adapt to their environment. Landraces are generally found in areas where cross-breeding has been able to occur, giving rise to many new genetic varieties with new genotypes constantly being introduced into the environment. Once an appropriate population is found, one can begin the process of selecting plants by removing those plants with undesired traits. This allows plants with the most desired traits to survive and give rise to a new line of crop that has been selectively chosen for its specific traits.

6

Biotechniques and the Alternative Agriculture

TECHNOLOGICAL CHANGE IN AGRICULTURE

Technological change in agriculture has been the foundation of economic growth and development. By raising yields and continually increasing the number of people who can be fed by the output of one worker in agriculture, change in agricultural technology has permitted an increasing proportion of labour and other productive resources to be devoted to non-agricultural production. Plant and animal breeding and chemical and mechanical technology plus improved husbandry have caused continual structural change in farming. These technological stimuli have arisen as a series of overlapping waves to create a process of technological change, the momentum of which looks likely to be maintained by intensified adoption of information and biotechnologies.

As a result of the cumulative processes to date, farms have become larger and more specialized, and farmers are more highly trained and have substituted machines for animal and human power; farming in the 'West' has become one of the most capital-intensive of industries. At the same time agriculture has become more dependent upon industry, and even more industrialized-the intensive production of eggs, broiler chickens, horticultural products and pigs has many characteristics which belong to industrial production rather than to traditional farming. Goodman *et al.* talk of 'the industrial appropriation of the rural processes'.

Processes of marketing farm products have passed off the farm to be performed by sophisticated distribution and retailing sectors which increasingly dictate details of production to the farm sector, while at the other end of the chain an increasing proportion of inputs is supplied by the chemical, pharmaceutical and engineering industries. The process of 'industrializing' agriculture has been transferred to the less-developed countries (LDCs) and is epitomized by the so-called 'Green Revolution'. This has entailed expansion of irrigation systems using modern pumps, engineered dams and canals, plus the increased use of inorganic fertilizer, insecticides, herbicides, fungicides and power machinery; although at its heart has been the 'old' biotechnology of breeding new cereal varieties which give high yields when supplied with chemical inputs. Because much of the impetus for this 'revolution' has been western technology and farming know-how, particularly that of the USA, there are those who are critical of what they see as the LDCs' increasing technological dependence upon the West, its industries and research institutions. What is undeniably the case is that technological advance has been faster, and public and private expenditure on research in western agriculture greater than in LDCs taken as a whole.

Thus exportable surpluses from the West have increased with generous public support, while many LDCs have had to resort increasingly to food imports to meet their needs. While at this stage it is difficult to foresee exactly when, and on what commercial scale, the new agricultural biotechnology 'revolution' which is now brewing will have its effect, most commentators agree that it will intensify the industrialization of agriculture, and that it will increase the technological dependence of most LDCs upon western firms.

It may cause considerable disruption to the economies of some countries and will cause farming operations to diverge increasingly from the pattern associated with the country yeoman. Rural life in many more remote farming areas will be further threatened by the continued deterioration in the economics of ruminant livestock farming on upland and low-productivity pastures. It is these redistributive and structural effects that are the central focus here rather than any attempt to estimate the scale and speed of uptake of specific biotechnology inputs and processes. Although prospective biotechnological change in agriculture is not different in many fundamental respects from other processes of technological change, in that it will operate through the provision of new or modified inputs (seeds, hormones, etc.) and the creation of new markets in industry, it does raise new issues.

Are there dangers from releasing life forms which are engineered using new methods which are quite different from those associated with varieties developed by traditional plant-and animal-breeding methods? Even if the dangers can be objectively assessed and turn out to be minimal, will what may be perceived as man-made and therefore 'unnatural' products be acceptable to society as a whole?

PRINCIPAL DIRECTIONS OF BIOTECHNOL OGICAL DEVELOPMENT

In a recent review of agricultural technology, the Office of Technology Assessment (OTA) of the US Congress defined biotechnology to include 'any technique that uses living organisms or processes to make or modify products, to improve plants or animals or to develop micro-organisms for specific uses-it focuses upon recombinant DNA and cell fusion technologies'. Longworth concurs with this definition, with the significant addition of tissue-culture techniques, an aspect of biotechnology which has great economic, and therefore social and political, potential impact. Despite debates about whether these or any other definitions are adequate, they do touch on the main aspects of biotechnology which deserve to be considered.

The aspect of this new biotechnology which most captures the imagination and stirs the greatest controversy is gene splicing or recombinant DNA techniques, which inspire researchers to consider the possibilities of producing reproducible animals and plants markedly different from current existing species, and referred to as transgenic species. Already in higher animals and plants recombinant DNA has produced transgenic forms which are being commercially exploited. Their agricultural significance is, however, so far limited, and commercial application is concentrated in highly profitable pharmaceutical and horticultural markets.

Genes have been introduced into several animal species which alter their protein synthesis to enable transgenic sheep to produce insulin in their milk, and rabbits to produce interferon. A completely different application has originated in Denmark for salmon, where it has proved possible to introduce germplasm which enables the salmon's physiology to handle heavy metals, which are normally toxic, so opening up new locations for farm fisheries.

Much current research is directed to conferring disease immunity on animals, and holds out the prospect of widespread commercial application. In plants one achievement has been the transfer of genetic resistance to antibiotics in the petunia, and another has been the introduction of storage-protein genes from French bean plants into tobacco plants. As yet commercial progress with recombinant-DNA technology in plants appears limited, but extensive opportunities beckon, particularly because plant research is less restricted by the ethical and animal-welfare concerns which apply to research on transgenic animals.

The most important commercial developments based on gene splicing have so far occurred with genetically much simpler microorganisms, and it is with these that the greatest short-to mediumrun commercial potential lies. Already genetically engineered micro-organisms are producing a variety of hormones, vaccines, enzymes and other proteins. Important examples are the production of insulin, vaccines for neo-natal diarrhoea in calves and piglets, and the bovine growth hormone BST identical to that produced in cows which can stimulate a 20 per cent increase in milk yield. BST is already a source of problems for legislators in the European Community, and has provoked strong reactions from the media and milk consumers. As far as larger animals, and cattle in particular, are concerned, it is developments in embryo transfer and in many processes for manipulating reproduction which hold out the prospect of continuing increases in yields, feed conversion efficiency and general economic efficiency. Already, apparently over 1 per cent of dairy calves in the USA are from embryo transplants, despite the high costs still associated with this procedure.

Widespread adoption of these sophisticated technologies would further distance livestock farming from its traditional rural simplicity, and from the natural mating of animals to produce offspring. It would place technical demands upon the operators which favour large-scale company farms capable of supporting a range of highly trained specialists. Longworth identifies new techniques of tissue and cell culture as having 'the potential for enormous advances in crop improvement in the next couple of decades'.

Those techniques 'can both increase the genetic diversity and greatly increase selection efficiency', and they permit innumerable plants to be reproduced asexually from single cells or small pieces of tissue. The particular technique known as 'callus culture' has been used for many years to clone highly valued horticultural plants such as orchids, and cloning of cuttings is widely practised for tree crops such as tea and palm oil as well as by millions of gardeners for garden plants. According to Longworth 'cell culture' of single cells has unexpectedly, and so far inexplicably, resulted in plants with different properties being regenerated from the same clump of parent tissue. Sugar cane, maize and potato plants regenerated in this way have been found which are resistant to important pathogens. This application of cell culture with the capacity to generate vast numbers of seedlings rapidly has the potential to simplify greatly the hitherto labourious procedures of plant breeding and selection.

However, in terms of current and immediate commercial importance, it is through micro-organisms that biotechnology has its greatest impact. Microbial fermentation processes have been of great commercial significance for centuries, for example in bread, wine and cheese production, and there is the prospect of considerable development. Two recent examples of new processes indicate the sorts of impact that such developments can have. In the late 1960s genetically engineered bacteria were developed which were able to digest corn starch to produce high-fructose corn syrup (HFCS) and which left as a residue corn gluten, which is now an important protein feed for livestock. HFCS has made considerable inroads in the United States and in Japan.

It has largely replaced sugar as a sweetener in Coca Cola as well as many other food and drink products. This has helped depress sugar prices to Third World growers. In the European Community steps have been taken to prevent HFCS and other new powerful sweeteners from being produced and from undermining the market for domestically grown beet sugar, which is supported by the Common Agricultural Policy. A second important application of microbial fermentation has been the production of ethanol from sugar cane in Brazil and from maize in the USA.

The commercial viability of these processes is critically dependent upon the price of oil as the main non-renewable source of fuel, and, to date, massive subsidies have been required to maintain the Brazilian and USA ethanol programmes. Longworth states, however, that a new biotechnology, Sucrotech, is being patented, which will not only reduce the cost of producing ethanol from sugar cane, but will simultaneously produce fructose at a cost which will be competitive with HFCS and may thereby reclaim part of the sweetener market for sugar cane.

That such possibilities are in prospect indicates how volatile the future might be; biotechnology has tilted the competitive balance from sugar cane to maize, causing economic pressure and even disruption to sugar-cane-dependent economies and may in future switch it back again. In the long run the capacity to ferment fuels microbially from renewable agricultural feedstocks points to an important long-term reorientation of agriculture if non-renewable oil becomes uncompetitively expensive as a fuel for cars and feedstock for certain chemicals.

It suggests that eventually there will be an increased emphasis on agricultural production of industrial feedstocks at the same time as continually increasing food output will be required to feed the expanding world population. All this will require considerable increases in agricultural productivity, to which biotechnology will increasingly contribute, but it will at the same time impose great strains on the natural environment and the structure of agriculture.

PRIVATE-SECTOR CONTROL OF BIOTECHNOLOGY DEVELOPMENT

It has been the role of the public sector to undertake research and development of agricultural technology as a public service to firms which might profit from translating that R and D into a commercial product or process, to farmers who might profit from adoption of the technology, and perhaps most importantly to consumers at home and abroad who benefited from the lower prices resulting from greater abundance. This was particularly true of the phase of change dominated by improvements in plant and animal breeding, the 'old' biotechnology. While the public sector had a role in basic research for chemical and mechanical technologies, the benefits of research expenditure in these areas were easier to capture by private companies investing in R and D, so that increasingly the public sector has taken a smaller role in these areas although maintaining a strong regulatory role in regard to agricultural chemicals in particular.

Where it is impossible to prevent others from escaping payment for the research costs, either because the product is easily copied (seeds which can be regenerated by farmers or other firms) or because proposals for new methods can be readily implemented, private firms are understandably unwilling to invest. Machines, insecticides, fungicides and other manufactured inputs do lend themselves more readily to private exploitation. Nevertheless the returns to R and D in these products do depend upon the degree of difficulty potential competitors would have in copying the product. In some cases there are inherent technical difficulties in copying the process, or it would be prohibitively expensive, but in other cases it is the ability to obtain patent protection which creates legal barriers to potential competitors' ability to become 'free-riders' and which protects incentives to private investment in R and D.

Traditionally, however, it has been impossible for plant and animal breeders to obtain patent rights for their products, which is one reason why public-sector R and D has remained so important in this area. For, as under the European Patent Convention (EPC) of 1973, it has been judged impossible for plant varieties to satisfy one of the key criteria to qualify for a patent, namely proof of 'an inventive step'. The application of standard breeding practices to generate new varieties by crossing existing plants or animal strains has not been deemed to be invention.

Thus under the EPC one set of exclusions from patentability is 'Plant or animal varieties or essentially biological processes for microbiological processes or the products thereof.' In the absence of patent rights plant-breeding firms in particular have worked hard to obtain other means of protection and royalties for their products. The history of the development of Plant Breeders' Rights (PER) is presented by Mooney, and is of particular interest because of the concerns he expresses about the consequences of allowing the basic genetic stock, which underpins agriculture and hence our whole society, from becoming private rather than public property. While Mooney's concerns are expressed in relation to 'old' biotechnology varieties of plants, they are of particular importance with respect to the products of the 'new' biotechnology.

For bio-engineered products are capable of meeting the inventive-step criterion of patentability, and both micro-organisms and transgenic animals have now been patented in the USA. Before examining developments in the seed industry it is worth touching upon the implications of changes in research policy which stress increasing reliance upon private R and D to develop and exploit the new biotechnology and other technologies.

In the UK government has decided that it should withdraw from funding what it terms 'near-market research', that is, research beyond the basic phase and which is preparation for commercial exploitation. This policy is based in part upon the arguments that industry should be investing more heavily in R and D, and that research strategies at the near-market stage should be driven by assessments of likely commercial success which can best be made by the firms involved, and will therefore lead to greater efficiency in allocating research funds. This has led to the closure of a number of government-financed agricultural research institutes, the scaling down of others and the sale of the National Seeds Organization by auction to Unilever; Unilever won against competition with BP and other major public companies. This deliberate attempt to switch an increasing proportion of agricultural R and D expenditure from the public to the private sector carries with it a number of risks.

In the first place there is a controversy about whether there is under-investment in R and D so that the returns to extra expenditure are high, or whether the converse is the case. If there is under-investment, then withdrawal of public support for applied research will exacerbate this, as the private sector will only undertake R and D expenditure on those products and processes from which exclusive benefits can be captured by the investor.

Thus in relation to biotechnology in the USA, where again public support for applied biotechnology research is relatively small, Stallman and Schmid (1987) argue that there will be emphasis on technologies which are applied in factory conditions where secrecy and control can be maintained. For technologies which cannot be confined to factories, such as seeds, Stallman and Schmid state that 'Firms are also considering mechanisms which "scramble" the genome of a plant in the second generation', in order to prevent farmers from reproducing seed with the enhanced, engineered characteristics. Clearly such actions are designed to frustrate the maximum spread of benefits from the technology and to maximize private profit for companies investing in research.

Another facet of commercially orientated biotechnology research is that it will aim at the most important crops and livestock products, those produced by the largest farming units, and those which are most heavily subsidized. In the latter case this will worsen the budgetary problems of adjusting agricultural policy in OECD countries. The probable neglect of minor crops, difficult habitats and small farmers means that the public sector will have a defined role in agricultural technology research, but one in which it will be relegated to the second division and where it is unlikely to prove successful in terms of the commercial yardstick of rates of return which is increasingly emphasized by public-research policy.

THE NEGLECT OF ALTERNATIVE AGRICULTURE

This chapter moves us from the past to the present, with an eye to the future. It makes the claim that, despite the emergence of alternatives, the trend towards genetic uniformity is likely to be perpetuated owing to the emphasis now placed upon new biotechniques. Biotechniques provide more powerful tools to achieve genetically uniform ideal plants, but they could also be deployed as part of strategies aimed at promoting deployment of *in situ* diversity. The argument presented here is that biotechniques are closing the door which alternative movements managed to prise open, if only briefly, in the 1980s. This could be made in various ways, but here, the argument is developed through analysis of the intellectual property rights (IPR) issue, since this has radically affected the issue of genetic resource control in recent years.

Support for biotechniques takes place against the backdrop of a global economy in which concern to maintain 'technological leadership' and 'competitiveness' has prevented serious consideration of important questions concerning technological options, and the need to consider the matching of ends with technological means. The more deep-seated this concern, the more likely it becomes that institutional change will underpin it, further legitimising the same concern. This is the domain of Arthur's self-reinforcing mechanisms. For this reason I suggest that we are becoming increasingly locked in to genetic uniformity only moments after it seemed that the door was possibly opening for (again, possibly) genetically more diverse alternatives.

GENETIC RESOURCE CONTROL

When southern corn leaf blight hit the US maize crop in 1970, the solution was found not through re-introducing genetic diversity, but in genes from Mayorbala maize found in a West African field. West African germplasm was used at virtually no cost to save US farmers from a disease that had cost them $1 billion in 1970. That US farmers should have been assisted so freely owed itself to the system of free exchange of germplasm operating at the time, based on the principle that biodiversity was 'the common heritage' of humankind.

The UN's Food and Agriculture Organisation (FAO) had discussed genetic resources at its founding conference in Quebec in 1946, and with the International Biological Programme (IBP) hosted a conference on plant genetics in 1961.

Little was being done to address the issue of genetic erosion. Most collecting of germplasm was done by academics in universities, and virtually all was done at a sub-national level, an exception being that conducted on maize. The issue of crop germplasm conservation was highlighted by the work of Erna Bennett, who in 1967 organised the second IBP/FAO Conference on genetic resources and subsequently set up the Crop Ecology Unit at the FAO in Rome.

Only a year after southern leaf blight destroyed much of the US crop, a cold winter followed by a dry spring drastically affected the wheat crop of the Soviet Union. Much of the land was planted to the Besostaja variety which was neither cold tolerant nor drought resistant. Wheat prices shot up, and when famine hit the Sahel region in Africa, Malthusianism became fashionable once more. Against this backdrop, Sir Otto Frankel, Bennett's co-editor for a path-breaking book on the subject, persuaded the 1972 Stockholm Conference on the Human Environment to adopt a resolution calling for concerted action on genetic resources. The need to conserve genetic resources was at last being taken seriously as the advancing BCM mode hastened their disappearance.

Hambridge and Bressman recognised long ago the tensions between free exchange in germplasm, and unfree exchange of the product derived from it:

From its rivals a nation may get the wheat germ plasm that enables it to supply its own needs or overwhelm those rivals in international trade... Will nations have the wisdom to deal with this situation, or will it lead to more bitter rivalries and more deadly conflicts, as the beneficent science of chemistry has enormously increased the deadliness of war? In his use of modern science, man has proved again and again that he is a bright child playing with fire.

The Wardian case, invented in 1829, facilitated an exodus of germplasm to one of a growing number of botanical gardens. The 'botanical chess game' has played an important role in shaping the international division of labour, a fact recognised by Marx:

You perhaps believe, gentlemen, that the production of coffee and sugar is the natural destiny of the West Indies. Two centuries ago, nature, which does not trouble herself about commerce, had planted neither sugarcane nor coffee trees there.

Metropolitan powers appreciated that control over commodity trade depended on restrictions on the movement of germplasm. History is therefore replete with examples of the heroic efforts of plant explorers in overcoming embargoes on the movement of seeds, the breaking of which was in many cases punishable by death.

Only in the post-colonial world did 'free exchange' reach its truly international apogee. Some interpret this as allowing gene-poor northern countries in the developed world to maintain access to germplasm residing in the gene-rich, financially poor, countries of the developing south. Yet if, as Galeano writes, the international division of labour was organised 'not by the Holy Ghost but by men', so it was with the system of germplasm exchange. As such, it could be changed by men.

The Seed Industry

Unless they were crossed inbreds, farmers could save seed for planting the following year without appreciable yield loss. The first claims for plant patents were made in 1885. In 1922, lawyers met in London to discuss patent protection for plant varieties but no action followed. It was the nursery industry which was primarily responsible for the passage of the Plant Patent Act of 1930 in the United States. For nurserymen, the obstacle to proprietary ownership of varieties lay in competition from other nurserymen, not farmers, since trees could easily be propagated in competing nurseries.

The Plant Patent Act of 1930 made it possible for asexually reproduced plants to be patented, with the exceptions of potatoes and Jerusalem artichokes. The rhetoric used in support of the act, that breeding had made such significant advances over the past decades, was actually completely irrelevant as far as asexually reproduced plants were concerned. Most of the work done by nurserymen lay in multiplying varieties that had been discovered by chance, and that were the product of insect or wind pollination, raising issues as to whether they should have been eligible for patents. Fowler concludes that: 'The PPA did not recognise the individual inventor or the creative act as much as it recognised and rewarded the system that produced the new variety, whether by luck or by designs'.

In France, ever since the turn of the century, rose breeders had been seeking the same recognition as inventors of machines. Early attempts were rejected by lawyers on grounds that even full disclosure would not make it possible for breeders to reproduce a variety. By 1928, however, there existed in the Ministry of Agriculture *de facto* protection of breeders' rights through an 'identity and purity service'.

The Italian High Court declared plant varieties patentable in 1948, but confusion led to calls for a special plant patent law. By 1957, with the view that plants and animals should not be patented in the ascendancy, the International Association of Plant Breeders for the Protection of Plant Varieties accepted an invitation to host a conference in Paris on plant breeders' rights (PBR), leading to the establishment of the Union for the Protection of New Varieties of Plants (UPOV) in 1961, whose International Convention was revised in 1972, 1978 and 1991.

For the most part, PBR legislation has been true to the UPOV Convention, requiring that plants pass the DUS test. The 1960s and 1970s saw several countries either joining UPOV or implementing a system of PBR of their own. Key to the passage of these acts was the belief that sexually reproduced plants could breed true, which, for sexually reproducing plants, is only the case for pure-line varieties (those having undergone four to nine generations of selfing). Also critical was the definition (thought by many to be impossible) of the term 'variety'.

The UPOV resorted to 'nothing other than a description of the steps of the method of breeding', or more accurately, pure-line (Mendelian) breeding. Hence, the extension of IPR to plants through PBR was an institutional innovation shaped by, and made possible by, changes in breeding techniques and technology respectively. However, these were not institutional changes waiting to be implemented as soon as these techniques emerged. Just as there were technical options open to breeders, so the institutional changes made represented a choice from myriad possibilities.

PBR facilitated an increasingly international outlook on the part of the seed industry. Modern varieties were spreading across the globe, and notwithstanding some efforts to improve disease resistance in new varieties, new seeds made increased use of other inputs more likely. In the 1980s, policy-related or structural adjustment lending undertaken by the World Bank advised privatisation of input supply industries, and an expanded role for private sector seed research and distribution, especially in the development of hybrids. The emphasis began to shift, as it had done in the developed world, away from the public system in favour of reduced public sector involvement.

As late as the 1960s, there were few multinational companies in the seed industry. A wave of acquisitions occurred in the 1970s as seed companies were bought up by transnational corporations, mainly food trading and petrochemical companies. Food traders, seeking to open up new export markets in the era of US 'food power', sought to extend their activities upstream. The development of high-input seeds by international agricultural research centres had also led (agro-) chemical companies to seek new markets in the developing world, so these companies sought to market seeds through the same channels. With PBR legislation in place in many developed countries, seeds were no longer a weak link in the input supply industry. UPOV, by creating a degree of harmonisation in PBR legislation, fostered the emergence of a global seed industry, whilst the horizontal integration of agricultural input supply has deepened the inter-relatedness of inputs over time.

The Challenge to the BCM Paradigm

The BCM mode has come under, and continues to operate in the face of, considerable pressures for change.

These are due to its:

- Ecological impact;
- Impact on food quality and health (in farming, and in consumption);
- Impact on rural communities; and
- Being supported by state policies, and high levels of farm support, which effectively exacerbate the problems mentioned above.

It is beyond the scope of this book to address each of these in detail. Suffice to say that, in the words of Almas and Nygard, the BCM mode has produced 'some of its own executioners'.

Aims to reform agricultural and farm support policies seek, increasingly, to re-direct support towards environmentally sound practices. These are generally believed to imply reduced use of agrochemicals, and also seeds. To the extent that it continues to be allowed, seed-saving becomes more economically attractive in times of low output prices. In the spirit of challenges to the BCM mode, it is to the issue of seeds and germplasm that we now turn.

SOWING THE SEEDS OF DISCONTENT

As the Rockefeller and Ford Foundations expanded their efforts in international agricultural research, they began to seek public funds to support their work. Following a meeting in Belaggio in April 1969 organised by the Rockefeller Foundation, 15 governments attended the first meeting of the Consultative Group for International Agricultural Research (CGIAR) in January 1971. The World Bank would provide a secretariat and administer finances.

International agricultural research, the year after Norman Borlaug received a Nobel Prize for his work on dwarf wheats in Mexico, had come of age.

In 1972, the CGIAR's Technical Assistance Committee (TAC) convened a meeting in Beltsville, USA, to formulate an international strategy for genetic resources conservation. After much debate, the International Board for Plant Genetic Resources (IBPGR) was set up as part of the CGIAR network. Under the IBPGR's system, the majority of the world's genebank accessions, mostly from developing countries given to believe that genetic resources were the common heritage of humankind, were stored in genebanks in developed nations increasingly predisposed to the notion of IPR over germplasm. In the late 1970s, developing countries were increasingly concerned by IBPGR statistics showing that of more than 1.9 million samples stored, 55 per cent were in developed countries and another 14 per cent were held in the Northern (donor) dominated CGIAR system. Collections were clearly biased towards crops of interest to the developed countries, the top cereal crops representing more than 75 per cent of accessions in the pre-1980 period. At a 1981 IBPGR/FAO/UNEP conference, Latin American countries pressurised the IBPGR, successfully, to increase collections of crops less prominent in international trade.

By the end of 1981, despite opposition from US and UK representatives, a Resolution tabled by the Mexican delegation had been passed at the FAO calling for the FAO Director General to draft elements of an international convention on plant genetic resources, and investigate the feasibility of establishing a new international gene bank. Two years later, at the FAO's biennial conference, Jose Ramon Lopez Portillo, son of the former Mexican President, forced another vote which led to an International Undertaking on Plant Genetic Resources (IUPGR) and the creation of an International Commission on Plant Genetic Resources (ICPGR). The aim of these moves was transfer of control over genetic resources from the developed countries, and IBPGR in particular, to the United Nations.

Under the IUPGR, the notion of free exchange was to be respected, and it was not just landraces that were to be 'available without restriction', but also 'special genetic stocks (including elite and current breeders' lines and mutants)'. This angered the American Seed Trade Association (ASTA), who were represented in the US FAO delegation.

They charged that the IUPGR struck: at the heart of free enterprise and intellectual property rights... The definition includes unimproved and obsolete varieties, land races, wild and weedy species, all of which the seed industry believes appropriate to be preserved and freely exchanged. However, it also includes improved elite varieties and breeding lines within the definition of plant genetic resources... This puts the Undertaking in direct conflict with the rights of holders of private property... The anti-private business bias of the Undertaking is clear.

The IUPGR also proposed establishing a network of base collections under the jurisdiction of the FAO. Yet the Undertaking was a mild and voluntary agreement rather than a legally binding convention.

The Birth of Biotechnology

In the second half of the twentieth century, enormous strides have been made in the life sciences, particularly in the discipline which has come to be known as molecular biology. As a result of this growing body of knowledge, new commercial opportunities appeared on the horizon based on the use of tools developed through new discoveries in this field. The idea that plants could be made resistant to pesticides was no new idea. Wiebe and Hayes discussed it decades ago with regard to the reaction of barley varieties to the application of DDT. Yet work in biotechnology brought such a strategy closer to hand, raising the possibility of breeding plants designed to tolerate applications of proprietary chemicals.

The use of plasmids, in 1973 by Cohen and Boyer, to mediate gene transfer made possible a new alchemy. In the immediate aftermath, biotechnologists in the US showed awareness of public unease regarding this new technology by proposing a moratorium on certain types of research. Since 1977, guidelines laid down by the National Institutes of Health (NIH) have been progressively relaxed. The desire to regulate the industry has dwindled as authorities were persuaded of the commercial significance of the new technologies.

Reduced regulation of the biotechnology industry began to be perceived, and not just in the United States, as a means through which a country could maintain or improve its position in emerging bioindustries. Field reported that 'industrial competitiveness appears to represent the central and overriding concern of national strategies'. From a different perspective, the United Nations Centre on Transnational Corporations opined that comparative advantage and the international division of labour were increasingly being shaped by technological prowess. Increasingly, regulation shied away from determining which technologies should be allowed for use, and the imperative of allowing new technologies to develop began to shape which regulations were considered acceptable.

The emergence of biotechniques for technology generation makes it possible to circumvent the constraints imposed upon genetic recombination by species incompatability. I argue below that formal agricultural research is undergoing a transition from the BCM mode to a biotechniques-mechanisation-legislative (BML) mode. This is not to imply that traditional plant breeding and chemical inputs are about to disappear from view, either now or in the near future.

The role of breeders, where they do not disappear altogether, is likely to undergo a change such that their work complements that of the biotechnologists, whilst the fact that to date, herbicide tolerance in crops is the most widely tested trait to date testifies to the likelihood of continued use of chemicals into the future. Nevertheless, the genetic determinants of interactions between the plant and various chemical and biological inputs are likely to become the focus of innovation in crop (and livestock) agriculture. The pivotal institutional innovation in enabling such a strategy to become privately profitable is IPR legislation.

Patenting Genetic Materials

In 1976, the first of the new biotechnology companies, Genentech, was formed by Herbert Boyer and venture capitalist Robert Swanson. In 1980, Genentech placed a share offering on the New York Stock Exchange, the prices of which shot up from $35 to $89 per share in twenty minutes, a record rate of increase. This was due to the fact that three months earlier, General Electric had successfully challenged an earlier decision by the US Patent and Trademark Office (PTO) which had ruled that an oil-degrading micro-organism developed by their scientist, Ananda Chakrabarty, was not patentable subject matter. The new ruling held that whether or not an invention was alive or dead was irrelevant to patent law.

In the PTO's ruling on Chakrabarty, the legal principle of 'pre-emption' disqualified materials protectable under the PPA or the PVPA from patent protection. But this ruling was also overturned in the 1985 *Ex parte Hibberd* case, in which Hibberd was granted patents on the tissue culture, seed, and whole plant of a corn line selected from tissue culture. Breeders could now choose the form of protection most suitable to them, including utility patents. The gene was being commodified.

Anxious to preserve its lead in the biotechnological race, the United States has moved fastest in bringing institutions into line with industry's desires. Employing both bilateral and multilateral channels, it has sought to harmonise standards across nations in line with its own structures, thus opening the way for global marketing of proprietary products of biotechniques. In November of 1982, at a ministerial level meeting held at the GATT at US insistence, the US proposed that the new round debate issues never before considered in earlier GATT rounds.

One such issue was trade in counterfeit goods, such as 'fake' Rolex watches, but the scope of this particular area was widened at the behest of the US and others to include the issue of IPR. Raghavan notes that this was 'thanks mainly to the negligence of the disorganised Third World countries, most of whom thought that it did not affect them'.

This would not have been so critical had it not been for the fact that what many countries saw as a preparatory discussion was subsequently proposed by the US and others as the agenda for a new round. The inclusion of many new issues was given justification through addition of the prefix 'trade-related'. Many countries hoped that by stifling their objections to the inclusion of new issues such as 'services' and IPR, they would be rewarded with concessions on 'old' issues, such as tariffs on tropical products (and escalating tariffs on processed products thereof), textiles, and a continuation of benefits under the Generalised System of Prefences (GSP). A compromise text was agreed at Punta del Este in Uruguay at the end of 1986 which included IPR. Even then, developing countries refused to negotiate on the subject before the mid-term review in Montreal in December 1988, where agreement was reached concerning the negotiating agenda.

The GATT agreement was finally signed at Marrakesh in April 1994 despite the fact that market access negotiations had not been verified, and with many developing country negotiators complaining that they had seen the texts only weeks before. Ratification in many countries was rushed through with little debate, and on 1 January 1995, the new World Trade Organisation (WTO) came into existence. This would co-exist with the GATT until the end of 1995. The final text of the TRIPs (Trade Related Intellectual Property Rights) agreement establishes new multilateral rules on IPR based on uniform minimum standards for their protection and enforcement, including their availability, use and scope.

As regards plant materials, the treaty does allow for exemptions on grounds of perceived environmental or public order impacts, yet at the same time, the treaty states that plant varieties shall be protected by patents 'or by an effective *sui generis* system or by any combination thereof'. Although developing countries and least developed countries are allowed, respectively, five and ten years to implement the agreement, the '*sui generis*' clause is to be reviewed four years after the date of entry into force of the WTO agreement. The TRIPs agreement offers little encouragement to communities that might seek to protect innovations which are the property of, as it were, the collective. The agreement recognises IPR as private rights, and also requires products to be 'capable of industrial application'. There is no mention of communities and their rights.

Before the GATT negotiations even began, the US had made its intentions in respect of IPR abundantly clear through applying pressure bilaterally. Mexico was targeted as early as the 1970s, but little progress was made until, in the mid-1980s, the US began to link the issue of IPR reform to expansion of GSP concessions. By the end of 1986, Mexico had adopted a revised Patent and Trademark Law, though as a result of the efforts of domestic lobbying, this was deemed inadequate by the US administration.

In 1984, the US Trade and Tariff Act had been revised, S.301 of Title III of which invested the prevailing Administration with coercive powers aimed at righting 'unfair' trading practices. In 1985, cases against Brazil and South Korea were initiated, the former concerning, *inter alia*, copyright issues regarding software, the latter concerning failure to protect intellectual property. The same issue led to talks with China, whilst India has also come under pressure to reform its IPR legislation in the past.

In January 1987, Mexico was informed of President Reagan's intention to withdraw $200 million of GSP benefits unless the perceived inadequacies of its new legislation were corrected. In 1988, the Omnibus Trade and Competitiveness Act was passed in the US. This included sections which came to be known as Super 301 and Special 301 respectively, retrospectively strengthening the coercive powers vested in the administration by the Trade and Tariff Act's 1974 revision. Under Special 301, a procedure was set up whereby the US Trade Representative could identify and initiate proceedings against countries considered to be offering inadequate IPR protection.

Within this list of countries, a Priority Watch List of countries was to be specified annually, and Mexico was on that first list in May 1989. When President Salinas de Gortari began his programme of liberalisation, and plans for a North American Free Trade Agreement were materialising, Mexico's stance on the IPR issue altered quite radically, and in 1990, when Mexico introduced a proposal for a TRIPs agreement at the GATT, it slipped off the 301 lists. Mexico's patent law was revised in June 1991 to explicitly allow for the patenting of plant varieties. It specifically addresses innovations likely to arise from the deployment of biotechnology.

As Fowler has made clear, GATT and the growing concerns of the US over intellectual property issues generally, enabled transnational corporations involved in agricultural biotechnology to have their concerns *vis-à-vis* the patenting of life addressed in new fora. In the case of the GATT, it became possible for the issue to be bundled up not only with concerns over patents and trademarks as they related to mechanical innovations, but also, since this was a take-all-of-it-or-leave-all-of-it package, with fourteen other areas with which the Uruguay Round was concerned. Significantly, Watkins notes, 'The major actors in [the TRIPs] exercise have been the US-based Intellectual Property Coalition – a grouping of 13 major companies, including IBM, DuPont and General Motors – *and European agro-chemical giants such as Unilever, Hoechst and Ciba Geigy*.' Thus, IPR was being simultaneously harmonised and extended across the globe.

UPOV

Paralleling the moves to enhance intellectual property protection under the auspices of the GATT were moves on the part of UPOV to bring the Convention into line with developments elsewhere, and particularly with respect to biotechnology. PBRs' research exemption made them inadequate for protecting biotechnically engineered plants since they offered protection at the level of the whole plant when what was required was protection at the level of the gene. But by 1987, it was clear that UPOV would be strengthened. According to Fowler *et al.*, UPOV's members had been divided between small seed houses and the integrated genetics supply industry, the former fearing gene patenting, the latter favouring new initiatives in this respect. UPOV was revised in March 1991.

Note that the right of farmers to save seed from one harvest for planting in the next, what the American Seed Trade Association had referred to as the 'farmers' right' in hearings on the PVPA, had become known as the 'farmers' privilege' and was no longer secure. Section 15.2 of the new Convention allows, as an optional exception, seed saving 'subject to the safeguarding of the legitimate interests of the breeder', implying that royalties should be paid to breeders where seed is saved.

On the other hand, there is a compulsory exemption for breeding other varieties. However, the interests of the breeder are, in general, strengthened since protection applies to 'Essentially derived and certain other varieties' as defined under Article 14 (5) (b) and (c). Lesser expresses concern that, since the definition is unclear, this will lead to quasi-IPR being granted to a breeder over thousands of attributes of a variety which he/she did nothing to create. The 'essentially derived' clause would appear to apply to genetic insertion, giving the owner of PBR the right to demand royalties from innovations based on insertion of one or two genes into a plant over which the right is held.

Increasingly, developed countries are bringing their PBR legislation into line with the 1991 UPOV Convention. Furthermore, although the TRIPs '*sui generis*' clause appears to offer room for manoeuvre in designing IPR for plant varieties, most believe that this translates into UPOV type standards and nothing less. Since accession to UPOV 1978 was closed at the end of 1995, this now implies the standards of UPOV 1991. Increasingly today, companies can choose which combination of protection they prefer, although European patent law currently forbids patenting of plant varieties.

BIOTECHNOLOGY FOR SOCIETAL DEVELOPMENT IN INDIA

The application of biotechnology for societal development has always received special attention since it benefits socially and economically disadvantaged sections of the society. Through this programme a number of projects are implemented for economic and technical empowerment of the beneficiaries. Efforts are made through training, demonstration and appropriate R andD inputs for providing tailor made solutions for location specific problems of SC/ST, women and rural population using modern biology and biotechnology. The following pages provide the highlights of the achievements.

PROGRAMMES FOR SC AND ST POPULATION

Biotechnology-based Programme for SC/ST population has benefited around 15,000 people directly and about 45,000 indirectly through 54 projects during the year. The projects have been supported in the areas of cultivation of medicinal and aromatic plants, biofuels, production of biofertilizers and biopesticides, organic farming, aquaculture, mushroom cultivation as well as human health care interventions etc. The universities, public funded institutions, Krishi Vigyan Kendras and voluntary/non-governmental organizations have been involved in the implementation of the projects. The target groups have received the benefit of hands-on training and field demonstrations.

Vermicomposting: MR Morarka GDC Rural Research Foundation Jaipur implemented a project in Jalgaon district of Maharashtra, to benefit 2600 SC/ ST target population on economic utilization of waste, agriculture residues and cow dung. Over 1780 MT of waste, agricultural residues and cow dung was utilized for vermicomposting through establishment of 28 community facility centres.

Beneficiaries are utilizing vermicompost on crops like onion, sugarcane, cotton, banana, papaya and wheat. Some farmers switched over from conventional crops to cash crops such as colorful capsicum, gerbera as well as inter cropping of safed musli, turmeric and ginger in banana plantation. Farmers have been trained to use bio-fertilizer, foliar sprays and mulching to use the agro-waste for fertility and management. Marketing linkages have been established with "Organic Shoppe", through which farmers are getting 10-20 poe cent premium for their organic produce. About 20 entrepreneurs have taken up vermiculture as commercial activity. extension Material in the local language was distributed to the farmers.

Mushroom cultivation: A project on promotion and production of essential oil, mushroom and vermicompost was implemented by RRL, Jorhat for revenue generation of the ST population of Arunachal Pradesh. Motivation and demonstrationcum-training programmes have been conducted in 13 villages of five districts on cultivation of citronella, production of mushroom and vermicompost. Demonstration units have been established at RRL Branch Naharlagun farm. Mushroom spawn, earthworms and citronella saplings were distributed to the beneficiaries. Several beneficiaries established vermicompost units on commercial basis and are selling vermicompost. The beneficiaries at Ziro, Jairampur, Doimukh, Manmao, Hetlong and Tengman have established oyster mushroom cultivation units. So far, 10 acres of land have been planted with citronella and 7 acres of land with patchouli at Sonajuli village of Papum Pare district.

Medicinal plants: A project on strengthening health care practices and conservation of medicinal and nutritious plants was continued at Herbal Folklore Research Centre, Tirupati for implementation in 100 villages of two Mandals in Andhra Pradesh. Training workshops were organized with the help of Yanadi tribal people for community health volunteers on identification and sustainable harvesting methods of 20 species of locally used medicinal and aromatic species.

The reputed healers were encouraged to prepare medicines needed for PHC from the identified plants. About 6000 seedlings were distributed to the beneficiaries for planting in kitchen gardens. Green health kits (samples of herbal medicines) and educational material in local language containing identification, cultivation techniques and preparation of medicines with their therapeutic properties were distributed to the trainees.

A project on creating awareness and promotion of medicinal plants for the production and marketing was implemented by RRL, Jorhat to benefit ST population of Arunachal Pradesh. Awareness and training programmes were organized on the importance, scientific cultivation and marketing of Piper longum, Kaempferia galanga, Andrographis paniculata, Stevia rebaudiana, Rauvolfia serpentina and Mentha piperita. Three nursery-cumdemonstration centers were established at Roing (1000 ft msl), New Bomdila (8160 ft msl) and Ziro (4350 ft msl) and 19000 saplings were distributed to ST farmers/NGOs for introduction and propagation in their fields. At high altitude, villagers were advised to cultivate Aconitum heterophyllym, Coptis teeta and Picrorrhiza kurroa. Beneficiaries have stopped collecting wild Homalonema aromatica, an important but rare aromatic plant of Arunachal Pradesh, and have started its cultivation in their farms. Ten community nurseries have been established with the help of Vivekananda Trust, Roing at ten villages of Lower Dibang Valley District. Twenty farmers were trained on cultivation and post-harvest management of medicinal plants and marketing tie-up has been established. Around 1266 local ST people participated in the awareness programmes.

A project on organic cultivation of medicinal and aromatic plants for income generation and sustainable development in Uttarkashi District of Uttaranchal was undertaken through Forests Conservation Programme of World Wide Fund for Nature, New Delhi.

Demonstration-cum-Resource Material Production Centre established for training on vermicomposting and cultivation of species.

Construction of poly house, vermi-composting units and planting of seeds of Rauvolfia serpentine completed. Awareness and motivation camps were organized for the farmers. Extension material brought out in the local language for cultivation of Rauvolfia serpentina, Coleus barbatus and Pelargonium graveolens were distributed to the farmers.

A project on developing skills and economically sound vermicompost units utilizing distillation waste of Mentha arvensis (menthol mint) was implemented for the benefit of SC/ST communities of Lucknow and Barabanki Districts of Uttar Pradesh. Around 200 farmers were trained on menthol mint nursery raising, transplanting, distillation and practices followed in vermicomposting. Demonstrations were carried out in farmers' field using high yielding variety 'Kushal'. Farmers trained on organic farming started vermicompost production using distillation and other agro-waste. The yield data obtained from the demonstrations showed the possibility of reduction of chemical fertilizers by 75 poe cent without affecting the yield.

BIOFUELS

A project was continued on cultivation of Jatropha in arid lands of Namakkal District in Tamil Nadu for production of biofuel and other products for SC/ST community by Centre for Research in Social Sciences, Coimbatore. A high tech nursery was established and plantlets of Jatropha were raised. Training programmes were organized for 223 members on cultivation of Jatropha in 31 acres of land. About 30,000 plants were supplied for cultivation. Awareness camps, training and demonstrations were conducted on Jatropha cultivation, oil extraction and soap making. SHG groups were trained in nursery techniques. Women self help groups were trained in various activities of jatropha cultivation and soap making. Extension material bought out in the local language was distributed to the beneficiaries.

Biocontrol agents: A project was implemented on development of eco-preneurship among the coastal community of Pondicherry by MS Swaminathan Research Foundation, Chennai. Sixteen SHGs involving women and youths were formed in Ariyankuppam, non-ankuppam, Kakayanthope and Veerampattinam villages. Demonstrations and trainings were conducted to educate trainees on basic aspects of IPM, importance of biocontrol agents, production and marketing of Trichogramma, importance of human nutrition, role of edible mushrooms and production and marketing of oyster mushroom. Micro-enterprise was developed for mushroom and Trichogramma production. Technologies were disseminated on biopesticide (Trichogramma), oyster mushroom production and vermicomposting to recycle the spent waste of oyster mushroom. Extension material brought out in the local language was distributed to the villagers.

AQUACULTURE

A project was undertaken on enhancing livelihood of SC/ST people in Orissa and Tamil Nadu by MS Swaminathan Research Foundation, Chennai for introducing various aquaculture enterprises to 250 participants in backyard ornamental fish culture and carp seed production. Fingerlings produced were distributed to the entrepreneurs for stocking. Various aspects on ornamental fish breeding, carp nursery and gold fish culture were disseminated to the self help groups, which helped them in earning a revenue of ₹ 200/month/household and an income of about₹ 3200/from carp nursery and ₹ 8500/from vegetables and flowers as bund crops. Community ponds were effectively utilized by stocking with rohu, catla, mirgali, Chinese and Indian carps and marketable size fish was harvested. Technologies transferred to the entrepreneurs on community pond aquaculture, carp nursery and ornamental fish culture and extension material in local language were distributed to the beneficiaries.

A project on utilization of integrated marine living resources by SC youth from Tuticorin region of Tamil Nadu was undertaken by College of Fisheries, Tuticorin. Lobster fattening and crab culture techniques were standardized and demonstrated to the beneficiaries, who have grown lobsters in cages for about 4 months and produced 250-350gms size for marketing. The beneficiaries were helped in availing credit facilities and bank loan. In the project around 120 people of target population have been trained. A manual on crab and lobster fattening technology was brought out for the distribution to the beneficiaries.

A project on freshwater prawn culture was implemented in village Naubasta Kalan in Allahabad District by University of Allahabad. Demonstrations were conducted on prawn hatchery technology and standardized methods for biocontrol of prawn disease for integrated giant freshwater prawn (Macrobrachium rosenbergii) culture. Trainings were conducted on scampi culture, breeding and seed production and microbial inputs were utilized for feed silage preparation using indigenous technology. Apiculture activity was introduced and disseminated to the target population and extension material was distributed to the beneficiaries.

GENETIC COUNSELING

Tribal population of northeast region was screened for anaemia and nutritional deficiencies through a programme implemented at Vivekananda Institute of Medical Sciences, Kolkata. The tribes of Arunachal Pradesh, Assam and Tripura displayed high incidence of anaemia. Awareness programme was organized for school children on inherited anaemia in pre-marital and pre-natal population. More than 1500 families from 80 villages were educated on anaemia, thalassaemia and hemoglobinopathies and counseling was provided to them. Leaflets were distributed for awareness of anaemia with special reference to Thalassaemia in local languages.

A project on molecular epideomiology of alcoholism was implemented jointly by Centre for Cellular and Molecular Biology, Hyderabad and Tribal Research Centre, Udagamandalam, Nilgiri District to study Kota tribe population in Nilgiri hills. General health awareness camps were arranged in seven villages for periodical medical check up. Blood samples were collected from 620 individuals and clinical data on health profile revealed that the disease morbidity is more for hypertension, heart related problems and diabetes. Genealogies indicated that the alcohol drinking habit is running in the families for the past two to three generations. Study undertaken on the health awareness camps and medical camps were found beneficial in counselling the tribal population.

FLORICULTURE

Over 180 women from Shillong, Itanagar and Chandigarh were trained in cultivation of commercially important and hybrid species of orchids. A total of 118 women of Shillong have raised 6 species *viz.* Dendrobium fimbriatum var.

oculatum, D. longicornu, D.lituiflorum, C. pendulum, C. aloifolium and C. giganteum in their backyard. These plants have started blooming. The beneficiaries have started selling their harvest in local market. Through S andT Council, Bhopal 200 women were trained in cultivation of gladiolus, marigold, mogra, rose and tuberose.

FOOD PROCESSING

At College of Horticulture, Vellanikara, Thrissur, 88 women were trained in various aspects of entrepreneurship in preparation of cocoa products including molding, packaging and marketing. Twenty-two women have already set up their production unit and are selling their products in local market.

In another project at Fisheries College, Tuticorin, 500 women were trained in cook-chill technology for seafood and various parameters for quality assurance *viz.* estimation of moisture content, bacteria load, presence of bacterial species, the nutritive value, shelf-life etc. The trained women were grouped in 48 SHGs for establishing market for seafood products.

At the Central Food Technological Research Institute, Mysore data on biochemical and microbial analyses for nutritionally important factors like protein, amino acids, isoflavones, genistein, diadzein and methods for improvement of shelf life were collected on for 'Tofu' preparation. Also a byproduct of Tofu preparation, Okara, was found to be rich in protein and amino acids has been used as a food supplement. In another project at CFTRI, 17 women were trained in preparation of various food products of high nutritive value, methods of removal of anti-nutritional factors like phytic acid and trypsin inhibitor found in soya bean, better texture and improved shelf life. The food items included fermented products of soya bean and fruit pulp. Also 34 women were trained in spirulina production and its application as health food. Through CFTRI, technologies for rural bakery are being disseminated to rural women in three villages.

POULTRY AND LIVESTOCK FARMING

The 'Action Research on Health and Socio Economic Development', Bolangir, Orissa, identified 100 women and trained them in poultry rearing. The beneficiaries were provided with parent birds of Carigold (dual purpose) and Vanaraja varieties. The beneficiaries were able to earn ₹ 1000/per month through the sale of eggs and birds. The beneficiaries were also trained in feed formulation using local ingredients, management practices and entrepreneurship skill. The trained women formed SHGs and are approaching a bank for setting up their poultry farms. The market is established through state owned apex cooperative known as OPOLFED.

Through three projects supported at Tamil Nadu Veterinary and Agricultural Sciences University and its Extension Centres at Namakkal,

Chennai and Tiruchirapally, over 1200 women were trained in poultry farming including feed formulation, farm management / health care and preparation of various food items such as quail egg pickle and ready-to-eat food products etc.

Aquaculture: Under a project at the North Eastern Hill University, Shillong, techniques were standardized for induced breeding for common carp and 4,00,000 fish seeds were reared up-to fries for distribution to the beneficiaries. Selected 92 women were trained in fish farming techniques including water quality testing, pond maintenance, induced breeding, harvesting and feed formulation using slaughter house waste. They were provided with fish fries for rearing in their own pond.

Over 470 women from Bhopal, Shimla, Solan and Bhubaneshwar were trained in polyculture fish farming. Interested women were also provided advance training in composite carp culture, integrated fish farming with duck/ poultry, pond management and value added products including fish processing. These women are selling their products in local market and are earning ₹ 3000/ per harvest/family.

Considering the opportunities in ornamental fish, a project was supported at the University of North Bengal. The identified women, 57 from Darjeeling and 54 from Jalpaigudi district were trained in rearing and breeding of guppy (Poecilia reticulate) in cemented cisterns and angel (Pterophyllum scalaer) in ponds. The women are being trained in management skill and marketing.

ORGANIC FARMING / VERMICOMPOST

At the Sri Sathya Sai Institute of Higher Learning, Anantapur Campus, Andhra Pradesh, neem varieties with higher level of azadirachtin were identified in 5 villages. Thirty trees were selected and 40 women from two villages were trained in fruit collection, drying and storage. Specialised training on maintenance of neem processing gadgets, processing of neem fruits for production of cake, oil or powder, maintenance of aza-rich plantlets, hardening, harvesting and application of neem based biopesticide was provided to five resource persons

At the University of Agricultural Sciences, GKVK Campus, Bangalore, five villages of Bangalore and Kolar were surveyed for the availability of organic waste. The waste materials were mostly sugarcane trash, coconut coir pith and vegetable wastes. Demonstration programmes for 20 to 25 farmwomen from each identified village were organized. Six interested women were given onfarm training. The technique of Trichoderma enriched vermicomposting was perfected and demonstrated at University campus.

At MSSRF, Chennai, phosphate-solubilising local strains were collected and screened. Simultaneously several awareness programmes on application of biofertilizer were organized for the villagers of Rediarchatram block. About 300 interested farmers were trained in large-scale production of biofertilizer.

HORTICULTURE

Through the Tamil Nadu Agricultural University, Coimbatore, 54 progressive farm women were identified for training on cultivation of hybrid rice. The women were provided with quality seed their cultivation over an area of approximately four acre. The yield for the first harvest (December to May) was recorded at 2000 kg/ha. The beneficiaries were able to earn an additional income of ₹20,000/per acre/year through three crops in an year. At Supi farms, Uttaranchal non-traditional vegetable crops were grown on farmers' land in a small way. An herbal garden was established. Incidence of pests to various crops was evaluated and farmers were trained on production of vermicompost and application of neem based biopesticide in horticultural crops.

MUSHROOM CULTIVATION

In an attempt to empower rural women through transfer of mushroom cultivation technology 93, women were trained at Himachal Pradesh Krishi Pradesh Vishwavidyalaya, Palampur and 750 at Tropical Botanic Garden and Research Institute, Kerala in mushroom cultivation. In addition, 15 women were trained in spawn production at TBGRI. The beneficiaries were given 15 days training in cultivation of Oyster and Agaricus species using wheat straw. The first harvest of about 1000 kg was sold @ ₹ 16/kg. The beneficiaries were also trained in various recipies of mushroom. Market linkage has been established for their products.

Under the project support at Dangoria Charitable Trust Hyderabad, apart from training women in processing various items of food, 100 women from 13 villages were trained in production of vermicompost, neem-based biopesticide and their application in organic farming, About 100 vermicompost units were set up using the ring method. The beneficiaries produced over 130 quintals of compost. The beneficiaries were encouraged to raise nurseries of drumstick, papaya and creeper spinach at their backyard using vermicompost. The beneficiaries were able to sell excess vermicompost and planting material.

SERICULTURE

A total of 550 women were trained for various activities including the organic cultivation of mulberry and rearing of cocoon and value added products from waste at the Central Sericultural Research and Training Institute, Mysore and at Vidyasagar University, Midnapore, West Bengal. The beneficiaries are getting an additional income through the sale of value added products and surplus vermicompost.

HEALTH CARE

Kidwai Memorial Institute of Oncology, Bangalore, Karnataka organized over 50 awareness camps on Carcinoma cervix to educate 2071 women for its prevention. Blood samples collected from 1830 women based on their medical history, 1812 were found negative and 18 positive cases were referred for treatment. A project on mental retardation including awareness generation among rual women, screening for early detection and genetic counseling was supported to Manovikas Bio-Medical Research. Over 7,000 families from three Gram Panchayats were surveyed. Ninety-one of these showing signs of mental retardation were shortlisted. The families were informed about the possible risk factors in future. Analysis of blood samples from these for genetic carrier showed only 4 cases of low level of fragile X chromosomes.

A project is supported at Saraswati Dental College, Lucknow for dental fluorosis. Water samples from 12 villages around Lucknow and Kanpur were analysed for fluoride content. Health camps were organized in 4 villages with a team of dental surgeon. A total of 2860 people were examined for dental fluorosis. The severely affected persons were advised for regular dental examination at SGPGI, Lucknow. They were also provided with calcium and cholicalciferol supplement. The blood samples from the severely affected boys and girls within the age group of 15 years and pregnant women were examined for their blood calcium level. Under another project supported at Mahavir Hospital And Research Centre, Hyderabad, Andhra Pradesh, blood lead levels in children and adults including men and women were measured to study environmental genotoxicity in women and children occupationally exposed to lead, a heavy metal.

RECOGNITION FOR FARMERS' RIGHTS

The aim to resist IPR strengthening has been closely related to attempts to encourage the use of diversity in agriculture. Whilst the agricultural biotechnology companies had their sights on institutional changes allowing for the granting of IPR over life through the GATT, the FAO was debating the issue of the rights of farmers over germplasm, especially from 1987 onward. Farmers' Rights would be the counterbalance to the spread of PBR. In 1987, an International Fund for Plant Genetic Resources was set up, and was legally established in 1989. The Fund was designed for genetic conservation and utilisation work, and administered by the ICPGR. Mexico argued that donations to such a Fund should be mandatory in the same way as are royalties to a patent holder, but no agreement was forthcoming.

By 1989, developing country governments had let it be known through negotiations at the FAO that failure on the part of developed country governments to acknowledge the concept of Farmers' Rights would result in those countries being denied access to developing country genetic resources. Similarly, developed country patents would not be honoured. In what was effectively an exercise in horse-trading, at the 1989 meeting, developed countries insisted on an additional Resolution modifying the International

Undertaking such that it recognised PBR. 'Free accesses to germplasm explicitly would not mean 'free of charge'. In return, a Resolution on Farmers' Rights was passed recognising the rights of farmers in respect of their work conserving, improving, and making available genetic resources.

In the midst of an increasingly polarised debate, a notable event was the Keystone Dialogue held at the Keystone Center in Colorado in 1988, at Madras in 1990, and Oslo in 1991. Major transnational corporations, NGOs, IBPGR, national genetic resource programmes, the academic community, and the Rockefeller Foundation reached the following consensus conclusions after the second meeting:

- Ipr were credited with encouraging development of new varieties, but also with encouraging genetic uniformity and erosion;
- Attempts to include ipr for plants under the auspices of the gatt negotiations were criticised; and
- Recognition of farmers' rights, and commendation of the idea of a fund such as that extant at the fao as a means of providing a form of concrete recognition thereof.

Corporate delegates refused to sanction a compensation mechanism, merely a fund recognising Farmers' Rights, a position which some participants would have found unacceptable were it not for the fact that it was agreed the fund should be mandatory. Furthermore, rather than the figure of $150 million proposed by NGOs at the FAO, the consensus figure arrived at was $500 million.

In November 1991, another Resolution concerning genetic resources was passed at the FAO. This amendment to the Undertaking upheld 'that nations have sovereign rights over their plant genetic resources and that breeders' lines and farmers' breeding material should only be available at the discretion of their developers during the period of development'. Although Farmers' Rights were recognised, they were given no substance. The International Fund, legally established in 1989, failed to materialise. Only in India have attempts been made to give substance to the concept through taxation of seed industry income.

Farmers' Rights still amount to little more than a polite thank you to farmers who have conserved genetic diversity *in situ*. This was most clearly illustrated in discussions at the June 1996 Leipzig Conference where the issue of Farmers' Rights showed that developed country donors were reluctant to support *in situ* conservation, partly, one suspects, because of issues related to sovereignty in respect of genetic resources (see next section).

The Convention on Biological Diversity

National sovereignty over genetic resources was a feature of the Convention on Biological Diversity (CBD), which entered into force as a legally binding international treaty at the end of 1993. The CBD is a framework convention whose objectives, stated in Article 1, are:

The conservation of biological diversity, the sustainable use of its components and *the fair and equitable sharing of the benefits arising out of the utilization of genetic resources, including by appropriate access to genetic resources and by appropriate transfer of relevant technologies*, taking into account all rights over those resources and to technologies, and by appropriate funding.

Article 3 of the CBD lays down the principle that: States have, in accordance with the Charter of the United Nations and the principle of international law, the *sovereign right to exploit their own resources* pursuant to their own environmental policies, and the responsibility to ensure that activities within their own jurisdiction or control do not cause damage to the environment of other states or of areas beyond the limits of national jurisdiction.

This effectively confirms that the free exchange principle is something of the past. For corporations, sovereignty appeared to cede too much control to governments over genetic resources which were of increasing value to biotechnology companies. For non-government organisations, this debate seemed to miss the point that it was not states who were really responsible for maintaining biological diversity within their borders, but local communities. The CBD offers little for local communities, and does not explicitly recognise Farmers' Rights, though the role of indigenous communities in conserving biodiversity is recognised in the preamble.

Articles 12, 17, 18 and 19 each refer to aspects of the biotechnology debate, but Article 16 of the CBD, dealing with transfer of technology, was the most heavily negotiated. Essentially, the debate centred around the fact that developing countries would most likely be providing raw materials for a biotechnology industry seeking to patent innovations. Article 16.2, suggests that: In the case of technology subject to patents and other intellectual property rights, such access and transfer shall be provided on terms which recognise and are consistent with the adequate and effective protection of intellectual property rights. The application of this paragraph shall be consistent with paragraphs 3, 4 and 5 below.

This Article was essentially a compromise meant to defuse the situation as regards the way in which technology transfer should account for IPR. These latter paragraphs make provision for the transfer of technology to developing countries on 'mutually agreed terms'. Article 16.5 suggests that the Contracting parties: recognizing that patents and other intellectual property rights may have an influence on then implementation of this Convention, shall cooperate in this regard subject to national legislation and international law in order to ensure that such rights are supportive of and do not run counter to its objectives.

The CBD's equivocation on IPR issues was perhaps the main reason why George Bush's US delegation felt unable to support the CBD in Rio. Ever since the late 1980s, when methods of screening increased in sophistication and fell in cost, plant research has acquired new significance for pharmaceutical companies. Thus, head of the US delegation, William Reilly, stated: 'We have negotiated in the Uruguay Round of GATT to try to protect intellectual property rights. We're not about to trade away here in an environmental treaty what we worked so hard to protect there'.

The CBD does not apply retrospectively. Thus, the legal status of *ex-situ* collections of resources donated by developing countries but housed in the developed world was left unclear. In 1994, the World Bank appears to have sought to prevent the CGIAR's *ex situ* collections from falling under intergovernmental control by taking control of these itself in exchange for new funds for the CGIAR.

However, 112 governments unanimously, and ultimately successfully, called for establishment of intergovernmental control over the CGIAR *ex situ* collections. The status of other collections held in developed country genebanks was discussed at the FAO's Leipzig Conference, but the outcome gave little encouragement to those countries which donated germplasm in the first place, as exemplified by the attempts of the company PHYTERA to gain access to genetic resources collected in developed country botanical gardens.

Indigenous Peoples

As so often in the past, indigenous peoples have been forgotten in the bulk of negotiations which affect a resource maintained largely by them. Their concerns span both the prime focus of the CBD and the issue of Farmers' Rights, as well as the TRIPs negotiations. Increasingly, the contributions made by indigenous peoples in terms not just of germplasm, but also knowledge regarding its use, are recognised.

Yet this recognition has led to little concrete action to protect their interests.In response principally to the heated debate generated by issues related to the patenting of indigenous people's cell lines, the Indigenous Peoples Biodiversity Network was formed with the objective of safeguarding their interests with respect to biodiversity and their knowledge thereof. Out of the debate concerning Farmers' Rights and IPR as they affect indigenous peoples has come awareness that current formal systems do not adequately recognise indigenous knowledge systems, in which knowledge is often held at the level of the community. More and more companies are screening plant materials for useful products, yet it is estimated that the chances of finding useful products are at least doubled if indigenous, or folk knowledge is utilised. An authority on the issue of IPR and indigenous peoples, Darrell Posey, notes that IPR pose seven problems for indigenous peoples:

- They do not grant rights to collective entities;
- They protect unique acts of discovery rather than transgenerational knowledge (from, for example, spirits, vision quests, or oral transmission) which tends to be public;

- They do not recognise non-western systems of ownership, access and tenure;
- They aim to promote commercialisation whereas the aims of indigenous peoples may be to prevent such activity;
- They recognise market values only and not spiritual, aesthetic, or cultural value;
- They are, as is clear from the above, intimately bound up with power relations; and
- They are expensive to obtain and difficult to defend.

Posey goes on to cite a number of examples where indigenous peoples have displayed almost uniform hostility to what they perceive as an insidious trend in IPR legislation which seeks to deepen the exploitation of their resources and their knowledge. They perceive simple recognition of their contributions, as exemplified by the Farmers' Rights issue, as typically patronising in the face of a continued absence of legal mechanisms adapted to meet their needs and concerns.

RAFI explores a number of ways in which indigenous communities could find space within existing legislation to protect their innovations. They argue that:

There is a strong case to be made that the uncompensated appropriation of farmers' varieties and medicinal plants constitutes real theft and that the parties responsible should be pursued under criminal law at the expense of national law enforcement agencies in the country where the theft occurs (the patenting country).

In the face of the forces mentioned above, it seems unlikely that much room for manoeuvre exists for those who would seek to place Farmers' Rights on the same level as IPR. Some countries have, however, been exploring the potential for exploiting the '*sui generis*' clause mentioned above in designing alternative IPR regimes which allow, for example, for communities, and not just individuals, to make IPR claims. As pointed out by Allen (1983), much innovation is the product of collective rather than individual efforts. In India, the concept of Collective Intellectual Property Rights is gaining credibility, whilst the Andean Pact is committed to developing a regime on collective rights of indigenous peoples.

This chapter has sought to show, through examining evolving IPR regimes, that a new mode of development of agricultural technologies is emerging. Some, indeed most of its most vocal and powerful supporters are drawn from the leaders of the BCM mode. The commercial possibilities presented by biotechniques saw pressure to extend IPR schemes. Following the 1980s, a decade which saw recognition gained in an international forum for the concept of Farmers' Rights, the rights of farmers to save seed first underwent conversion to a 'privilege', and were then consigned to history.

The old BCM mode still prevails despite the attacks of environmentalists and the growing awareness of available alternatives. If the possibilities for alternatives to thrive alongside the BCM mode seem limited, they are likely to be more so as the BCM mode is superseded by BML techniques. Although the BML mode's supporters have often appeared as keen as environmentalists to see the back of the old BCM mode, their agenda is not an alternative based on bringing diversity back into the picture, but the ushering in of new techniques aimed at increasing the potential for achieving a genetically uniform ideal plant.

BIOTECHNOLOGICAL APPLICATIONS RELEVANCE TO THE INDIAN FARMER

Increased food production must keep pace with the country's population increase. Not only the quantity of food made available should improve. To keep pace with the present population growth and consumption pattern, India's food requirement has been estimated to cross 225 million tons by 2005 AD. This would mean an annual agricultural growth rate of 6.7 percent, a daunting task considering the rapidly shrinking resource-base and fast declining input-use efficiency in major cropping systems.

Not withstanding the impressive gain in the agricultural production, the vast agricultural potential still remains highly under-utilized. There are serious gaps both in farm yield realization and technology transfer, as the national average yield of most crop varieties of most crop varieties is low.

Not many superior varieties is mot pulses, many oilseeds and vegetable crops have been evolved in recent past. In case of vegetables the varieties which were imported or developed many decades ago are still being grown by the farmers. Their yields are low and their inherent ability to respond to improve production technology is also limited. Farmers have accepted bybrids in my vegetable crops such as cabbage, tomato and capsicum.

RESEARCH AND TECHNOLOGY TRANSFER

The Central and State governments continued to invest heavilyu in research, hence the Indian public sector research system has increased in size in real terms but by mid 90s levelled off. The number of hybrids and varieties developed and released by ICAR Crop Improvement Projects and State Agricultural Universities in major and minor crops are enormous and have been well documented.

Private research to develop superior hybrids was initiated by few seed companies in mid 60s. By 1987-88, the number increased to over 12 and in 1997-98 it is estimated at 34. With the governments recognition of provate sector's performance capabilities and with "New Seed Policy" announced in 1988, the number of private companies involved in crop improvement programmes have been steadily increasing (Table below).

NUMBER OF PRIVATE COMPANIES ENGAGED IN CROP IMPROVEMENTPROGRAMME (1987-1998)

For want of financial, infrastructural and trained human resources, most private companies specialize in breeding superior hybrids in few crops characterized by low volume, high value and amenable to hybrid technology. In addition to the ICAR and Agriculture University research system, about 13 private sector seed companies are actively involved in vegetable R andD. Some of them have joint ventures with multinational or companies specializing in vegetable crop research and utilize research facilities available with their overseas partners.

Table : Number of Companies with Crop Improvement programmes

Crop	*1987-88*	*1993-94*	*1997-98*
Pearl Millet	12	23	25
Sorghum	10	12	15
maize	06	15	20
Sunflower	10	30	20
Cotton	09	21	34
Hibrid Rice	-	4	6

Few have created facilities for field testing imported and public bred varieties. The role of private sector in vegetable research is primarily in the development of F hybrids and few open pollinated varieties.

Table : Hybrid Vegetable and Melon seeds of Private R andD available in the market (1996-97)

Crop	*No. of Companies*	*No. of hybrid Varieties*
Tomatoes	11	45
Brinjal	10	38
Chilli	8	27
Bhendi	13	25
Cabbage	11	40
Cauliflower	8	21
Ridgegourd	6	8
Watermelon	5	13
Muskmelon	2	7

Number of private/public hybrids in major crops in market during 1995 are compared with the number in 1997.

In pearl millet, sorghum, maize and cotton number of hybrids introduced by private, companies has been steadily increasing. Many hybrids especially in paddy and cotton have regional adaptability. The number of hybrids which are accepted by the farmers and may be grown on 2 percent or more area is much lesser, so also the number evaluated in coordinated project trails or identified for release. Many hybrids recently supplied by the private sector embodies new technology which is either imported or developed by private companies. Private R andD's real investment in research has quadrupled between 1986 and 1997. Subsidiaries or joint ventures with multinational companies account for 30 percent of all private seed industry research in 1997. Both domestic companies and multinationals companies have started research programmes on rice and mustard. Recently, Monsanto has also focused on hybrid wheat.

Table : Number of Private and Public Bred Hybrids in Major Crops in the Market during 1995 and 1997

	Number of Hybrids in Market				
	1995		*1997*		
Crop	*Private*	*Public*	*Private*	*Public*	*No of Private Hybrids grown on 2% plus area*
Pearl Millet	50	04	60	06	14
Sorghum	22	04	41	05	6
Maize	57	03	67	03	12
Sunflower	47	05	35	06	10
Cotton	73	15	150	15	19
Hybrid Rice	4	4	8	4	-

Few private sector companies have also initiated joint ventures (*e.g.* Mahyco-Monsanto Biotech, ProAgro-Agro-Evo) to transfer specific genes to their product lines to provide effective protection against pests. Some of them have taken initiative to treat their seed with high-tech chemicals to provide protection against sucking pests (aphids, jassids, thrips) during seedling and grand growth crop stages and diseases (downy mildew in bajra) during flowering.

FARMING TECHNIQUES IN SUSTAINABLE AGRICULTURE

Sustainable agriculture is a special kind of agriculture along with farming technique which makes maximum utilization of the environment without causing any form of harm to it. The outputs from this Sustainable Agriculture are devoid of any types of inorganic chemicals such as pesticides and insecticides. This farming technique uses the organic way of farming. All these factors produce the products in a more environment friendly manner and are thus healthier for the consumers to consume.

- Sustainable Agriculture Practice is regarded as the need of the hour by most of the countries over the world keeping in view environmental concern.
- Sustainable Agriculture Washington is an organization, which has been rendering promotional methods in the field of agriculture. Find more about Sustainable Agriculture in Washington.

- National Sustainable Agriculture is one of the leading national level programme in United States of America. Get more on National Sustainable Agriculture.

The other characteristic feature of Sustainable Agriculture are given in the bullet form below:

- The treatment towards the animals in such kind of farming are more humane because they are given enough space to graze and are not bred like factory products.
- The workers who are associated with the Sustainable Agriculture receive more humane behavior from the employers and it has been observed that the wages offered to these farm workers are fairer than the ones working in the so called industrial farms.
- The one of the most important feature of Sustainable Agriculture is its extended support towards the small farming community farmers. It is essentially being emphasized in this type of farming that the organic food items that have been produced by the small farmers of the rural areas are bought and marketed to the consumers through this Sustainable Agriculture route.

Sustainable Agriculture got its first legal recognition in the United States of America through the Public Law 101-624, Title XVI, Subtitle A, Section 1683 which was introduced in the year 1990. This law defined this farming practice type to be more Eco-friendly and having a sight for the long run satisfaction of the nutrient values in accordance with the nutritional requirements of the human body. It was especially emphasized in the law that the non renewable resources that are used in the farming process needs to be used at a minimal level in Sustainable Agriculture. It emphasizes that the viability conditions of the small farmers in the rural areas needs to be enhanced through Sustainable Agriculture through proper disposal of wage and proper distribution of their produced products.

Information on Sustainable Agriculture would be incomplete if the mention of the Organic Farming is not considered in the same. For more development in this field, research on sustainable agriculture is going on and with every passing day, this farming type is given more emphasis by the American as well as the French government.

FARMING AND NATURAL RESOURCES

Water. When the production of food and fibre degrades the natural resource base, the ability of future generations to produce and flourish decreases. The decline of ancient civilizations in Mesopotamia, the Mediterranean region, Pre-Columbian southwest U.S. and Central America is believed to have been strongly influenced by natural resource degradation from non-sustainable farming and forestry practices. Water is the principal resource that has helped agriculture and society to prosper, and it has been a major limiting factor when mismanaged.

Water supply and use. In California, an extensive water storage and transfer system has been established which has allowed crop production to expand to very arid regions. In drought years, limited surface water supplies have prompted overdraft of groundwater and consequent intrusion of salt water, or permanent collapse of aquifers. Periodic droughts, some lasting up to 50 years, have occurred in California.

Several steps should be taken to develop drought-resistant farming systems even in "normal" years, including both policy and management actions:

- Improving water conservation and storage measures,
- Providing incentives for selection of drought-tolerant crop species,
- Using reduced-volume irrigation systems,
- Managing crops to reduce water loss, or
- Not planting at all.

Water quality: The most important issues related to water quality involve salinization and contamination of ground and surface waters by pesticides, nitrates and selenium. Salinity has become a problem wherever water of even relatively low salt content is used on shallow soils in arid regions and/or where the water table is near the root zone of crops. Tile drainage can remove the water and salts, but the disposal of the salts and other contaminants may negatively affect the environment depending upon where they are deposited.

Temporary solutions include the use of salt-tolerant crops, low-volume irrigation, and various management techniques to minimize the effects of salts on crops. In the long-term, some farmland may need to be removed from production or converted to other uses. Other uses include conversion of row crop land to production of drought-tolerant forages, the restoration of wildlife habitat or the use of agroforestry to minimize the impacts of salinity and high water tables. Pesticide and nitrate contamination of water can be reduced using many of the practices discussed later in the Plant Production Practices.

Wildlife: Another way in which agriculture affects water resources is through the destruction of riparian habitats within watersheds. The conversion of wild habitat to agricultural land reduces fish and wildlife through erosion and sedimentation, the effects of pesticides, removal of riparian plants, and the diversion of water. The plant diversity in and around both riparian and agricultural areas should be maintained in order to support a diversity of wildlife. This diversity will enhance natural ecosystems and could aid in agricultural pest management.

Energy: Modern agriculture is heavily dependent on non-renewable energy sources, especially petroleum. The continued use of these energy sources cannot be sustained indefinitely, yet to abruptly abandon our reliance on them would be economically catastrophic. However, a sudden cutoff in energy supply would be equally disruptive. In sustainable agricultural systems, there is reduced reliance on non-renewable energy sources and a substitution of renewable sources or labour to the extent that is economically feasible.

Air: Many agricultural activities affect air quality. These include smoke from agricultural burning; dust from tillage, traffic and harvest; pesticide drift from spraying; and nitrous oxide emissions from the use of nitrogen fertilizer. Options to improve air quality include incorporating crop residue into the soil, using appropriate levels of tillage, and planting wind breaks, cover crops or strips of native perennial grasses to reduce dust.

Soil: Soil erosion continues to be a serious threat to our continued ability to produce adequate food. Numerous practices have been developed to keep soil in place, which include reducing or eliminating tillage, managing irrigation to reduce runoff.

CONCEPT THEMES

Sustainable agriculture integrates three main goals—environmental health, economic profitability, and social and economic equity. A variety of philosophies, policies and practices have contributed to these goals. People in many different capacities, from farmers to consumers, have shared this vision and contributed to it. Despite the diversity of people and perspectives, the following themes commonly weave through definitions of sustainable agriculture.

Sustainability rests on the principle that we must meet the needs of the present without compromising the ability of future generations to meet their own needs. Therefore, *stewardship of both natural and human resources* is of prime importance. Stewardship of human resources includes consideration of social responsibilities such as working and living conditions of laborers, the needs of rural communities, and consumer health and safety both in the present and the future. Stewardship of land and natural resources involves maintaining or enhancing this vital resource base for the long term.

A *systems perspective* is essential to understanding sustainability. The system is envisioned in its broadest sense, from the individual farm, to the local ecosystem, *and* to communities affected by this farming system both locally and globally. An emphasis on the system allows a larger and more thorough view of the consequences of farming practices on both human communities and the environment. A systems approach gives us the tools to explore the interconnections between farming and other aspects of our environment.

A systems approach also implies *interdisciplinary efforts in research and education*. This requires not only the input of researchers from various disciplines, but also farmers, farmworkers, consumers, policymakers and others.

Making the transition to sustainable agriculture is a process. For farmers, the transition to sustainable agriculture normally requires a series of small, realistic steps. Family economics and personal goals influence how fast or how far participants can go in the transition. It is important to realize that each small decision can make a difference and contribute to advancing the entire system further on the "sustainable agriculture continuum." The key to moving forward is the will to take the next step.

Finally, it is important to point out that *reaching toward the goal of sustainable agriculture is the responsibility of all participants in the system*, including farmers, laborers, policymakers, researchers, retailers, and consumers. Each group has its own part to play, its own unique contribution to make to strengthen the sustainable agriculture community.

The remainder of this document considers specific strategies for realizing these broad themes or goals. The strategies are grouped according to three separate though related areas of concern: Farming and Natural Resources, Plant and Animal Production Practices, and the Economic, Social and Political Context. They represent a range of potential ideas for individuals committed to interpreting the vision of sustainable agriculture within their own circumstances.

PLANT PRODUCTION PRACTICES

Sustainable production practices involve a variety of approaches. Specific strategies must take into account topography, soil characteristics, climate, pests, local availability of inputs and the individual grower's goals.

Despite the site-specific and individual nature of sustainable agriculture, several general principles can be applied to help growers select appropriate management practices:

- Selection of species and varieties that are well suited to the site and to conditions on the farm;
- Diversification of crops and cultural practices to enhance the biological and economic stability of the farm;
- Management of the soil to enhance and protect soil quality;
- Efficient and humane use of inputs; and
- Consideration of farmers' goals and lifestyle choices.

Selection of site, species and variety: Preventive strategies, adopted early, can reduce inputs and help establish a sustainable production system. When possible, pest-resistant crops should be selected which are tolerant of existing soil or site conditions. When site selection is an option, factors such as soil type and depth and location should be taken into account before planting.

Diversity: Diversified farms are usually more economically and ecologically resilient. While monoculture farming has advantages in terms of efficiency and ease of management, the loss of the crop in any one year could put a farm out of business and/or seriously disrupt the stability of a community dependent on that crop. By growing a variety of crops, farmers spread economic risk and are less susceptible to the radical price fluctuations associated with changes in supply and demand.

Properly managed, diversity can also buffer a farm in a biological sense. In annual cropping systems, crop rotation can be used to suppress weeds, pathogens and insect pests. Also, cover crops can have stabilizing effects on the agroecosystem by holding soil and nutrients in place, conserving soil moisture with mowed or standing dead mulches, and by increasing the water infiltration rate and soil water holding capacity. Cover crops in orchards and vineyards can buffer the system against pest infestations by increasing beneficial arthropod populations and can therefore reduce the need for chemical inputs. Using a variety of cover crops is also important in order to protect against the failure of a particular species to grow and to attract and sustain a wide range of beneficial arthropods.

Optimum diversity may be obtained by integrating both crops and livestock in the same farming operation. This was the common practice for centuries until the mid-1900s when technology, government policy and economics compelled farms to become more specialized. Mixed crop and livestock operations have several advantages.

First, growing row crops only on more level land and pasture or forages on steeper slopes will reduce soil erosion. Second, pasture and forage crops in rotation enhance soil quality and reduce erosion; livestock manure, in turn, contributes to soil fertility. Third, livestock can buffer the negative impacts of low rainfall periods by consuming crop residue that in "plant only" systems would have been considered crop failures. Finally, feeding and marketing are flexible in animal production systems. This can help cushion farmers against trade and price fluctuations and, in conjunction with cropping operations, make more efficient use of farm labour.

Soil management: A common philosophy among sustainable agriculture practitioners is that a "healthy" soil is a key component of sustainability; that is, a healthy soil will produce healthy crop plants that have optimum vigour and are less susceptible to pests. While many crops have key pests that attack even the healthiest of plants, proper soil, water and nutrient management can help prevent some pest problems brought on by crop stress or nutrient imbalance. Furthermore, crop management systems that impair soil quality often result in greater inputs of water, nutrients, pesticides, and/or energy for tillage to maintain yields.

In sustainable systems, the soil is viewed as a fragile and living medium that must be protected and nurtured to ensure its long-term productivity and stability. Methods to protect and enhance the productivity of the soil include using cover crops, compost and/or manures, reducing tillage, avoiding traffic on wet soils, and maintaining soil cover with plants and/or mulches. Conditions in most California soils do not defence the buildup of organic matter. Regular additions of organic matter or the use of cover crops can increase soil aggregate stability, soil tilth, and diversity of soil microbial life.

Efficient use of inputs: Many inputs and practices used by conventional farmers are also used in sustainable agriculture. Sustainable farmers, however, maximize reliance on natural, renewable, and on-farm inputs. Equally important are the environmental, social, and economic impacts of a particular strategy. Converting to sustainable practices does not mean simple input substitution. Frequently, it substitutes enhanced management and scientific knowledge for conventional inputs, especially chemical inputs that harm the environment on farms and in rural communities. The goal is to develop efficient, biological systems which do not need high levels of material inputs.

Growers frequently ask if synthetic chemicals are appropriate in a sustainable farming system. Sustainable approaches are those that are the least toxic and least energy intensive, and yet maintain productivity and profitability. Preventive strategies and other alternatives should be employed before using chemical inputs from any source. However, there may be situations where the use of synthetic chemicals would be more "sustainable" than a strictly non-chemical approach or an approach using toxic "organic" chemicals. One grape grower switched from tillage to a few applications of a broad spectrum contact herbicide in the vine row. This approach may use less energy and may compact the soil less than numerous passes with a cultivator or mower.

Consideration of farmer goals and lifestyle choices. Management decisions should reflect not only environmental and broad social considerations, but also individual goals and lifestyle choices. Adoption of some technologies or practices that promise profitability may also require such intensive management that one's lifestyle actually deteriorates. Management decisions that promote sustainability, nourish the environment, the community *and* the individual.

ANIMAL PRODUCTION PRACTICES

In the early part of this century, most farms integrated both crop and livestock operations. Indeed, the two were highly complementary both biologically and economically. The current picture has changed quite drastically since then. Crop and animal producers now are still dependent on one another to some degree, but the integration now most commonly takes place at a higher level—*between* farmers, through intermediaries, rather than *within* the farm itself.

This is the result of a trend towards separation and specialization of crop and animal production systems. Despite this trend, there are still many farmers, particularly in the Midwest and Northeastern U.S. that integrate crop and animal systems—either on dairy farms, or with range cattle, sheep or hog operations. Even with the growing specialization of livestock and crop producers, many of the principles outlined in the crop production section apply to both groups. The actual management practices will, of course, be quite different. Some of the specific points that livestock producers need to address are listed below.

Management Planning: Including livestock in the farming system increases the complexity of biological and economic relationships. The mobility of the stock, daily feeding, health concerns, breeding operations, seasonal feed and forage sources, and complex marketing are sources of this complexity. Therefore, a successful ranch plan should include enterprise calendars of operations, stock flows, forage flows, labour needs, herd production records and land use plans to give the manager control and a means of monitoring progress towards goals.

Animal Selection: The animal enterprise must be appropriate for the farm or ranch resources. Farm capabilities and constraints such as feed and forage sources, landscape, climate and skill of the manager must be considered in selecting which animals to produce. Ruminant animals can be raised on a variety of feed sources including range and pasture, cultivated forage, cover crops, shrubs, weeds, and crop residues. There is a wide range of breeds available in each of the major ruminant species, *i.e.*, cattle, sheep and goats. Hardier breeds that, in general, have lower growth and milk production potential, are better adapted to less favorable environments with sparse or highly seasonal forage growth.

Animal nutrition: Feed costs are the largest single variable cost in any livestock operation. While most of the feed may come from other enterprises on the ranch, some purchased feed is usually imported from off the farm. Feed costs can be kept to a minimum by monitoring animal condition and performance and understanding seasonal variations in feed and forage quality on the farm. Determining the optimal use of farm-generated by-products is an important challenge of diversified farming.

Reproduction: Use of quality germplasm to improve herd performance is another key to sustainability. In combination with good genetic stock, adapting the reproduction season to fit the climate and sources of feed and forage reduce health problems and feed costs.

Herd Health: Animal health greatly influences reproductive success and weight gains, two key aspects of successful livestock production. Unhealthy stock waste feed and require additional labour. A herd health programme is critical to sustainable livestock production.

Grazing Management: Most adverse environmental impacts associated with grazing can be prevented or mitigated with proper grazing management. First, the number of stock per unit area must be correct for the landscape and the forage sources. There will need to be compromises between the convenience of tilling large, unfenced fields and the fencing needs of livestock operations.. Second, the long term carrying capacity and the stocking rate must take into account short and long-term droughts. Especially in Mediterranean climates such as in California, properly managed grazing significantly reduces fire hazards by reducing fuel build-up in grasslands and brushlands. Finally, the manager must achieve sufficient control to reduce overuse in some areas while other areas go unused. Prolonged concentration of stock that results in permanent loss of vegetative cover on uplands or in riparian zones should be avoided. However, small scale loss of vegetative cover around water or feed troughs may be tolerated if surrounding vegetative cover is adequate.

Confined Livestock Production: Animal health and waste management are key issues in confined livestock operations. The moral and ethical debate taking place today regarding animal welfare is particularly intense for confined livestock production systems. The issues raised in this debate need to be addressed.

Confinement livestock production is increasingly a source of surface and ground water pollutants, particularly where there are large numbers of animals per unit area. Expensive waste management facilities are now a necessary cost of confined production systems. Waste is a problem of almost all operations and must be managed with respect to both the environment and the quality of life in nearby communities. Livestock production systems that disperse stock in pastures so the wastes are not concentrated and do not overwhelm natural nutrient cycling processes have become a subject of renewed interest.

7

Applications of Genetic Biotechnology in Horticulture

The requirement of fruits and vegetables is increasing proportionally with the increasing population in the country. How do we keep horticultural production on par with the burgeoning population? Although conventional plant breeding techniques have made considerable progress in the development of improved varieties, they have not been able to keep pace with the increasing demand for vegetables and fruits in the developing countries. Therefore an immediate need is felt to integrate biotechnology to speed up the crop improvement programmes. Biotechnological tools have revolutionized the entire crop improvement programmes by providing new strains of plants, supply of planting material, more efficient and selective pesticides and improved fertilizers. Many genetically modified fruits and vegetables are already in the market in developed countries. Modern biotechnology encompasses broad areas of biology from utilization of living organisms or substances from those organisms to make or to modify a product, to improve plant or animal or to develop micro-organisms for specific use. It is a new aspect of biological and agricultural science which provides new tools and strategies in the struggle against world's food production problem. The major areas of biotechnology which can be adopted for improvement of horticultural crops are –

GENETIC ENGINEERING OF PLANTS

Using techniques of genetic engineering many useful genes have been introduced into plants and many transgenic plants have been developed in which the foreign DNA has been stably integrated and resulted in the synthesis of appropriate gene product. Transgenic plants have covered about 52.6 m hectares in the Industrial and developing countries upto 2001. Genes for the following traits have been introduced to the crop plants.

Herbicide tolerance: Transgenic plants are developed that are resistant to herbicides allowing farmers to spray crops so as to kill only weeds but not their crops. Many herbicide tolerant plants have been developed in tomato, tobacco, potato, soybean, cotton, corn oilseed rape, petunia, etc. Glyphosate is one of the most potent broad spectrum environment friendly herbicide known, it is marketed under the trade name Round up. Glyphosate kills plants by blocking the action of an enzyme (5-enolpyruvyl shikimate-3-phosphate synthase) (EPSPS) an essential enzyme in the biosynthesis of aromatic amino acids, tyrosine, phenylalanine and tryptophan. Amino acids are building blocks of protein.

Transgenic plants resistant to Glyphosate have been developed by transferring gene of EPSPS that over prodoce this enzyme thus inhibiting the effect of Glyphosate. A number of detoxifying enzymes have been identified in plants as well as in microbes. Some of these include glutahthione-s-transferase or GST in maize and other plants which detoxifies the herbicidebromoxynil and phosphinothricin acetyl transferase (PAT) which detoxifies the herbiside PPT (L-phosphinothricine). Transgenic plants using *bxn* gene from Klebsiella and *bar* gene from Strepotomyces have been obtained in potato, oilseed, sugarbeet, soybean, cotton and corn and are found to be herbicide resistance. These transgenic plants reduce the use of weeding labour, farmers cost and increase yield.

Engineering pathogen resistance: Viruses are the major pests of crop plants which cause considerable yield losses. Many strategies have been applied to control virus infection using coat protein and satellite RNA. Viruses are submicroscopic pockets of nucleic acid (DNA or RNA) enclosed in a protein coat and can multiply within a host cell. Use of viral coat protein as a transgene for producing virus resistant plants is one of the most spectacular successes achieved in plant biotechnology. Coat protein gene from tobacco mosaic virus (TMV) classified as a positive strand RNA virus has been transferred to tobacco, making it nearly resistant against TMV. Using gene for nucelocapsid protein resistance has been introduced in crops like tomato, tobacco, lettuce, groundnut, pepper and in ornaments like *Impatiens*, *Ageratum* and *Crysnathemum* against tomato spotted wilt virus. Use of satellite RNA (SATRNA) makes many transgenic plants resistant toCucumber Mosaic Virus (CMV). Transgenic resistant plants have also been developed against alfalfa mosaic virus, potato virus X, Rice tungro virus, tobacco rattle virus and Papaya ring spot virus.

During the last decade many resistance genes whose products are involved in recognizing the invading pathogens have been identified and cloned. A number of signaling pathways which follow the pathogen infection have been dissected. Many of the antifungal compounds synthesized by plants which combat fungal infections have been identified. The major strategies for developing fungal resistance have been production of transgenic plants with antifungal molecules like proteins and toxins, and generation of hypersensitive response through R genes or by manipulating genes of SAR pathway. A chitinase gene from bean plants in tobacco and *Brassica napus* showed enhanced resistance to *Rhizoctonia solani*. In another case chitinase gene obtained from *Serratia marcescens*(soil bacterium) is introduced in tobacco making it resistant to *Alternaria longipes* which causes brown spot diseases. Acetyl transferase gene is introduced in tobacco making it resistant to*Pseudomonas syringea*, a causal agent of wild fire disease.

Stress resistance : A number of genes responsible for providing resistance against stresses such as to water stress heat, cold, salt, heavy metals and phytohormones have been identified. Studies are also being conducted on metabolites like proteins and betains that have been implicated in stress tolerance. Resistance against chilling was introduced into tobacco plants by introducing gene for glycerol-1-phosphate acyl-transferase enzyme from *Arabidopsis*. Many plants respond to drought stress by synthesizing a group of sugar derivatives called polyols (Mannitol, Sorbitol and Sion). Plants that have more polyols are more resistant to stress. Using a bacterial gene capable of synthesizing mannitols it is possible to raise the level of mannitol very high making plants resistant to drought.

Fruit Quality: Tomatoes which ripen slowly are helpful in transportation process. Transgenic tomato with reduced pectin methyl esterase activity and increased level of soluble solids and higher pH increases processing quality. Tomatoes exhibiting delayed ripening have been produced either by using antisense RNA against enzymes involved in ethylene production (Eg ACC synthase) or by using gene for deaminase which degraded l-aminocyclopropane-l-carboxylic acid (ACC) an immediate precursor of ethylene. This increases the shelf life of tomatoes. These tomatoes can also stay on the plant long giving more time for accumulation of sugars and acids for improving flavour. It is produced at commercial level in European and American countries. Tomatoes with elevated sucrose and reduced starch could also be produced using sucrose phosphate synthase gene. Starch content in potatoes has been increased by 20-40 poe cent by using a bacterial ADP glucose pyrophosphorylase gene.

Pest resistance: The insecticidal beta endotoxin gene (bt gene) has been isolated from *Bacillus thuringiensis* the commonly occurring soil bacteria and transferred to number of plants like cotton, tobacco, tomato, soybean, potato, etc. to make them resistant to attack by insects. These genes produce insecticidal crystal proteins which affect a range of lepidopteran, coleopteran, dipteran insects. These crystals upon ingestion by the insect larva are solubilised in the highly alkaline midgut into individual protoxins which vary from 133 to 136 kDa in molecular weight. Insecticidal crystal protein produced during vegetative growth of the cells (VIP)are also found to be highly effective against insect control. Bt resistant plants are already in the market.

Male sterility and Fertility restoration: This is helpful in hybrid seed production. Transgenic plants with male sterility and fertility restoration genes have become available in *Brassica napus*. It facilitates production of hybrid seed without manual emasculation and controlled pollination as often done in maize. In 1990, Mariani and others from Belgium have successfully used a gene construct having another specific promoter from TA29 gene of tobacco and bacterial coding sequence for a ribonuclease gene from Bacillus Sp. (barnase gene) for production of transgenic plants in *Brassica napus*. Here the translated gene prevented normal pollen development leading to male sterilily.

TISSUE CULTURE

One of the widest applications of biotechnology has been in the area of tissue culture and micro propagation in particular. It is one of the most widely used techniques for rapid asexual in vitro propagation. This technique is economical in time and space affords greater output and provides disease free and elite propagules. It also facilitates safer and quarantined movements of germplasm across nations. When the traditional methods are unable to meet the demand for propagation material this technique can produce millions of uniformly flowering and yielding plants. Micropropagation of almost all the fruit crops and vegetables is possible now. Production of virus free planting material using meristem culture has been made possible in many horticultural crops. Embryo rescue is another area where plant breeders are able to rescue their crosses which would otherwise abort.

Culture of excised embryos of suitable stages of development can circumvent problems encountered in post zygotic incompatibility. This technique is highly significant in intractable and long duration horticultural species. Many of the dry land legume species have been successfully regenerated from cotyledons, hypocotyls, leaf, ovary, protoplast, petiole root, anthers, etc., Haploid generation through anther/pollen culture is recognized as another important area in crop improvement. It is useful in being rapid and economically feasible. Complete homozygosity of the offspring helps in phenotype selection for quantitative characters and particularly for qualitatively inherited characters making breeding much easier successful isolation, culture and fusion of plant protoplasts has been very useful in transferring cytoplasmic male sterility for obtaining hybrid vigour through mitochondrial recombination and for genetic transformation in plants.

In vitro germplasm conservation is of great significance in providing solutions and alternative approaches to overcoming constrains in management of genetic resources. In crops which are propagated vegetatively and which produce recalcitrant seeds and perennial crops which are highly heterozygous seed storage is not suitable. In such crops especially, *in vitro* storage is of great practical importance. These techniques have successfully been demonstrated in a number of horticultural crops and there are now various germplasm collection centers. In vitro germplasm also assures the exchange of pest and disease free material and helps in better quarantine.

Plant breeders are continually searching for new genetic variability that is potentially useful in cultivar improvement. A portion of plants regenerated by tissue culture often exhibits phenotypic variation atypical of the original phenotype. Such variation, termed somaclonal variation may be heritable i.e. genetically stable and passed on to the next generation. Alternatively, the variation may be epigenetic and disappear following sexual reproduction. These heritable variation are potentially useful to plant breeders.

MOLECULAR DIAGNOSTICS

Nucleic acid probes:- It is now possible to detect the plant diseases even before onset of symptoms by using cDNA probes. Probes are nucleic acid sequences of pathogen causing organisms labeled with certain markers. cDNA probes corresponding to specific regions of the pathogens can be generated using standard recombinant DNA technique.

Monoclonal antibodies (McAb): Immunochemical techniques are extremely useful for the rapid and accurate routine detection of plant pathogens and ultimately the diagnosis of plant disease and their relatedness, The introduction of hybridoma technology has provided methods for the production of homologous and biochemically defined immunological reagents of identical specificity which are produced by a single cell line and are directed against a unique epitope of the immunizing antigen. The great potential of McAbs in phytopathological diagnostics is essential because of homogeneous antibody preparations with defined activity and specificity can be produced in large quantities over long periods. Even though hybridoma technology is a laborious and expensive enterprise compared to standard immunization procedures it is going to be widely used for large scale diagnosis.

MOLECULAR MARKERS

The possibilities of using gene tags of molecular makers for selecting agronomic traits has made the job of breeder easier. It has been possible to score the plants for different traits or disease resistance at the seedling stage itself. The use of RFLP (Restriction Fragment Length polymorphism), RAPD (Random Amplified Polymorphic DNA), AFLP (Amplified Fragment Length Polymorphism) and isozyme markers in plant breeding are numerous. RFLPs are advantageous over morphological and isozyme markers primarily because their number is limited only by genome size and they are not environmentally or developmentally influenced. Molecular maps now exist for a number of crop plants including corn, tomato, potato, rice, lettuce, wheat, Brassica species and barley. RFLPs have wide ranging applications including cultivar finger printing, identification of quantitative trait loci, analysis of genome organization, germplasm introgression and map-based cloning. AFLP is becoming the tool of choice for fingerprinting because of its reproducibility compared to RAPD.

Microsatellile or simple sequence repeats (SSRS) markers have also become the choice for a wide range of applications in genotyping, genome mapping and genome analysis.

DEVELOPMENT OF MICROBIAL INOCULAN

Indiscriminate and injudicious use of chemical fertilizers and pesticides for the crop production and control of insect-pests has resulted in pollution of the environment deterioration of soil health and development of resistance by many insects and residue problems. Hence there is a great concern world wide to use safer biofertilisers and biopesticdies in the integrated nutrient management and pest management systems.

Biofertilizers are micro-organisms which fix atmospheric nitrogen or solubilise fixed phosphorus in the soil and make more nutrients available to the plant. Some of the organisms providing major inputs are the biological nitrogen fixing organisms like *Rhizobium, Azotobacter, Azospirillum* and phosphate solubilising organisms like *Bacillus polymyxa, B. magaterium, Pseudomonas striata* and certain fungal species of Aspergillus and *Penicillium.*

The benefits of using micro-organisms as fertilizers are many fold. They are less expensive, non-toxic to plants, do not pollute the ground water nor render the soil acidic and unfit for growth of plants. *Rhizobium* forms nodules on the roots of leguminous plants and help in fixing nitrogen from the atmosphere to ammonium irons which get converted to amino acids in the plant system. Inoculation with this bacteria helps in reducing addition of nitrogenous fertilizers to the soil. *Azospirillum* is also found colonizing inter cellular spaces inside the root system. These bacteria also contribute substantially to the nitrogen requirement of the plant.

Phosphate solubilising bacteria are another group of micro-organisms which solubilise the insoluble phosphorus in the soil and make them readily available to the crop.

Mycorrhiza is the symbiotic association of the roots of crop plants with non-pathogenic fungus. They provide nutrients absorbed from deeper layers of soil to the plants. They help the plants in better plant establishment and growth when inoculated. Many fruit crops like papaya, mango, banana, citrus, pomegranate are found to be dependent on this association and are greatly benefited by its inoculation in procuring higher phosphate and other nutrient from the soil. These mycorrhizal associations help the plants in overcoming pathogen attack also. They improve soil characters too.

Genetic modification of microbes: By using DNA recombination technique it has been possible to genetically manipulate different strains of these bacteria suitable to different environmental conditions and to develop strains with traits with capacity for better competitiveness and nodulation. Biopesticides are biological organisms which can be formulated as that of the pesticides for the control of pests. Biopesticides are gaining importance in agriculture, horticulture and in public heatlh programmes for the control of pests. The advantages of using biopesticides are many.

They are specific to target pests and do not harm the non target organisms such as bees, butterflies and are safe to humans and live stocks, they do not disturb the food-chain nor leave behind toxic residues. Some of the microbial pesticides used to control insect pests are *Bacillus thuringiensis* species to control various insect pests. Insecticidal property of these bacteria are due to crystals of insecticidal proteins produced during sporulation. These proteins are stomach poisons and are highly insect specific. Bt toxins could kill plant parasitic nematode too. Number of baculoviruses (BV) nuclear polyhedrosis virus (NPV) is being developed as microbial pesticides both nationally and internationally, A few examples of these are *Heliothis, Spodoptera, Plusia, Agrotis, Trichoplusia*, etc.

Biocontrol agents: These are other microbes which are antagonistic to several pathogenic fungus and are good substitutes to fungicides or insecticide. These are Bacillus sps. *Pseudomonas fluorescens, Trichoderma, Verticillium sp., Streptromyces* sps. etc. These organisms are commercially available.

ORGANIC HORTICULTURE

Organic horticulture is the science and art of growing fruits, vegetables, flowers, or ornamental plants by following the essential principles of organic agriculture in soil building and conservation, pest management, and heirloom variety preservation.

The Latin words hortus and cultura together form horticulture, classically defined as the culture or growing of garden plants. Horticulture is also sometimes defined simply as "agriculture minus the plough." Instead of the plough, horticulture makes use of human labour and gardener's hand tools, although some small machine tools like rotary tillers are commonly employed now.

GENERAL

Double Digging, Vermicompost, Mulches, cover crops, compost, manures, and mineral supplements are soil-building mainstays that distinguish this type of farming from its commercial counterpart. Through attention to good healthy soil condition, it is expected that insect, fungal, or other problems that sometimes plague plants can be minimized. However, pheromone traps, insecticidal soap sprays, and other pest-control methods available to organic farmers are also sometimes utilized by organic horticulturists.

Horticulture involves five areas of study. These areas are floriculture, landscape horticulture, olericulture and postharvest physiology. All of these can be, and sometimes are, pursued according to the principles of organic cultivation. Organic horticulture is based on knowledge and techniques gathered over thousands of years. In general terms, organic horticulture involves natural processes, often taking place over extended periods of time, and a sustainable, holistic approach - while chemical-based horticulture focuses on immediate, isolated effects and reductionist strategies.

ORGANIC GARDENING SYSTEMS

There are a number of formal organic gardening and farming systems that prescribe specific techniques. They tend to be more specific than, and fit within, general organic standards. Biodynamic farming is an approach based on the esoteric teachings of Rudolf Steiner. The Japanese farmer and writer Masanobu Fukuoka invented a no-till system for small-scale grain production that he called Natural Farming. French intensive and biointensive methods and SPIN Farming are all small scale gardening techniques.

A garden is more than just a means of providing food, it is a model of what is possible in a community - everyone could have a garden of some kind and produce healthy, nutritious organic food, a farmers market, a place to pass on gardening experience, and a sharing of bounty, promoting a more sustainable way of living that would encourage their local economy. A simple 4' x 8' raised bed garden based on the principles of bio-intensive planting and square foot gardening uses fewer nutrients and less water, and could keep a family, or community, supplied with an abundance of healthy, nutritious organic greens, while promoting a more sustainable way of living.

Other methods can also be used to supplement an existing garden. Methods such as composting, or vermicomposting. These practices are ways of recycling organic matter into some of the best organic fertilizers and soil conditioner. Vermicompost is especially easy. The byproduct is also an excellent source of nutrients for an organic garden.

SOIL

Managing the soil is very important. If your garden is healthy then insects will not attack the plants. Insects only attack plants that are unhealthy. As an example illustrating the relative health of plants, insects will not feed on plants which have been genetically modified so as to reduce foliar damage; thus, genetically modified plants are healthier than unmodified ones. To keep your garden healthy give it organic matter and humus to survive. The most important thing is to give your garden lots of attention and your energy.

PEST CONTROL APPROACHES

Differing approaches to pest control are equally notable. In chemical horticulture, a specific insecticide may be applied to quickly kill off a particular insect pest. Chemical controls can dramatically reduce pest populations in the short term, yet by unavoidably killing natural control insects and animals, cause an increase in the pest population in the long term, thereby creating an ever increasing problem. Repeated use of insecticides and herbicides also encourages rapid natural selection of resistant insects, plants and other organisms, necessitating increased use, or requiring new, more powerful controls.In contrast, organic horticulture tends to tolerate some pest populations while taking the long view.

Organic pest control requires a thorough understanding of pest life cycles and interactions, and involves the cumulative effect of many techniques, including:

- Allowing for an acceptable level of pest damage
- Encouraging predatory beneficial insects to flourish and eat pests
- Encouraging beneficial microorganisms
- Careful plant selection, choosing disease-resistant varieties
- Planting companion crops that discourage or divert pests
- Using row covers to protect crop plants during pest migration periods
- Rotating crops to different locations from year to year to interrupt pest reproduction cycles
- Using insect traps to monitor and control insect populations

Each of these techniques also provides other benefits, such as soil protection and improvement, fertilization, pollination, water conservation and season extension. These benefits are both complementary and cumulative in overall effect on site health. Organic pest control and biological pest control can be used as part of integrated pest management. However, IPM can include the use of chemical pesticides that are not part of organic or biological techniques.

HORTICULTURE AND ANTHROPOLOGY

The origins of horticulture lie in the transition of human communities from nomadic hunter gatherers to sedentary or semi-sedentary horticultural communities, cultivating a variety of crops on a small scale around their dwellings or in specialized plots at some remove (such as the "milpa" or maize field of mesoamerican cultures). In forest areas such horticulture is often carried out in swiddens ("slash and burn" areas). A characteristic of horticultural communities is that useful trees are often to be found planted around communities or specially retained from the natural ecosytem.

Horticultural communities may be distinguished from agricultural ones by:

(1) the small scale of the cultivation, using small plots of mixed crops rather than large field of single crops
(2) the use of a variety of crops, often including fruit trees
(3) the encouragement of useful native plants alongside direct cultivation
(4) continued use of other forms of livelihood.

In pre-contact North America the semi-sedentary horticultural communities of the eastern woodlands (growing maize, squash and sunflower) contrasted markedly with the mobile hunter gatherer communities of the Plains people.

In central America, Mayan horticulture involved augmentation of the forest with useful trees such as papaya, avocado, cacao, ceiba and sapodilla. In the cornfields, multiple crops were grown such as beans (using cornstalks as supports), squash, pumpkins and chili peppers, in some cultures tended mainly or exclusively by women.

PESTS AND DISEASES OF ROSES

Roses Rosa sp. are susceptible to a number of pests, diseases and disorders. A large number of the problems affecting roses are seasonal and climatic. Certain varieties of roses are naturally more resistant or immune than others to certain pests and diseases.

Cultivation requirements of individual rose species and cultivars, when observed, often assist in the prevention of certain pests, diseases and disorders.

Pests

Pests are often considered to be the insects that affect roses:

- *Aphids (Greenfly) (Order Hemiptera: Family Aphididae) Macrosiphum Rosae:* Likely to be found on new shoots and buds, aphids are soft bodied insects 1-2mm long. Often green but occasionally light-brown, and sometimes with wings, they may cover (in a colony) the complete growing tip of the plant. Aphids are most active in spring and summer and multiply at a prodigeous rate feeding on the sap of the plant by piercing the plant cells via a proboscis. In large quantities they may seriously retard the growth of the plant and ruin buds. They are particularly damaging to the new shoots with subsequent damage to the emerging leaves which become malformed with much the same appearance as leaf-curl in peaches.
- *Two-spotted Mite (Spider-mites or Red spider mite) (Order Acari: Family Tetranychidae) Tetranychus Urticae:* Previously known as red-spider mite these arachnids prefer the underside of leaves and are difficult to see with an unaided eye. Evidence of their presence is silvering of leaves where the mites have destroyed individual leaf cells. Fine webbing and eggs on the undersides of leaves is further evidence of the presence of Tetranychus urticae.
- *Thrips (Order Thysanoptera):* Thrips are slim-winged insects 1mm in length, resembling fine black slivers of wood. Preferring light-coloured blooms and often appearing in plague numbers flowers are often left looking bruised and lustreless.

- *Caterpillars (Order Lepidoptera) See also List of Lepidoptera which Feed on Roses:* The tortryx (tortrix) moth Lozotaenia forsterana is a prominent pest of roses, although not the sole pest. The caterpillars are green, up to 15mm long, and can be found boring into buds or within curled leaves. When disturbed the caterpillars move swiftly, dropping to the ground on a fine thread. Damage is chewn leaves and flowers and buds with "shot holes".

Cottony Cushion Scale (Order Hemiptera: Family Coccoidea) Icerya purchasi

This scale infests twigs and branches. The mature female is oval in shape, reddish-brown with black hairs, 5 mm long. When mature the insect remains stationary and produces an egg sac in grooves, by extrusion, in the body which encases hundreds of red eggs. The insect causes little damage but produces copious honeydew (frass) that can cause damaging sooty mould.

California Red Scale (Order Hemiptera: Family Coccoidea) Aonidiella Aurantii

A hard scale, orange to orange-pink, the female covering being less than 1.5mm across. Often in plague numbers this scale infests upper surfaces of foliage causing yellowing, leaf fall, and twig and branch dieback. Serious infestations can cause plant death.

Rose Scale (Order Hemiptera: Family Coccoidea) Aulacaspis rosae

Mainly found on the stems and branches of the plant, lack of control will allow the pest to spread to flower stalks and petioles. At this point the plant would be stunted, spindly and with a white, flaky crust of scales on the bark. Female Aulacaspis rosaemay live for 1 year and may lay 80 eggs each with several overlapping generations living within milliimetres of the original parent.

- *Leaf cutting bee (Order Hymenoptera: Family Megachilidae) Megachile spp.*

Leafcutter bees are 6-16mm long and mostly black with bands of light-coloured hair. They chew pieces from the edges of leaves. The pieces are regular in shape, circular or oval. Damage is not often significant.

- *Nematodes (Eelworms)(Order Tylenchida: Family Heteroderidae)*

Root-knot nematode Meloidogyne spp.

See - Root-knot nematode - symptoms of Meloidogyne infestation in roses is stunting, slow-growth, pale green leaves and wilting in mild weather.

- *Metallic flea-beetles (Order Coleoptera: Family Chrysomelidae) Altica spp.*

The small, shiny and metallic Altica beetles have thickened hindlegs adapted to jumping, similar to fleas. The insects are 3mm long and chew holes of irregular shapes in young leaves and buds. As the leaves enlarge so do the holes.

Diseases

Fungal, bacterial and viral diseases affect roses:

Fungal Disease

- *Black Spot (Class Leotiomycetes: Family Helotiales) Marssonina rosae syn. Diplocarpon rosae:* Marssonina rosae causes black spots on leaves. The spots, which may be as much as 12mm across, are generally circular and have an irregular edge often with a yellow halo. Leaves frequently turn yellow and fall early. Sometimes new leaves are produced, and these may also become affected.
  Continual defoliation will cause weakness, die-back or death of the plant. Some very susceptible species may have stems affected with a considerable reduction in plant vigour.
- *Powdery Mildew Oidium sp.:* Oidium produces a very fine, powdery coating on the surface of buds and leaves. Significant cases have stems and particularly thorns, infected. Attacks on young leaves and buds will cause deformity with retardation of growth. Infected buds will fail to open. The disease is likely in hot, humid weather, with fungal spores overwintering on the stems and fallen leaves.
- *Downy mildew (Class Oomycetes: Family Peronosporaceae) Peronospora sparsa:* Peronospora causes purple-red to dark-brown spots on the leaves with irregular margins, however, often angular. Stems, petioles and flower stalks can split and spotted with purple marks. Buds, sepals, petals and calyces can be affected and will present purple spots. New growth affected will be deformed. The disease is spread by wind.
- *Rust Phragmidium mucronatum:* Rose rust appears as yellow patches on the surface of leaves, with orange pustules of spores underneath the leaf. The fungus is spread by wind. Affected leaves fall prior to healthy ones and plants may be defoliated in serious infections.
- *Anthracnose Sphaceloma rosarum:* Spots caused by this fungus originate from a point where leaves are watersoaked, ususally unnoticeable at first, until they turn black with a very distinct defined edge. As the spots enlarge the centre becomes gray and may fall out resulting in a shot-hole appearance. Defoliation may occur but is often not serious.
- *Grey mould (Class Leotiomycetes: Family Sclerotiniceae) Botrytis cineria:* On roses grey mould is prmiarily a disease of the flowers and buds, leaves are infrequently attacked. Infected buds rot on the stem and infection may progress down the stem. On petals botrytis cineria produces pink rings.
- *Verticillium wilt (Class Incertae sedis: Family Verticillium) Verticillium dahliae*

- *Sooty moulds Alternaria spp.:* Sooty mould appears as black, dry powder on leaves similar to chimney soot. Many sooty moulds grow on the honeydew (frass) produced by sap-sucking insect such as aphids and soft scales. Alternaria does no direct damage to plants but surface cover of leaves will reduce the plants capacity to photosynthesise and may create an unsatisfactory plant appearance.
- *Canker Leptosphaeria coniothyrium and Cryptosporella umbrina:* Cankers present as small yellowish or reddish spots on bark slowly increasing in size. Leptosphaeria coniothyrium turns brown, increases in size, and may eventually girdle the stem. The tissue within the infection begins to dry out and shrink, presenting a shrivelled appearance. If the disease infects only part of the stem, growth above the canker will continue. If it girdles the stem, however, growth will cease and the stem will die.

Viral Disease

- *Rose Mosaic:* This diesase is caused by a complex or viruses and is charaterised by yello patterns on the leaves. The patterns vary considerably, ranging between all-over fine blotches to patterns of lines in waves. The patterns may appear on a few or many leaves. Plants are infected by this virus at propagation using infected plant material.
- *Rose Wilt:* Rose wilt is a complex of viruses and is referred to as dieback in some areas. The disease can be spread by vectors such as aphids. Symptoms are variable and range from stunted growth to curled young leaves. The soft tissue symptoms are more evident in spring and new leaves will reflex towards their own petioles. The affected leaves are brittle and easily fall from the plant. Fully formed leaves will 'wilt' as if the plant were water stressed.

Bacterial Disease

- *Crown gall rot (Class Alpha Proteobacteria: Family Rhizobiaceae) Agrobacterium rhizogenes:* This disease is characterised by large lumps at the base of the plant stem or on roots. Galls may appear higher on stems as the disease progresses. Galls are soft compared to surrounding plant tissues. The pathogenic bacteria enter the plant via a wound. If the disease affects the plant whilst it is young the plant may be affected to the degree where it will not produce blooms. All affected plants wilt readily and grow poorly.

Environmental Disorders

- *Frost:* Frost will destroy fresh growth causing stems and leaves to wilt, turn black and fall away from the plant. Timing pruning to promote growth after the threat of frost is a means to avoid frost damage.

- *Salinity:* Salinity will present in roses as limp and light brown leaves with dry leaf margins. Soil may require testing to determine salinity levels. Symptoms will present if salinity is greater than 1200 parts per million.
- *Herbicide Damage:* Overspray or soil leaching of herbicidal sprays can present with several symptoms: Prolonged exposure to overspray of glyphosatewill cause yellow leaves and new leaves will be small and elongated.

Hormone weedsprays (*e.g.* 24-D and 245-T) may cause grotesque new growth with thin twisted leaves and distorted buds. Plants may die in severe cases. Pre-emergent herbicides contacting the plants' rootsystem via the soil will cause yellowing foliage. Effects of soil borne herbicide may take several years to clear.

APPLICATIONS IN HORTICULTURE AND FORESTRY

MICROPROPAGATION

In recent years, the application of micropropagation techniques as an alternative mean of asexual propagation of important plants has increased the interest of workers in various field. The micropropagation techniques are preferred over the conventional asexual propagation methods because of the following reasons: In this method only a small amount of tissue is needed as the initial explant for regeneration of millions of clonal plants in a year, *(b)* this method provides a possible alternative method for developing resistance in many species; it provides a mean for international exchange of plant materials, hence the problem for introduction of disease can be solved in quarantine; *in vitro* stock can be quickly proliferated as it is not season dependent, and valuable germplasm can be stored for a long time (Hu and Want, 1983; Mascarenhas and Muralidharan, 1989).

Regeneration of plantlets in cultured plant cell and tissues has been achieved in many trees of high economic value. Many of the studies are aimed at large scale micropropagation of important trees yielding fuel, pulp, timber, oils or fruits. Therefore, clonal forestry and horticultureare gaining an increasing recognition as an alternative for tree improvement. However, strategies for transferring cultured plants from*in vitro* to field conditions are based on relatively higher priced horticultural species rather than agricultural and forestry species (Fossard, 1987). Regeneration of plantlets in cultured tissue has been described (Murashige, 1974) to be accomplished into 3 stages *(see* shoot culture)'. Fossard (1987) has given a detailed account of stages of micropropagation

At planting out stage the plantlets fail to survive because of sudden change in the environment and invasion by soil microbes. Jagannathan (1987) has described that the regenerates should be transferred first to green house and then to field. The humidity should be controlled by covering the plants with transparent polyethylene sheets. This acclimatization requires several weeks whichshould be followed by potting into sterile peat or soil

In recent years, the interest has, aroused in commercializing the *in vitro* propagation of forest trees. This will bring about refinement in the existing procedures to make micropropagation more cost effective. Mascarenhas, Muralidharan and coworkers are making efforts to commercialize this biotechnology with respect to forest trees. However, development of automated procedure, plant delivery systems using somatic embryos and artificial seeds are also in progress. For betterment and improvement of tree plants of high economic value a break through in forestry research has come with production of artificial seeds in *Eucalyptus* (Muralidharan and Mascarenhas, 1989), and genetic transformation and *in vitro* regeneration in conifers (Gupta, 1989). Moreover, micropropagation has been successfully done in many trees (Gupta *et al.* 1980, 1981; Jaiswal and Pratap Narayan, 1985; Amin and Jaiswal, 1988; Mascarenhas and Muralidharan, 1989).

Mascarenhas and Muralidharan (1989) reviewed the tissue culture studies carried out on forest trees in India. Some of the important plants are: Acacia nilotica, Albizia lebbeck, A. procera, Azadirachta indica, Bauhinia purpurea, Butea monosperma, Dalbergia sp., Dendrocalmusstrictus, Eucalyptus sp. Ficus religiosa, Morus sp., Populus sp., Shorea robusta, Tectona grandis (all angiosperms), Biota oriental's, Cedrus deodara, Cryptomena japonica, Picea smithiana, Pinus sp. (all gymnosperms)

In Vitro Establishment of Mycorrhiza

Mycorrhizal fungi show highest level of specialization of parasitism. But the major problems with them is their failure to grow on an artificial medium in laboratory. Therefore, establishment and multiplication of mycorrhizal fungi on cultured tissue of the same host plant, if successfully developed, may be a good tool for handling mycorrhizal fungi, production of high potential inoculum and their establishment in root systems of nursery plants in horticulture and forestry, and plantation of mycorrhiza-infested seedlings into field. Only one report is available on this work. Kiernan *et al.* (1984) successfully produced strawberry plants by tissue culture which was infected by a mycorrhizal fungus, *Glomus* sp.

Many attempts have been made to establish Vesicular Arbuscular Mycorrhizal (VAM) fungi in axenic culture but unfortunately none of them got success. It was assumed that self inhibition of hyphal growth occurs in the growing germ tubes and the self inhibition compounds were recovered by adding activated charcoal into an agar medium that absorbs inhibitory compounds produced by germ tubes into medium. Cultures of mycorrhizas synthesized aseptically are grouped into two: the whole plant cultures and excised root cultures. Both the types of cultures are known as genetobiotic or monogenic systems (Rhodes, 1983). Due to presence of two organisms, it is also known as two member culture.

Mosse (1962) for the first time, reported the establishment of two member cultures. Appressorium formation and root penetration were much more likely to occur if a *Pseudomonas* sp. was present in culture. It is, therefore, suggested that 3 organisms *i.e.* fungus-plant-bacterium might be necessary for the development of symbiosis. Moreover, mycorrhizal fungi have been cultured only on cortical tissues of roots which were separated from the whole plant, as in root organ cultures where it acted as food base. VAM fungi have very high degree of specialization for food base on root cortex (Rhodes, 1983).

NURSERY (HORTICULTURE)

A nursery is a place where plants are propagated and grown to usable size. There are retail nurseries which sell to the general public, wholesale nurseries which sell only to other nurseries and to commercial landscape gardeners, and private nurseries which supply the needs of institutions or private estates. Some retail and wholesale nurseries sell by mail.

Nurseries grow annuals, perennials, and woody plants (trees and shrubs). These have a variety of uses: decorative plants for flower gardening and landscaping, garden vegetable plants, and agricultural plants. Nurseries often grow plants in a greenhouse, a building of glass or in plastic tunnels, designed to protect young plants from harsh weather (especially frost), while allowing access to light and ventilation. Modern greenhouses allow automated control of temperature, ventilation and light and semi-automated watering and feeding. Some also have fold-back roofs to allow "hardening-off" of plants without the need for manual transfer to outdoor beds.

Some nurseries specialize in one phase of the process: propagation, growing out, or retail sale; or in one type of plant: groundcovers, shade plants, fruit trees, or rock garden plants. Nurseries remain highly labour-intensive. Although some processes have been mechanised and automated, others have not. It remains highly unlikely that all plants treated in the same way at the same time will arrive at the same condition together, so plant care requires observation, judgement and manual dexterity; selection for sale requires comparison and judgement. A UK nurseryman has estimated that manpower accounts for 70 poe cent of his production costs.

Business is highly seasonal, concentrated in spring and autumn. There is no guarantee that there will be demand for the product - this will be affected by temperature, drought, cheaper foreign competition, fashion, etc. A nursery carries these risks and fluctuations. Annuals are sold in trays (undivided containers with multiple plants), flats (trays with built-in cells), peat pots, or plastic pots. Perennials and woody plants are sold either in pots, bare-root or balled and burlaped and in a variety of sizes, from liners to mature trees. Plants may be propagated by seeds, but often desirable cultivars are propagated asexually by budding, grafting, layering, or other nursery techniques.

CLASSIFICATION OF PLANTS

Plants are classified in several different ways, and the further away from the garden we get, the more the name indicates a plant's relationship to other plants, and tells us about its place in the plant world rather than in the garden. Usually, only the Family, Genus and species are of concern to the gardener, but we sometimes include subspecies, variety or cultivar to identify a particular plant.

Starting from the top, the highest category, plants are classified as follows. Each group has the characteristics of the level above it, but has some distinguishing features. The further down the scale you go, the more minor the differences become, until you end up with a classification which applies to only one plant.

EMERGING MODEL SYSTEMS IN PLANT BIOLOGY

The economic and ecological importance of forest trees provide the impetus for developing model systems to study tree biology. Wood is one of the most valuable commodities in the industrialized world, exceeding the weight of all other structural materials combined. More than half of the world's annual wood harvest is used as fuel, primarily in less-developed countries. In nature, forest trees are the principal form of terrestrial biomass.

Wild forests provide irreplaceable environmental benefits such as carbon sequestration, watershed protection, and habitat for endangered wildlife. Trees themselves are intrinsically interesting, since they represent the pinnacle of evolutionary refinement in the competition among plants for access to sunlight, thereby dominating many of the Earth's most awe-inspiring landscapes.

While many aspects of tree biology are common to all plants, and hence can be studied in very tractable model species such as *Arabidopsis*, some unique facets of tree anatomy and physiology must be investigated in trees themselves. A large number of these unique characters are related to their perennial growth habit.

Extensive formation of secondary xylem (wood) is perhaps the most obvious, but other characters include leaf and flower phenology, seasonal reallocation of nutrients, cold hardiness, iterative development of a complex crown form, and juvenile-mature phase change. The great lifespan and large size of forest trees pose special challenges to biologists, but are fundamental to understanding the role of trees in natural and managed ecosystems.

The experimental power of genetics, including molecular biology and genomics, has now permeated every discipline in biology. Clearly, an ideal model forest tree must have the potential to be manipulated using standard genetic approaches, such as mutagenesis and the creation of transgenic plants. In spite of the fact that other forest trees have been put forward as model systems, including *Salix* (willow), *Eucalyptus*, and *Pinus* (pine), for the reasons discussed below *Populus* has led the way in making use of genetic methods to study tree structure and function.

The knowledge gained from model forest tree systems will be put to use in the protection of forest ecosystems and in practical tree breeding. By defining, quantifying, and understanding tree physiological processes and morphological structures we hope to mathematically model, predict, and get a conceptual grasp of tree growth and development. In contrast to agronomic crops, relatively few physiologically based growth models have been developed for trees, largely due to a lack of fundamental physiological information and the difficulty of working with complex perennial plants (Ford 1985; Amaro and Tomé 1999). Just as in forest genetics, poplars have been adopted by tree physiologists as a model system.

The interplay between poplar genetics and physiology, at scales as large as the landscape and as small as a molecule, is the subject we now explore in greater depth. Ultimately, we wish to trace tree physiology and anatomy to the level of individual genes, understanding and manipulating their interactions with each other and the environment.

POPULUS TAXONOMY, DISTRIBUTION, AND GENERAL CHARACTERISTICS

Poplars are dioecious, wind-pollinated, and produce large amounts of pollen and small, cotton-tufted seed that is dispersed by wind and water in early summer.

Capable of rapidly invading disturbed sites, many species occupy habitats in the dynamic environment of riverine floodplains where they form a key component of riparian forests (Braatne *et al.* 1996). Others, such as the aspens, commonly colonize upland areas after intense, stand-replacing fires (Burns and Honkala 1990).

All poplars also have the capacity to reproduce asexually, mostly by sprouting from the root collar of killed trees, or from abscised or broken branches that become embedded in the soil. Aspen and white poplars propagate through sucker shoots that arise from horizontal roots, often after a fire, typically leading to clonal stands that may cover several hectares. Recurrent fires can maintain genets of such clones for hundreds of years.

Strengths of Populus as a Model System

Well established collaboration among poplar biologists. Extensive collaboration between poplar geneticists, physiologists, and pathologists has set a solid scientific foundation for joint efforts in the future. Shared genetic materials, genetically informative two- and three-generation pedigrees, DNA-based genetic markers, common field measurement protocols, clonal plantation trials supported by industry/government/academic partnerships, and funding from multi-agency grants have established a poplar research network of proven productivity. Working groups under the aegis of the International Poplar Commission (IPC) of the Food and Agriculture Organization (FAO) of the United Nations, and of the International Union of Forestry Research Organizations (IUFRO), as well as several university/industrial research cooperatives, help to coordinate the work and the exchange of information. Two books synthesizing a large body of data on poplar biology have been published recently, reviewing and solidifying the status of poplar as a model forest tree (Klopfenstein *et al.* 1997; Stettler *et al.* 1996).

Abundant genetic variation in natural populations. Adaptations to the diverse conditions inherent in large distribution ranges, as well as prominent genetic polymorphisms in local populations, offer poplar researchers a rich source of variation in tree morphology, anatomy, physiology, phenology, and response to biotic and abiotic stress. Much of this variation is under moderate to strong genetic control (Farmer 1996). Extensive natural populations remain for many species, especially in North America.

Ease of sexual propagation and interspecific hybridization. Poplars are bred in the greenhouse on detached female branches with pollen that can be stored for several years. Each pollination can yield hundreds of seeds within 4 to 8 weeks. Seeds germinate within 24 hours and give rise to 1-2 m tall seedlings by the end of the same year. Few if any trees can match such efficiency. Poplar species within the same section, and many of the species from different sections, can be hybridized. Because all members of the genus are diploid (2n = 38), hybrids are fertile and can generate F2 and backcross progenies that segregate for a wide range of traits. F1 hybrids often show heterosis in growth and associated characteristics which make them attractive for commercial use (Zsuffa *et al.* 1996).

Physiological responses to environmental variables are rapid and pronounced. As a consequence of the rapid juvenile growth of poplars, it is possible to measure short-term responses to biotic (*e.g.*, disease) and abiotic (*e.g.*, drought, elevated CO_2, ozone) factors. Opportunities are being explored for using this rapid growth as a means of carbon sequestration, particularly in Europe. The short turnaround time for physiological studies makes it possible to use molecular genetic approaches requiring very large sample sizes (hundreds or thousands of physiological measurements). High-throughput poplar physiology has led to the genetic mapping of quantitative trait loci (QTLs) controlling morphological (Bradshaw and Stettler 1995), phenological (Frewen *et al.* 2000), and pathological (Cervera *et al.* 1996; Newcombe and Bradshaw 1996; Newcombe *et al.* 1996; Villar *et al.* 1996) traits. Candidate genes for some of these QTLs have been identified (Frewen *et al.* 2000). Close coupling of physiological traits and biomass productivity. Critical components of productivity in *Populus* have been examined from a physiological perspective (Ceulemans 1990).

Both structural and functional components were identified at different organizational levels: the individual leaf, the branch, and the whole tree. Among the process-related components, stomatal morphology and behavior (Tschaplinski *et al.* 1994), leaf morphology and leaf growth physiology (Ridge *et al.* 1986), leaf and whole tree photosynthesis (Isebrands *et al.* 1988), and root development (Friend *et al.* 1991) showed the most significant genetic variation. Key anatomical characters include those that determine whole tree leaf area and its duration, such as leaf demography (Chen *et al.* 1994; Hinckley *et al.* 1989), leaf size and distribution (Dunlap *et al.* 1992), branch size and distribution (Ceulemans *et al.* 1990), and leaf and branch orientation (Isebrands and Michael 1986). A favorable combination of several leaf physiological as well as whole tree and canopy structural traits explains the superior growth of selected hybrid poplar clones. QTLs for many of these traits have been mapped genetically (Bradshaw and Stettler 1995; Wu *et al.* 1997), but the genes controlling these traits remain to be identified and characterized.

Well-characterized molecular physiology. The large physical size of poplars, coupled with the well-understood movement of carbon assimilates (Vogelmann *et al.* 1982) and co-transported systemic wound signals between source and sink leaves (Davis *et al.* 1991; Clarke *et al.* 1998; Constabel and Ryan 1998) is an advantage over small plant models such as *Arabidopsis* in studies involving systemic signal movement. Induced resistance to insect herbivory is acquired systemically, and is phenocopied in part by mechanical injury (Havill and Raffa 1999).

This positions poplar as an excellent model for synthesizing ecological and molecular perspectives of induced resistance to insect herbivory. Dramatic patterns of nitrogen allocation, utilization, and storage. Riparian ecosystems receive nutrient inputs episodically, which may help explain why poplars tissues store high levels of nitrogen in the form of vegetative storage proteins for subsequent utilization (Coleman *et al.* 1994; Lawrence *et al.* 1997). We have observed healthy poplar leaves with nitrogen concentrations exceeding 8 poe cent of dry weight (J. Cooke, K. Brown, J. Davis, unpublished), *vs.* the 1-1.5 poe cent maximum levels found even in fertilized conifer needles. Nitrogen levels fluctuate dynamically among organs within the same tree (roots, stem, and leaves) in concordance with seasonal rhythms of active growth and dormancy. The stem anatomy of poplar trees is particularly suited to studying the role of phloem-transmissible substances such as glutamine in regulating nitrogen allocation, as phloem can be specifically perturbed by girdling while xylem transport remains intact (*e.g.*, Sauter and Neumann 1994).

Physiological process models for poplar growth and development. Large databases of anatomical, physiological, and silvicultural traits are available for a modest number of *Populus* hybrids and clones. Several physiologically-based growth and productivity models have been developed from these data. The models include basic information on carbon uptake and allocation in poplar, as well as components and parameters of leaf display and crown structure. Most of the models simulate carbon uptake, carbon allocation, growth, and/or light interception in poplar and incorporate some specific parameters of leaf display, position in the tree, and branch structure (Chen *et al.* 1994; Host *et al.* 1996; Isebrands *et al.* 1996). Data on the physiological and structural growth determinants at the leaf, branch, and whole tree level indicate that differences in clonal productivity can be incorporated into the ideotype concept developed for poplar tree breeding under short rotation intensive culture (Dickmann 1985; Dickmann and Keathley 1996).

Cloning of individual tree genotypes. The ease with which most materials can be vegetatively propagated is one of poplar's premier assets. Cloning captures genetic variation and allows it to be replicated in space and time in separate experiments. Cloning 'freezes' genetic variation in hybrids and permits the side-by-side growth of multiple generations of a pedigree. Cloning permits the growth of abnormal plants under field conditions that in the competitive environment of a seedling population would be impossible. Cloning also allows destructive sampling for physiological studies, the sharing of materials among laboratories, and the buildup of cumulative knowledge on selected genotypes.

Closely related to other angiosperm model plants. Unlike the pines and other gymnosperms, poplars are relatively recently diverged from other angiosperms, such as *Arabidopsis*, which serve as models for integrating genetics into the study of plant biology (Fig. 1).

Small genome size. The haploid genome size of *Populus* is 550 million base pairs (bp) (Bradshaw and Stettler 1993), only 4 times larger than the genome of the model plant *Arabidopsis*, and 40 times smaller than the genomes of conifers such as loblolly pine. The small poplar genome simplifies gene cloning, Southern blotting, and other standard molecular genetics techniques. The physical:genetic distance ratio for poplar is close to 200 kb/centiMorgan, almost identical to *Arabidopsis*, making *Populus* an attractive target for map-based (positional) cloning of genes.

Basic molecular genetics toolkit. Linkage maps of the poplar genome have been made with a variety of marker types, including allozymes, RFLPs, STS/CAPs, RAPDs, AFLPs, and microsatellites (Liu and Furnier 1993; Bradshaw *et al.* 1994; Cervera *et al.* 1996; Frewen *et al.* 2000). Bacterial artificial chromosome (BAC) genomic libraries containing 10 haploid poplar genome equivalents (50,000 clones) have been constructed for *P. deltoides* (W. Boerjan, pers. comm.) and for *P. trichocarpa* (J. Vrebalov, pers. comm.).

These libraries have been arrayed into microtiter plates and replicated for sharing with other researchers. Expressed sequence tag (EST) databases of modest size are in the public domain (Sterky *et al.* 1998). Facile transformation and regeneration to create transgenic poplars. The ease of creating transgenic *Populus* is unmatched by any other forest tree. Some poplars, such as the hybrid aspen (*P. alba* x *tremula*) clone 717-1B4, can be genetically transformed with *Agrobacterium* and regenerated efficiently into intact transgenic trees within 6-10 months (Jouanin *et al.* 1993). Transgenesis is the 'gold standard' for demonstration of gene function, and so is crucial for basic research into physiological processes.

Obviously, there are also potential commercial applications for transgenic poplars. The ease with which genes are cloned and transgenic trees are produced in *Populus*, set against the backdrop of a poplar EST database of wood-forming genes (Sterky *et al.* 1998), has created a fertile scientific field for understanding and manipulating wood composition in dramatic ways (Hu *et al.* 1999). Progress is likely to accelerate in this important research area as the tools of genomics, biochemistry, and cell biology are collectively brought to bear on manipulating wood quality and other important physiological or morphological traits.

Weaknesses of Populus as a Model System

Modest commercial importance. Poplar is grown commercially on a wide geographic scale and many wild populations, especially aspen in the boreal forest, are harvested by industry. However, from a corporate viewpoint, poplars are much less important than pines and other softwoods, or eucalypts. It can be difficult to persuade companies to invest in (or lobby for) poplar research, either as a model tree, or as a component of fibre/energy plantations. For a variety of historical, political, and economic reasons, most of the forest products industry still relies on harvests from natural or extensively managed stands and has yet to turn to the intensive agricultural paradigm for wood production for which poplar is ideally suited.

Long generation interval. Many willows and some eucalypts can be induced to flower within a year or two from seed. Most poplars will not flower earlier than 4 years of age, and many will take twice that long. The long generation interval is an impediment to practical breeding and selection and the development of informative pedigrees. It limits the applicability of conventional experimental genetic techniques such as induced mutagenesis, since producing homozygotes from heterozygous mutants is impracticably slow. Outcrossing mating system. Poplars are dioecious, so self-pollinations cannot be done (with a very few hermaphroditic exceptions). Classical genetic tools such as inbred lines cannot be produced rapidly enough to be useful. Large physical size of trees. There is no such thing as a 'small experiment' when one works with poplars! Beyond one year of age the physiologically active crown is above human reach, requiring towers or remote sensing. Lack of well-planned breeding programmes.

The advantages of cloning have been pointed out, but there is a hidden disadvantage. Most poplar breeding programmes have relied upon more-or-less random interspecific hybridization to produce large progeny arrays, followed by very intense clonal selection to identify the best clones out of perhaps 10,000 seedlings. While demonstrably effective at producing remarkable genetic gains in short order, the scarcity of well-planned, long-term, sustained breeding programmes and associated breeding materials hampers genetic studies. For example, multi-generation pedigrees are the basis for genetic mapping experiments designed to identify QTLs affecting important tree phenotypes, yet few such pedigrees exist.

What is needed to increase the value of populus as a model system?

Long-term support for germplasm collection and maintenance. Nothing is more destructive to sustained progress on long-lived plants than inconsistently funding basic genetic infrastructure such as breeding, maintenance of living clone banks, and annual measurement of experimental plantations. Unfortunately, these routine but important activities are difficult to justify to most funding agencies and usually are jettisoned at the first dip in finances. A solution to this problem might be to establish at least one public genetic stock center on each continent with an active interest in poplar research. This would promote the centuries-old practice of exchanging pollen, seed, and cuttings from genetic stocks with value for research and commercial use. The poplar research community's tradition of accumulating and integrating knowledge about shared pedigrees would be encouraged.

Shorter generation interval; early or inducible flowering. Delayed flowering represents a lost opportunity for tree geneticists to provide suitable pedigrees for physiological studies, and is currently the single most limiting factor in the conventional genetic improvement of forest trees. The development of methods for inducible early flowering, either by chemical/physical treatment or genetic engineering, will make it possible to decrease the generation interval for purposes of breeding, while also preventing excessive flowering from siphoning photosynthate away from wood formation in production plantations. While some progress has been made in this area with poplars (Weigel and Nilsson 1995), much remains to be done (Rottman *et al.* 2000). Complete genomics toolkit. Probably no other group of organisms will benefit more from the development of genomics tools than forest trees, where conventional genetic approaches are so limited. Since most of the interesting traits in trees are inherited quantitatively, a dense genetic map composed of 2500-5000 genetic markers which identify homologous loci in all *Populus* species and have a high information content, such as microsatellites or single-nucleotide polymorphisms (SNPs), should be produced specifically for QTL mapping.

The genome of *Populus* is small enough to invite the construction of a complete physical map of overlapping BACs. The physical map could be tied to the genetic map either by mapping genetic markers onto the physical map, or by developing genetic markers directly from BAC sequences having known physical positions. The physical and genetic maps would serve several important purposes: as a starting point for map-based gene cloning, as a means to determine the extent of collinearity between the *Populus* genome and that of the completely sequenced genome of *Arabidopsis* (thus making effective use of the growing knowledge of *Arabidopsis* genome structure and function), and as a scaffold for the eventual sequencing of the entire *Populus* genome.

Until the *Populus* genome is completely sequenced, perhaps in the next 5-10 years, a comprehensive EST database will provide a source of SNPs, and become the foundation for microarray development and use for studying global patterns of gene expression. EST microarrays will allow poplar researchers to couple physiological measurements with gene expression profiles, illuminating the genes, biochemical pathways, and cellular processes that are affected by a given environmental perturbation or transgene.

Steve Strauss (pers. comm.) has suggested an ambitious strategy of discovering genes with novel functions in poplar by generating 50,000 dominant gain-of-function mutants in transgenic poplars by activation tagging (Walden *et al.* 1994), which is an insertional mutagenesis strategy that is more suitable for trees with their outcrossing mating systems than traditional mutagenesis approaches that yield mostly recessive mutants. The incorporation of such tools will be important if we are to carry out forward and reverse genetics with equal facility in forest trees.

IMPORTANT FEATURES OF FLOWERING PLANTS

Flowering plants (Angiospermae) represent one of the largest groups of primary producers. Their contribution to the production of oxygen as well as that to the nutriment of animals and man is consequently very large. All features reviewed in this chapter refer to seed-producing plants, also called spermatophytes.

Typically, flowering plants are organized into an underground root and a shoot above ground that consists itself of a stem and leaves. The organs of a plant that serve sexual reproduction are the flowers. Part of the pollinated flower ripens and becomes the fruit.

In contrast to many other plant groups, flowering plants are striking, numerous and common. They are the most important group of the so-called primary producers that generate the prerequisite for life on earth: oxygen. Green plants have the ability to convert solar energy into chemical energy (photosynthesis) producing the oxygen necessary for all other organisms as a by-product. The useable plants among the flowering plants are - directly or indirectly - the basis of human existence; they are, too, an important economical factor. A basic knowledge of flowering plants should therefore be among everybody's general knowledge.

Much has been written about flowering plants and every reader of this chapter will miss something that he regards worth knowing, while he might find other information trivial. But everybody will understand that it is impossible to review in a few lines a theme about which an extensive, partly popular scientific literature exists. And although this term may sometimes be used in a disparaging way, most of the popular scientific literature is scientifically correct, lucid and, above all, very well illustrated.

To get an idea of the variety of existing plants and to get to know special species, it is necessary to identify them. Many books on classification with different approaches exist. In many popular books, colour photos or drawings are used and often is the colour of the flower a primary feature of recognition. Most of the so-called scientific books on classification work on dichotomic keys, *i.e.* the user is asked a lot of questions in succession and has at each one to decide between two answers. This procedure is continued until the plant has been identified.

The "scientific nature" is mostly based on the completeness of the varieties present in a book, since almost all of the illustrated books contain only the most common or most striking plants. Recently, electronic tools have been used to figure out identification keys for plants.

The research into the flora of Central Europe has a tradition that goes back for centuries. It is mirrored in the nearly complete modern floras and books on classification. Incomplete, if existing at all, are books on the classification of less well discovered regions, like tropics, subtropics and many mountain areas.The question of the origin of the wealth of forms (evolution) is discussed elsewhere where it is also shown that mountains with their rather small and isolated areas provide ideal conditions for the coming into being of new species. This is the reason, why even very experienced botanists equipped with renowned books on classification of the Central European flora will sometimes and in some places fail (Alps).

This chapter will deal with the characteristics of flowering plants that produce seed (phanerogames or spermatophytes) only. Many of the structures present in this plant group can be found with other non-flowering plants, too. But mosses, ferns and algae miss some features, like flowers or seeds while others, like roots or leaves exist in an incomplete way or are replaced by other organs. The body of vegetation of many-celled algae (and mosses) is called thallus, that of flowering plants, ferns and fern-like plants (pteridophytes) is called cormus. The latter are therefore summed up as cormophytes. The special features of the different plant groups. The body of vegetation of a "typical" flowering plant consists of an underground root and a shoot above ground. The shoot is organized into stem and leaves. Each of these basic organs can exist in many variations and these again can be combined in many different ways. The almost unlimited ability of combination is one of the main reasons of the existence of such a high number of species while at the same time, the identification of the relations of the species is aggravated.

If seemingly different organs with different functions can be traced back to the same basic organ, they are called homologous. It is also spoken of homology. Contrasting is analogy where organs with a similar look and function have descended from different organs.

THE STEM

Typically, flowering plants are organized into an underground root and a shoot above ground that itself consists of a stem and leaves. The stem is the main supporting axis of a flowering plant. It is made up of nodes, the part where one or more leaves are borne and of the usually longer, in-between lying internodes. Stems are characterized by their type of branching, by their surface and their symmetry. Special cases of stems are those of climbing plants, succulence, runners and rhizomes.

Shoot and root can be clearly distinguished within a seed. After germination, they do develop into the shoot that grows towards the sun and into the root which grows towards the earth's score. The shoot that develops during the first stage of development is called the main- or the primary shoot and depending on the plant, it does bear one, two, or more cotyledons. Growing and sprouted shoots are organized into nodes and the in-between lying internodes.

The nodes are the parts where the leaves are inserted. At the tip of a growing shoot sits the terminal bud of the shoot apex that is surrounded by the leaf (or flower) primordias. In the area of the nodes are axillary buds where the axillary shoots insert. Vegetable shoots are either annual or perennial. The realised alternative depends usually on the species and in some species on the quality of the environment.

The shoot of annual species is mostly fleshy and is usually called a stalk while perennial plants lignify and thus form trunks and branches. Some species, like Graminaceae or Poaceae have a hollow stalk that is called a blade of grass.

To characterize a shoot, its type of branching and its surface (smooth, hairy, with or without grooves, round or squared) is used, as well as the ratio of heights to diameter.

It can generally be distinguished between monopodial and sympodial branching. Monopodial branching exists, if a continuous main shoot has feeble apical shoots. Sympodial branching is given, if apical shoots of different generations exist at the same time since the apical meristem of each sympodial branch has either died or is differentiated into a flower, a thorn or a tendril and has thus lost its ability to grow. The respective shoot tip is nearer to the ground than that of the next apical shoot. If only one of the apical shoots has the same direction as the main shoot would have had, then a monochasium results which may sometimes be difficult to distinguish from monopodial branching.

But if two buds of a sympodial branch sprout at the same time, a more or less fork-shaped pattern, a dichasium, is formed. Hardly ever do several buds put forth at the same time (pleiochasium).

An extreme ratio of heights to diameter is found in climbing plants where the diameter (and the stability) is very small compared to the length so that upright growing is only possible with the support of other plants or growing aids. Succulence is another extreme, that is typical for plants living in arid or saline habitats.

Most shoots are of a radial symmetry but with a few species, the apical shoots may be changed into flat, leaf-like phylloclades. Often, especially with wooden plants, the axes are differentiated into long and short shoots. The internodes of the long shoots that cause elongation with trees and shrubs are longer than the short shoots. These again are often specialized apical shoots with shortened internodes. The thorns of white-horn (Crataegus), for example, or those of sloe (Prunus spinosa) correspond to such short shoots.

Apical shoots may, too, be changed into runners or rhizomes. Runners are horizontally growing shoots with greatly prolonged internodes that are either underground or above ground and are sometimes bearing leaves. The internodes produce sometimes roots at the part that is nearest to the main shoot (shoot-borne roots). Some parts of these runners can switch to an upright growth and thus behave like a normal shoot. They serve to propagate vegetatively (strawberries are a typical example). Rhizomes are mostly underground and grow in a horizontal (plagiotrop) manner. They are rather enduring shoots whose internodes usually stay short. They serve as survival organs and are often at the same time used for storage; sometimes, they end in specially developed storage tubers like in potatoes. Except for small, scale-like leaves, rhizomes do usually have no leaves. Shoots which happen to come into touch with a moist soil surface may start to develop leaves and flowers.

ROOTS

Roots, the lower, underground part of a plant are characterized by their lack of leaves. They are the structures that function to take up water, nutriments and minerals from the soil. The primary root is the main axis of a root system.

Mostly, it bears thinner side-roots. Special types of primary roots are tap roots or turnips. Adventitious roots are roots originating directly from the stem. The most striking morphological feature of roots is their lack of leaves. Roots are organized into primary roots and side-roots. The latter may develop a frequently branched and extensive root system. Among the prominent types of roots are the tap roots where the primary root is dominant while the side-roots are hardly developed at all. The primary root grows deep into the soil. Swollen roots that have developed into storage organs are called turnips. Their formation may involve the hypocotyl. Swollen side-roots can also become storage organs (tubers).

The root-system of many monocots (like, that of the poaceae,for example) is completely formed from adventitious roots. Adventitious roots are roots that originate directly at the shoot. Rhizomes and shoots that touch the soil are able to form adventive roots. This fact is used to propagate plants vegetatively through cuttings (parts with nodes in them).

Epiphytes do often develop air roots. They are, just like usual roots able to take up minerals and rain water and sometimes they can even perform photosynthesis.

The picture shows a two-dimensional model of a root seeking water in the soil during its development. The initial water distribution has been predetermined, forming an S-shaped zone of high concentration indicated by the light colour. The growing tips of the main root and rootlets absorb water that diffuses in the soil. The decreased water concentration is indicated by dark areas that emerge around the root system. In areas with insufficient water concentration the rootlets cease to grow before they have reached their potential full length.

A leaf is an aerial and lateral outgrowth of the stem of a usually flat and dorsiventral anatomy. It functions mainly to manufacture food by photosynthesis and consists typically of a stalk also called petiole, a flattened blade, the lamina, and the leaf base. Strands of conducting and strengthening tissues, the veins, run through it. Their pattern, also called leaf venation, is a feature of characterization. Leaves may be simple, *i.e.* undivided or compound (composed of several parts called leaflets). The blade margin and the leaf arrangement at the stem are further features of characterization.

While potentially unlimited growth is a feature of the shoot, that of leaves is usually limited. They are almost always flat and have a dorsiventral anatomy, *i.e.* their upper and their undersurface are constructed differently and have different functions. This morphological and anatomical peculiarity is strongly connected to the main function of leaves, photosynthesis. With most conifers, the leaves are very thin and often tough (needles).

A typical leaf is organized into blade (lamina), stalk or petiole and leaf base. In some plant species, the leaf base is plainly set off against the petiole while the transition is invisible in others. Sometimes, it is merely indicated by a faint widening. Occasionally, lateral excrescences develop. Most species have clearly recognizable petioles.

If they are missing, the leaves are said to be sessile in which case the leaf base may sometimes surround the shoot. The extremes are laminas with fused basal rims evoking the impression of a leave that the shoot grows through (Claytonia perfoliata). The basis of the leaf stalk may even grow down on the shoot a little bit further. With many monocots, like Poaceae or Graminaceae, the shoot is completely surrounded by a sheath-forming petiole.

The blade is radiated by leaf veins that contain the fibres. It is distinguished between parallel, arched, pinnated or netlike vein structures.

Just as variable as the vein structure is the shape of the leaves. The fundamental difference is that between simple and compound leaves. The compound leaves are built from several small leaves or from pinnations that sit in a regular organisation at the undivided or branched rhachis.

Cataphylls are simply organized, scale-like leaves. They can be found at the sprout, at the basis of sprouting shoots or underneath the normal leaves and in wooden plants also as a protection for the buds.

Finally, bracts have to be mentioned that live above the region of the usual leaves, for example at the inflorescence as subtending bracts for the flowers. But they can also have very different functions. Remarkable are the strikingly colored bracts of Bougainvillea spectabilis and Poinsettia pulcherrima (poinsettia) that add strongly to the attractivity of the rather inconspicuous flowers. Another example are the Cuckoo-pint and its relatives where the bract encloses the inflorescence and is at the same time a refined trap for pollinating insects (Arum maculatum).

Some species, especially those that live in water have differently shaped leaves (heterophylly). An example is Ranunculus aquatilis (water crowfoot) which has pinnation-shaped, submerse living leaves and flat swimming leaves.

Another feature of recognition is the blade margin that can be shaped very differently. It can be plain, serrated, double serrated, crenated or sinuated. Leaves with deep clefts are called pinnatified, pectinated, palmated or lobed.

Quite a few species have structures called leaf metamorphosis (variations), because they are homologous to leaves. Among these are thorns where the moulding of the blade does not occur and the leaf veins have been strengthened by the deposit of a though material, as well as leaf tendrils that also miss the lamina but have an extremely pliable and tension-proof petiole. There may, too, be succulence of leaves.

Leaves are always organized around the shoot in a regular pattern that can be recognized best when looked at from above. It is obvious in most cases that the leaves are arranged in the way of a screw and that there is always the same angle measurable between subsequent leaves (angle of divergence).

CONTRIBUTIONS OF MOLECULAR BIOLOGY IN HORTICULTURE

More and more pathologists are using molecular probes for early detection of diseases. However, the main contributions of modern biotechnology remain probably in the hands of the plant breeders. Indeed, molecular markers are of great use in detecting desirable new genes, or to identify important QTL. This marker-assisted selection is very useful for many vegetable plants. Other applications are DNA - fingerprinting and genetic engineering. This recent technology aims to insert and to express new specific genes into a selected plants.

MARKER-ASSISTED SELECTION

Molecular markers and genome mapping will reach a large extension in the next breeding programmes of tomatoes, peas, cabbages, melons, potatoes,...

It will be possible to improve the speed and the incorporation efficiency of desirable new genes by utilizing of closely linked selectable molecular markers genes by utilizing of closely linked selectable molecular markers. The number of backcross will be reduced. Tanksley *et al.* (1981) show that only with 12 markers, one on each tomato chromosome, the composition of the recurrent parent genome is similar to that observed in the absence of selection after the third backcross two years later. Therefore, it's easy to understand the high improvement that we can get with the 700 tomato markers already known in 1990 for this crop.

The utilization of molecular markers could also identify important quantitative trait loci (QTL). Recently Foolad and Chen (1998) have identified 13 RAPD markers at eight genomic regions, that were associated with QTLs affecting salt tolerance during germination in tomato.

DNA Fingerprinting

The DNA polymorphism observed by RFLP, AFLP, or RAPD, allows cultivar identification of fruit trees, apple, citrus,... or other vegetatively propagated plants. Although some mutants, as coloured fruits, escape these identification techniques, some people would evaluate the genetic uniformity of regenerated plants in this way.

Genetic Engineering

Genetic engineering consists of introducing a foreign gene into a plant genome to create a new function. Or at the contrary, it could be to reduce or suppress an existing gene function, as in the antisense strategy.

1. *Historic:* In this field, progress were very rapid. In 1974 the pathogenicity of Agrobacterium tumefaciens, due to a large plasmid called Ti, was demonstrated. In 1977, Chilton *et al.* demonstrate a stable incorporation of plasmid Ti-DNA into higher plant cells. In 1983,

the team of Van Montagu and Schell in Gent succeeded in producing a transgenic plant, and eleven years later the first transgenic cultivars were available in the market.

2. *Transgenic Plants with New Agronomic Traits:* Two different techniques are commonly used to introduce genetic information (DNA) inside cells, protected by pectocellulosic walls. A particle bombardment process, called biolistic, is used with success for monocots gynmosperm, squash, peas,...Today, this technique allows the bombardment of meristems and avoids the difficulties of regeneration. But the mediation of Agrobacterium remains till now the most routinely system to transform dicotyledonous plants.
   The main agronomic traits introduced in horticultural plants and already commercialised are Bt toxin and herbicide resistance. Other studies concern virus resistance, male sterility...
3. *Antisense Strategy:* Calgene created in 1994 the first commercial transgenic plant, a long shelf life tomato, by the suppression of polygalacturonase activity due to an antisense gene. However, t activity due to an antisense gene. However, this Flavr Savr tomato variety was removed from the trade 3 years later, because of its disease susceptibility and its lack of productivity !
   Later, other tomato varieties with long storage qualities were obtained by the utilization of an antisense RNA inhibition of ACC synthase or ACC oxydase, two ethylene precursors.
   The antisense technique was also used to reduce the lignification of woody plants, by blocking the enzymes involved in the precursor of lignin biosynthesis. Another interesting application was the induction of white flowers in petunia and in different other ornamental plants by the suppression of chalcone synthase activity.
4. *Transmission of the New Traits:* These traits induced by the transferred DNA are transmitted to the progeny as a dominant Mendelian character. Nevertheless in some cases, transgenic plants with a strong gene expression can generate progenies with only a faint no gene expression. This problem is not clearly elucidate.
5. *Marker, Reporter, Promotor and Expression Genes:* To select the transformed cells, some marker genes, very often resistant to antibiotics or to herbicides, are attached to the coding sequence, as a promoter or an expression gene. Those allow the new gene to express in the whole plant or in a specific plant tissue.
   Some reporter genes are used to follow the evolution one.
   Some reporter genes are used to follow the evolution of the transformed cells. The most frequent reporter genes are the gene "gus" or the gene lux (luciferase).

6. *Horticultural Transgenic Crops already released in USA:* The Flavr Savr tomato was the first genetically engineered whole food approved for commercial sale in 1994. Four other transgenic tomatoes with a delayed ripening were approved later (1995 and 1996).

In 1995, potatoes with Bt genes and a squash cultivar resistant to two viruses were released. Two years later, another squash cultivar resistant to 3 viruses and a papaya line also resistant to viruses were approved. Two other horticultural crops are waiting the authorities approval: red hearted cichory (Radicchio) with male sterility and resistance to herbicide, and a tomato with Bt toxin.

It there are so few commercial transgenic plants in horticulture, it's probably due to the relatively low acreage of these horticultural crops in comparison with other agricultural crops. In 1997, in the world 5 million ha were planted with glyphosate resistant soybeans released by Monsanto alone.

Nowadays, it's true also that many transgenic plants exist in research laboratories, awaiting authorization for field testing, as they are very interesting model plants for learning physiology: lettuce with less nitrate by increasing nitrate reductase gene active with less nitrate by increasing nitrate reductase gene activity, a yeast ribonuclease gene to reduce viroid infection on potato, modification of lignin synthesis in plant to produce trees adapted to paper industry, or timber, or biomases production, regulation genes to control tree architecture,....

Situation of Biotechnologies in the World: Development and Risks

Situation in the Young Countries or in Developing Countries: A general use of in vitro micropropagation in the developing countries is now a reality, as it was predicted by Albert Sasson (1993). African countries for instance are producing potatoe, banana, cassava...

Even the production of transgenic plants is no longer a prerogative of the Northern countries. Different experiments are now starting in the South, specially in the international laboratories from CGIAR. Although CGIAR's expenditures on biotechnology for 1997 raise only 24.2 million US$ on an annual budget of 345 million US$ this Consultative Group on International Agricultural Research plays a very important role as contributor to agricultural research for developing countries. The large international research centres IITA (Ibadan, Nigeria), CIP (Lima, Peru), CIAT (Cali, Columbia),... belong to the CGIAR. Their roles and NGO's are primordial. The genetic transformation their roles and NGO's are primordial. The genetic transformation of cassava is an excellent example. Its tuberous roots provide food for over 500 million people, mostly small-scale farmers. Unfortunately, till recently this integral plant for food security in developing countries has been recalcitrant to transformation approaches. However, thanks to the participation of international institutes located in cassava growing countries, CIAT and IITA with the help of four development associations from Swiss, UK, The Netherlands and USA, were able to initiate procotols for cassava transformation and regeneration.

Other networks, like the Cassava Biotechnology Network (CBN) previously described, exist. REDBIO, a technical cooperation network on plant biotechnology, is supported by FAO to promote a best use of scarce manpower, equipment and other resources in Latin America.

On the other hand, more and more private laboratories exist in the developing countries which have a high scientific and technical level (China, Singapore, Taiwan, India, Brasil, Mexico, Chile,...). In these countries, European, American or Japanese agrochemical companies are investing. Also these countries will have their chance to develop their own technologies, and to export to the Northern markets.

Meanwhile, genetic engineering techniques are being applied mainly to crops which are important for g applied mainly to crops which are important for the industrialized world, not crops on which the world's hungry depend. Therefore, it is unrealistic when Monsanto writes "the experts said that biotechnological innovation will increase the crop productivity without occupation of new lands, saving tropical forest of best quality and animal habitat". However, the World Bank was promising in a 1997 publication Bioengineering of crops that" transgenic crops could improve food yield by up to 25 per cent in the developing countries and could help to feed an estimated additional three billion people over the next 30 years".

But, not all people are confident in this prospective. To the contrary, some are afraid that the poorest countries, where more than 700 million people are chronically undernourished will miss the biotechnological revolution.

Nevertheless, there is some hope to transfer to the developing countries high technology, which don't undermine the environmental and social network. So, the Agricultural Biotechnology for Sustainable Productivity Project (ABSP project) is managed by Michigan University. Four American universities, two research centres from France and USA, and two American private companies are also involved in this project. The ABSP's objectives are the reduction of losses due to pathogens and the reduction of losses due to pathogens and pests by using transgenic plants (sweet potatoes, potatoes, cucurbits, tomatoes) and by cloning commercial value-plants (bananas, pineapple, coffee).

Situation in the Northern Countries

1. *Mass Propagation:* In 1988, Pierik (1991) predicted a European in vitro propagation of 212 million plantlets/year. He was also predicting an important development of this activity for the future. Following Pierik this quantity accounted for only 5 poe cent of the plants able to be propagated by vegetative means. However a few years later

Pierik's prediction was not confirmed. The world production was estimated at 600 million. In USA, Zimmerman (1997) was recording a production of 121 million plantlets for a total of 110 laboratories, half produced in 11 laboratories, and 6 labs were propagating each more than 6 million plants per year.

In Europe, only the Dutch production of micropropagated plants is precisely known. In 1990 the total production was 95.5 million. In 1995, the Dutch in country production reached only 53,7 million, whereas importations increased 77,3 million. (37 million from Poland and 17,1 million from India) (Pierik, data unpublished). These laboratory relocations to low-cost manpower countries are commonly observed, and put the question of the quality management.

2. *Transgenics:* The GMOs invade progressively all the US lands: 1.6 poe cent in 1996, 15 poe cent in 1997, 50 poe cent in 2000..... It will represent an estimated market of 3,9 million US$ in 2003 ! The largestt agrochemical companies are merging to become giants of agrobusiness such as Novartis, Monsanto, Zeneca,...

   Following Monsanto in 1996, about 950 T insecticides were saved by introduction of resistant cotton, id est a net savings of 81 US$/ha for the grower. Glyphosate-resistant soybeans (4 million ha in 1997) allow a savomgs of 44 to 49 US$/ha. (Information from cultivar, suppl. n 436, Feb. 16, pp. 24-37, 1998).

3. *Perspectives, Limitations and Environmental Risks*: Ecological impact related with the introduction of GMOs is always an open debate.

   The application to agriculture of these new technologies certainly opens interesting perspectives, but also raises potential problems. The risk of crop transgene spreading has been demonstrated. A researcher of Clemson University in South Carolina reported "that in a population of wild strawberries growing within 50 meters of a strawberry field, more than 50 poe cent of the wild plants contained marker genes from the cultivated strawberries".

A Danish team have shown a possible rapid spread of genes from oilseed rape to the weedy relative *Brassica campestris*.

There are other risks. *Brassica campestris*.

There are other risks. The introduction of Bt gene allows a drastic reduction in the use of toxic chemicals for crop protection. But a poorly controlled use of Bt-technology can destroy more effectively the predators than the pests.

Or when" many crop plants are transformed with similar effective traits, in such situation, many polyphagous pest species, which by nature are more flexible evolutionarily than those that have a narrower diet, are likely to overcome Bt resistance very quickly". Therefore, before releasing a transgenic plant, any risk has to be weighed against the benefit of the transgenic crops.

We must not forget that annually in the world 500.000 acute pesticide poisonings, with 5000 deaths, are observed.

BIOTECHNOLOGIES BASED ON VEGETATIVE PROPAGATION

Strategies supporting large-scale utilization of genetic material with a narrow genetic base must be appropriately integrated into tree improvement programmes. Vegetative propagation within such programmes allows for a fast release of new materials and for appropriate matching of clones to different local conditions. It also allows continued cultivation of given clones and to efficiently change the mixture of clones used in a given programme. Vegetative propagation also supports other currently available biotechnologies (*in vitro* storage and cryopreservation; *in vitro* selection).

IN VITRO STORAGE AND CRYOPRESERVATION

In vitro storage refers to the storage of germplasm in aseptic culture under laboratory conditions, while cryopreservation refers to the storage of cells, tissues, seeds, etc. at temperatures of liquid nitrogen (-196°C). The two techniques do not seem to be widely used in genetic conservation activities for forest trees, but they may serve as back-up strategies for species with seed storage problems.

Micropropagation

For crop and horticultural species, micropropagation (*in vitro* vegetative propagation of plants) is now the basis of a large commercial industry involving hundreds of laboratories around the world. Successful protocols now exist for a large number of forest tree species and the number of species for which successful use of somatic embryogenesis has been reported is increasing (somatic embryogenesis is a step in micropropagation where somatic cells are differentiated into somatic embryos).

So, in the future, it is likely that micropropagation in the forestry sector will become commercially more important. Compared to vegetative propagation through cuttings, the higher multiplication rates available through micropropagation seem to offer a quicker capture of genetic gains obtained in clonal forestry strategies.

One major factor impeding early application of micropropagation in many large-scale forest plantation programmes, is that breeding and selection of desired clones are not sufficiently advanced for clonal forestry to be contemplated.

Current high costs will also be an impediment to the direct use of micropropagation in many programmes. Technologies resembling those used commercially in horticulture are most likely to be affordable for a limited number of high-value forest tree species, particularly those for which propagation by cuttings is difficult. Micropropagation is unlikely to be used for the production of planting stock of non-industrial forest tree species.

In Vitro Selection

In vitro selection refers to the selection of germplasm based on test results using tissue culture under laboratory conditions. Many recent publications for crop plants have reported useful correlations between *in vitro* responses and the expression of desirable field traits, most commonly disease resistance. Positive results are available also for tolerance to herbicides, metals, salt and low temperatures.

For the selection criteria of major general importance in forest trees (in particular vigour, stem form and wood quality), poor correlations with field responses will limit the usefulness of *in vitro* selection. However, *in vitro* selection may be of possible interest in forestry programmes for screening disease resistance and tolerance to salt, frost and drought.

8

Forest Genetics and Biotechnology

THE DEVELOPMENT OF FORESTRY

A feature of fundamental importance, which distinguishes forests from some other resources, is that they are renewable. This means that the mature forest, as it is cut, may be replaced by a new crop of trees; thus a continuous yield of wood is provided from the same area of land. The application of this principle is the primary basis of forestry practice, which may be defined briefly as the growing and using of successive crops of timber.

As the practice of forestry develops, the forests often come to be looked upon less as a natural resource and more as a crop that is dependent in large measure on man's knowledge and skill. However, forestry is not concerned exclusively with growing timber as a crop; forests contribute in many other ways to the well-being of mankind and proper forestry practices assure the maintenance and enlargement of these benefits.

Man is not a conservationist by nature: history shows that as long as the existing forest appears to be adequate to the demand, it is subject to exploitation for immediate revenue and there is little concern for its future productivity. In other words, forest conservation, involving some investment of time and money, only begins when it is apparent that current practices are endangering the resource with critical effects upon the supply of forest products and services.

Down through the centuries, as man has increased in numbers and spread over the face of the earth, he has caused and is still causing widespread destruction of the forest cover, partly of necessity to provide land for agriculture, settlement and industrial development, partly through ignorance and greed and partly in the belief that the forest resource is inexhaustible.

This latter viewpoint is to some extent justified in the early development of countries and regions that have extensive forest wealth and relatively small populations. Here the forest is often looked upon as a hindrance to settlement and agricultural development, with the result that areas cleared may exceed requirements and include lands quite unsuitable for the purpose in mind.

Man has in fact retained under intensive agriculture only a small portion of the area that was originally cleared for this purpose. It is estimated that, as a result of his activities, man has destroyed about two thirds of the world's original forest cover; the remaining forests containing merchantable timber today occupy less than one fifth of the world's land area.

In some parts of the world forest destruction has been unchecked, despite the adverse effect that this has had on the general economy and on the living standards of the people. Asia, in particular, has suffered from the effects of forest clearance and forest destruction, with widespread soil deterioration, inadequate supplies of wood to meet basic domestic needs and low living standards for a large portion of the population. Elsewhere, as the danger became apparent, methods of managing the forests to ensure their continued productiveness were evolved and thus the science of forestry came into being. However, the philosophy of conservation has developed slowly and forestry has invariably been the child of necessity: only after periods of forest destruction and land clearance has the need for husbandry and planned forest economy been realized.

TRANSITION FROM EXPLOITATION TO MANAGEMENT

The transition from exploitation to management takes time and requires the solution of many problems, technical, economic and administrative. In the beginning it may be characterized by a strong reaction against spoliation aimed at preserving the forest intact — "locking it up" — for posterity. In general, however, this is the antithesis of sound management of a renewable resource, which for its fullest development requires periodic harvesting and regeneration; an exception would be the preservation of representative samples of natural forest for scientific study and as "wilderness" areas. Modern forestry originated in Germany and France and much of its development during the past two centuries has taken place in northwestern Europe. Attention to forest conservation in these countries came early owing to an increasing demand for timber and firewood and the difficulty of securing wood cheaply from other countries.

These conditions and the serious effects of deforestation on stream flow, particularly in montane regions, made evident an urgent need for forest management. During the nineteenth century expanding populations and the iron and coal economy of the industrial revolution created new demands for wood and gave added impetus to the development of forestry.

Slowly and usually only as the need became urgent, other countries have followed the lead of western Europe and developed forestry practices to meet their own requirements. Such developments usually begin with restriction on the use of the forest, as in the setting aside of timber reserves and with protection of the forest against its more serious enemies, particularly fire. These steps are followed by efforts to secure new growth after cutting, by encouraging natural reproduction and by planting: with this the practice of silviculture begins. Finally, the accessible, productive forests are managed under cutting regulations and suitable silvicultural techniques to ensure a continuous yield of timber in perpetuity. Such forests are said to be managed on a "sustained-yield" basis.

While the natural forest usually provides the starting point for a forestry programme, it may not represent the optimum conditions that an area is capable of and its perpetuation is not always looked upon as the ideal objective of forest management. The aim, rather, may be to improve the natural forest as far as possible within the limits set by the local environment, silvicultural knowledge and economic considerations and in view of the purpose for which the forest is to be used.

Attempts to go beyond these limits and completely change the natural forest pattern have sometimes failed. A well-known instance was the century-long attempt to convert the lowland hardwood forests of Saxony to pure Norway spruce. Despite early rapid growth, the spruce was often badly damaged by snow and frost as well as by fungal and insect attacks. By the end of the third crop of spruce, conditions had so deteriorated that hardwood species had to be reintroduced and it was generally conceded that the plan had been unsuccessful. On the other hand, there are many examples of the successful introduction of types and species of trees different from those in the natural forest. The success of these operations depends largely on careful study of the ecological factors involved.

The aim of forestry may vary from one country to another and even between regions within the same country, depending upon local conditions. Usually the primary purpose is to grow sufficient wood to meet a country's requirements — first for domestic consumption, then for export to foreign markets. Products other than wood, usually referred to as minor forest products, may sometimes be an objective of management. These vary in importance and kind from one country to another, *e.g.*, turpentine, resin, oils, fruits, maple syrup, bark for tanning, etc. The southeastern United States and the Landes in Gascony, France, are two regions where the production of resin and turpentine from species of pine is a major industry carried on in association with timber production.

While the production of wood and other commodities is often primary, in its broader aspects forestry is concerned with the interaction between the forest and its environment, as, for example, in the effects of the forest on the stabilization of soil, on the maintenance and restoration of soil fertility, on the regulation of stream flow and on the provision of a suitable environment for valuable fish and wild life. The protective influences of the forest are many and far-reaching and under certain conditions may be more important than the supplying of wood.

This is particularly true in some parts of the tropics and subtropics, in crucial drainage areas and in regions of light, sandy soil unsuitable for cultivation, where the removal of the forest cover often results in near-desert conditions. Forestry may also aim at the full development of the recreational value of an area; but even where recreation or protection are the main objectives, controlled cutting of wood can usually be carried on advantageously, if the cutting operations do not interfere with the primary purpose of management.

The term "multiple use" is widely used in referring to the management of a forest area for two or more purposes, one of which may be dominant, but not exclusive.

Bases of Forestry

Forestry may be defined as the scientific management of forests for the continuous production of goods and services. While its practice involves many problems of an economic and administrative nature, the solution of which is essential for success, its proper development must in the first place be based on a sound knowledge of the forests and of the laws that govern their development and on an understanding of the characteristics of the individual species that make up the forests and of the interrelationship of the forests with their environment.

The forests of the world include a great many species of trees, not necessarily having any close botanical relationship, with varying adaptations to climate, soil and topography and more or less distinct reproductive characteristics, forms and rates of growth, sizes and ages at maturity, light requirements and abilities to resist damage by insects and other destructive agencies. It is with an intimate knowledge of these features of the trees and tree communities that sound forestry practice must begin.

GENE SELECTION AND CLASSIFICATION OF MICROARRAY DATA USING RANDOM FOREST

Selection of relevant genes for sample classification (e.g., to differentiate between patients with and without cancer) is a common task in most gene expression studies e.g.,. When facing gene selection problems, biomedical researchers often show interest in one of the following objectives:

1. To identify relevant genes for subsequent research; this involves obtaining a (probably large) set of genes that are related to the outcome of interest, and this set should include genes even if they perform similar functions and are highly correlated.
2. To identify small sets of genes that could be used for diagnostic purposes in clinical practice; this involves obtaining the smallest possible set of genes that can still achieve good predictive performance (thus, "redundant" genes should not be selected).

We will focus here on the second objective. Most gene selection approaches in class prediction problems combine ranking genes (e.g., using an F-ratio or a Wilcoxon statistic) with a specific classifier (e.g., discriminant analysis, nearest neighbour). Selecting an optimal number of features to use for classification is a complicated task, although some preliminary guidelines, based on simulation studies by [4], are available.

Frequently an arbitrary decision as to the number of genes to retain is made e.g., keep the 50 best ranked genes and use them with a linear discriminant analysis as in; keep the best 150 genes as in. This approach, although it can be appropriate when the only objective is to classify samples, is not the most appropriate if the objective is to obtain the smaller possible sets of genes that will allow good predictive performance. Another common approach, with many variants e.g., is to repeatedly apply the same classifier over progressively smaller sets of genes (where we exclude genes based either on the ranking statistic or on the effect of the elimination of a gene on error rate) until a satisfactory solution is achieved (often the smallest error rate over all sets of genes tried).

A potential problem of this second approach, if the elimination is based on univariate rankings, is that the ranking of a gene is computed in isolation from all other genes, or at most in combinations of pairs of genes, and without any direct relation to the classification algorithm that will later be used to obtain the class predictions. Finally, the problem of gene selection is generally regarded as much more problematic in multi-class situations (where there are three or more classes to be differentiated), as evidence by recent papers in this area e.g.,. Therefore, classification algorithms that directly provide measures of variable importance (related to the relevance of the variable in the classification) are of great interest for gene selection, specially if the classification algorithm itself presents features that make it well suited for the types of problems frequently faced with microarray data. Random forest is one such algorithm.

Random forest is an algorithm for classification developed by Leo Breiman that uses an ensemble of classification trees [14-16]. Each of the classification trees is built using a bootstrap sample of the data, and at each split the candidate set of variables is a random subset of the variables. Thus, random forest uses both bagging (bootstrap aggregation), a successful approach for combining unstable learners [16,17], and random variable selection for tree building. Each tree is unpruned (grown fully), so as to obtain low-bias trees; at the same time, bagging and random variable selection result in low correlation of the individual trees. The algorithm yields an ensemble that can achieve both low bias and low variance (from averaging over a large ensemble of low-bias, high-variance but low correlation trees). Random forest has excellent performance in classification tasks, comparable to support vector machines. Although random forest is not widely used in the microarray literature, it has several characteristics that make it ideal for these data sets:

- Can be used when there are many more variables than observations.
- Can be used both for two-class and multi-class problems of more than two classes.
- Has good predictive performance even when most predictive variables are noise, and therefore it does not require a pre-selection of genes (i.e., "shows strong robustness with respect to large feature sets",*sensu*).
- Does not overfit.
- Can handle a mixture of categorical and continuous predictors.
- Incorporates interactions among predictor variables.
- The output is invariant to monotone transformations of the predictors.
- There are high quality and free implementations: the original Fortran code from L. Breiman and A. Cutler, and an R package from A. Liaw and M. Wiener.
- Returns measures of variable (gene) importance.
- There is little need to fine-tune parameters to achieve excellent performance. The most important parameter to choose is *mtry*, the number of input variables tried at each split, but it has been reported that the default value is often a good choice. In addition, the user needs to decide how many trees to grow for each forest (*ntree*) as well as the minimum size of the terminal nodes (*nodesize*). These three parameters will be thoroughly examined in this paper.

Given these promising features, it is important to understand the performance of random forest compared to alternative state-of-the-art prediction methods with microarray data, as well as the effects of changes in the parameters of random forest. In this paper we present, as necessary background for the main topic of the paper (gene selection), the first through examination of these issues, including evaluating the effects of *mtry*, *ntree* and *nodesize* on error rate using nine real microarray data sets and simulated data.

The main question addressed in this paper is gene selection using random forest. A few authors have previously used variable selection with random forest and use filtering approaches and, thus, do not take advantage of the measures of variable importance returned by random forest as part of the algorithm. Svetnik, Liaw, Tong and Wang propose a method that is somewhat similar to our approach.

The main difference is that first find the "best" dimension (p) of the model, and then choose the p most important variables. This is a sound strategy when the objective is to build accurate predictors, without any regards for model interpretability. But this might not be the most appropriate for our purposes as it shifts the emphasis away from selection of specific genes, and in genomic studies the identity of the selected genes is relevant (e.g., to understand molecular pathways or to find targets for drug development).

The last issue addressed in this paper is the multiplicity (or lack of uniqueness or lack of stability) problem. Variable selection with microarray data can lead to many solutions that are equally good from the point of view of prediction rates, but that share few common genes. This multiplicity problem has been emphasized by and and recent examples are shown in Although multiplicity of results is not a problem when the only objective of our method is prediction, it casts serious doubts on the biological interpretability of the results. Unfortunately most "methods papers" in bioinformatics do not evaluate the stability of the results obtained, leading to a false sense of trust on the biological interpretability of the output obtained. Our paper presents a through and critical evaluation of the stability of the lists of selected genes with the proposed (and two competing) methods.

In this paper we present the first comprehensive evaluation of random forest for classification problems with microarray data, including an assessment of the effects of changes in its parameters and we show it to be an excellent performer even in multi-class problems, and without any need to fine-tune parameters or pre-select relevant genes. We then propose a new method for gene selection in classification problems (for both two-class and multi-class problems) that uses random forest; the main advantage of this method is that it returns very small sets of genes that retain a high predictive accuracy, and is competitive with existing methods of gene selection.

EVALUATION OF PERFORMANCE AND COMPARISONS WITH ALTERNATIVE APPROACHES

We have used both simulated and real microarray data sets to evaluate the variable selection procedure. For the real data sets, original reference paper and main features and further details are provided in the supplementary material. To evaluate if the proposed procedure can recover the signal in the data and can eliminate redundant genes, we need to use simulated data, so that we know exactly which genes are relevant. Details on the simulated data are provided in the methods and in the supplementary material.

Estimation of error rates

To estimate the prediction error rate of all methods we have used the.632+ bootstrap method. The.632+ bootstrap method uses a weighted average of the resubstitution error (the error when a classifier is applied to the training data) and the error on samples not used to train the predictor (the "leave-one-out" bootstrap error); this average is weighted by a quantity that reflects the amount of overfitting. It must be emphasized that the error rate used when performing variable selection is not what we report in as prediction error rate. To calculate the prediction error rate as reported, for example, the.632+ bootstrap method is applied to the complete procedure, and thus the samples used to compute the leave-one-out bootstrap error used in the.632+ method are samples that are not used when fitting the random forest, or carrying out variable selection. The.632+ bootstrap method was also used when evaluating the competing methods.

Stability of variable (gene) selection evaluated using 200 bootstrap samples. "# Genes": number of genes selected on the original data set. "# Genes boot.": median (1st quartile, 3rd quartile) of number of genes selected from on the bootstrap samples....

Effects of parameters of random forest on prediction error rate

Before examining gene selection, we first evaluated the effect of changes in parameters of random forest on its classification performance. Random forest returns a measure of error rate based on the out-of-bag cases for each fitted tree, the OOB error, and this is the measure of error we will use here to assess the effects of parameters. We examined whether the OOB error rate is substantially affected by changes in*mtry*, *ntree*, and *nodesize*.

For both real and simulated data, the relation of OOB error rate with *mtry* is largely independent of *ntree* (for *ntree* between 1000 and 40000) and *nodesize* (nodesizes 1 and 5). In addition, the default setting of *mtry* (*mtryFactor* = 1 in the figures) is often a good choice in terms of OOB error rate. In some cases, increasing *mtry* can lead to small decreases in error rate, and decreases in *mtry* often lead to increases in the error rate. This is specially the case with simulated data with very few relevant genes (with very few relevant genes, small *mtry* results in many trees being built that do not incorporate any of the relevant genes). Since the OOB error and the relation between OOB error and *mtry* do not change whether we use *nodesize* of 1 or 5, and because the increase in computing speed from using *nodesize* of 5 is inconsequential, all further analyses will use only the default *nodesize* = 1. These results show the robustness of random forest to changes in its parameters; nevertheless, to re-examine robustness of gene selection to these parameters, in the rest of the paper we will report results for different settings of *ntree* and *mtry* (and these results will again show the robustness of the gene selection results to changes in *ntree* and *mtry*).

Gene selection using random forest

Random forest returns several measures of variable importance. The most reliable measure is based on the decrease of classification accuracy when values of a variable in a node of a tree are permuted randomly, and this is the measure of variable importance that we will use in the rest of the paper. (In the Supplementary material we show that this measure of variable importance is not the same as a non-parametric statistic of difference between groups, such as could be obtained with a Kruskal-Wallis test). Other measures of variable importance are available, however, and future research should compare the performance of different measures of importance.

To select genes we iteratively fit random forests, at each iteration building a new forest after discarding those variables (genes) with the smallest variable importances; the selected set of genes is the one that yields the smallest OOB error rate. Note that in this section we are using OOB error to choose the final set of genes, not to obtain unbiased estimates of the error rate of this rule. Because of the iterative approach, the OOB error is biased down and cannot be used to asses the overall error rate of the approach, for reasons analogous to those leading to "selection bias". To assess prediction error rates we will use the bootstrap, not OOB error. In our algorithm we examine all forests that result from eliminating, iteratively, a fraction,*fraction.dropped*, of the genes (the least important ones) used in the previous iteration. By default,*fraction.dropped* = 0.2 which allows for relatively fast operation, is coherent with the idea of an "aggressive variable selection" approach, and increases the resolution as the number of genes considered becomes smaller. We do not recalculate variable importances at each step as mention severe overfitting resulting from recalculating variable importances. After fitting all forests, we examine the OOB error rates from all the fitted random forests. We choose the solution with the smallest number of genes whose error rate is within u standard errors of the minimum error rate of all forests. Setting $u = 0$ is the same as selecting the set of genes that leads to the smallest error rate. Setting $u = 1$ is similar to the common "1 s.e. rule", used in the classification trees literature; this strategy can lead to solutions with fewer genes than selecting the solution with the smallest error rate, while achieving an error rate that is not different, within sampling error, from the "best solution". In this paper we will examine both the "1 s.e. rule" and the "0 s.e. rule".

On the simulated data sets backwards elimination often leads to very small sets of genes, often much smaller than the set of "true genes". The error rate of the variable selection procedure, estimated using the.632+ bootstrap method, indicates that the variable selection procedure does not lead to overfitting, and can achieve the objective of aggressively reducing the set of selected genes. In contrast, when the simplification procedure is applied to simulated data sets without signal, the number of genes selected is consistently much larger and, as should be the case, the estimated error rate using the bootstrap corresponds to that achieved by always betting on the most probable class.

For additional results using different combinations of *ntree* = {2000, 5000, 20000}, *mtryFactor* = {1, 13}, *se*= {0, 1},*fraction.dropped* = {0.2, 0.5}). Error rates when performing variable selection are in most cases comparable (within sampling error) to those from random forest without variable selection, and comparable also to the error rates from competing state-of-the-art prediction methods. The number of genes selected varies by data set, but generally the variable selection procedure leads to small (< 50) sets of predictor genes, often much smaller than those from competing approaches. There are no relevant differences in error rate related to differences in *mtry*, *ntree* or whether we use the "s.e. 1" or "s.e. 0" rules. The use of the "s.e. 1" rule, however, tends to result in smaller sets of selected genes.

Stability (uniqueness) of Results

We have evaluated the stability of the variable selection procedure using the bootstrap. This allows us to asses how often a given gene, selected when running the variable selection procedure in the original sample, is selected when running the procedure on bootstrap samples. The results here will focus on the real microarray data sets. For other combinations of*ntree*, *mtryFactor*, *fraction.dropped*, *se*) shows the variation in the number of genes selected in bootstrap samples, and the frequency with which the genes selected in the original sample appear among the genes selected from the bootstrap samples. In most cases, there is a wide range in the number of genes selected; more importantly, the genes selected in the original samples are rarely selected in more than 50 poe cent of the bootstrap samples. These results are not strongly affected by variations in *ntree* or *mtry*; using the "s.e. 1" rule can lead, in some cases, to increased stability of the results.

Discussion

We have first presented an exhaustive evaluation of the performance of random forest for classification problems with microarray data, and shown it to be competitive with alternative methods, without requiring any fine-tuning of parameters or pre-selection of variables. The performance of random forest without variable selection is also equivalent to that of alternative approaches that fine-tune the variable selection process. We have then examined the performance of an approach for gene selection using random forest, and compared it to alternative approaches. Our results, using both simulated and real microarray data sets, show that this method of gene selection accomplishes the proposed objectives. Our method returns very small sets of genes compared to alternative variable selection methods, while retaining predictive performance. Our method of gene selection will not return sets of genes that are highly correlated, because they are redundant. This method will be most useful under two scenarios: a) when considering the design of diagnostic tools, where having a small set of probes is often desirable; b) to help understand the results from other gene selection approaches that return many genes, so as to understand which ones of those genes have the largest signal to noise ratio and could be used as surrogates for complex processes involving many correlated genes. A backwards elimination method, precursor to the one used here, has been already used to predict breast tumor type based on chromosomic alterations.

We have also thoroughly examined the effects of changes in the parameters of random forest (specifically *mtry*, *ntree*, *nodesize*) and the variable selection algorithm (*se*, *fraction.dropped*). Changes in these parameters have in most cases negligible effects, suggesting that the default values are often good options, but we can make some general recommendations. Time of execution of the code increases H" linearly with *ntree*. Larger *ntree* values lead to slightly more stable values of variable importances, but for the data sets examined, *ntree* = 2000 or *ntree* = 5000 seem quite adequate, with further increases having negligible effects. The change in *nodesize* from 1 to 5 has negligible effects, and thus its default setting of 1 is appropriate. For the backwards elimination algorithm, the parameter *fraction.dropped* can be adjusted to modify the resolution of the number of variable selected; smaller values of*fraction.dropped* lead to finer resolution in the examination of number of genes, but to slower execution of the code. Finally, the parameter *se* has also minor effects on the results of the backwards variable selection algorithm but a value of *se* = 1 leads to slightly more stable results and smaller sets of selected genes.

In contrast to other procedures our procedure does not require to pre-specify the number of genes to be used, but rather adaptively chooses the number of genes have conducted an evaluation of several gene selection algorithms, including genetic algorithms and various ranking methods; these authors show results for the Leukemia and NCI60 data sets, but the Leukemia results are not directly comparable since focus on a three-class problem. They report the best results with the NCI60 data set estimated with the.632 bootstrap rule.

These best error rates are 0.408 for their evolutionary algorithm with 30 genes and 0.318 for 40 top-ranked genes. Using a number of genes slightly larger than us, these error rates are similar to ours; however, these are the best error rates achieved over a range of ranking methods and error rates, and not the result of a complete procedure that automatically determines the best number of genes and ranking scheme. A comparative study of feature selection and multi-class classification. Although they use four-fold cross-validation instead of the bootstrap to assess error rates, their results for three data sets common to both studies (Srbct, Lymphoma, NCI60) are similar to, or worse than, ours. In contrast to our method, their approach pre-selects a set of 150 genes for prediction and their best error rates are those over a set of seven different algorithms and eight different rank selection methods, where no algorithm or gene selection was consistently the best. In contrast, our results with one single algorithm and gene selection method (random forest) match or outperform their results.

Recently, several approaches that adaptively select the best number of genes or features have been reported. For the Leukemia data set our method consistently returns sets of two genes, similar to using an exhaustive search method, and lower than the numbers given by of 3 to 25 have proposed a Bayesian model averaging (BMA) approach for gene selection; comparing the results for the two common data sets between our study and theirs, in one case (Leukemia) our procedure returns a much smaller set of genes (2 vs. 15), whereas in another (Breast, 2 class) their BMA procedure returns 8 fewer genes (14 vs. 6); in contrast to BMA, however, our procedure does not require setting a limit in the maximum number of relevant genes to be selected have developed a method for gene selection and classification, LS Bound, related to least-squares SVMs; their method uses an initial pre-filtering (they choose 1000 initial genes) and is not clear how it could be applied to multi-class problems.

The performance of their procedure with the leukemia data set is better than that reported by our method, but they use a total of 72 samples (the original 38 training plus the 34 validation of [44]) thus making these results hard to compare. With the colon data sets, however, their best performing results are not better than ours with a number of features that is similar to ours. Proposed two Bayesian classification algorithms that incorporate gene selection (though it is not clear how their algorithms can be used in multi-class problems).

The results for the Leukemia data set are not comparable to ours (as they use the validation set of 34 samples), but their results for the colon data set show error rates of 0.167 to 0.242, slightly larger than ours (although these authors used random partitions with 50 training and 12 testing samples instead of the.632+ bootstrap to assess error rate), with between 8 and 15 features selected (somewhat larger than those from random forest). Finally, [31], applied both shrunken centroids and a genetic algorithm + KNN technique to the NCI60 and Srcbt data sets; their results with shrunken centroids are similar to ours with that technique, but the genetic algorithm + KNN technique used larger sets of genes (155 and 72 for the NCI60 and Srbct, respectively) than variable selection with random forest using the suggested parameters.

In summary, then, our proposed procedure matches or outperforms alternative approaches for gene selection in terms of error rate and number of genes selected, without any need to fine-tune parameters or preselect genes; in addition, this method is equally applicable to two-class and multi-class problems, and has software readily available. Thus, the newly proposed method is an ideal candidate for gene selection in classification problems with microarray data.

A reviewer has alerted us to the paper by Jiang et al., previously unknown to us. In fact, our approach is virtually the same as the one used by Jiang et al., with the exception that these authors recompute variable importances at each step (we do not do this in this paper, although the option is available in our code) and, more importantly, that their gene selection is based both in the OOB error, as well as the prediction error when the forest trained with one data set is applied to a second, independent, data set; thus, this approach for gene selection is not feasible when we only have one data set. Jiang et al. [45] also show the excellent performance of variable selection using random forest when applied to their data sets. The final issue addressed in this paper is instability or multiplicity of the selected sets of genes.

From this point of view, the results are slightly disappointing. But so are the results of the competing methods. And so are the results of most examined methods so far with microarray data, as shown in [29] and [30] and discussed thoroughly by [27] for classification and by [28] for the related problem of the effect of threshold choice in gene selection. However, and except for the above cited papers and [6,46] and [5], this is an issue that still seems largely ignored in the microarray literature. As these papers and the statistical literature on variable selection (e.g., [40,47]) discusses, the causes of the problem are small sample sizes and the extremely small ratio of samples to variables (i.e., number of arrays to number of genes). Thus, we might need to learn to live with the problem, and try to assess the stability and robustness of our results by using a variety of gene selection features, and examining whether there is a subset of features that tends to be repeatedly selected. This concern is explicitly taken into account in our results, and facilities for examining this problem are part of our R code.

The multiplicity problem, however, does not need to result in large prediction errors. That the very different classifiers often lead to comparable and successful error rates with a variety of microarray data sets. Thus, although improving prediction rates is important, when trying to address questions of biological mechanism or discover therapeutic targets, probably a more challenging and relevant issue is to identify sets of genes with biological relevance. Two areas of future research are using random forest for the selection of potentially large sets of genes that include correlated genes, and improving the computational efficiency of these approaches; in the present work, we have used parallelization of the "embarrassingly parallelizable" tasks using MPI with the Rmpi and Snow packages [50,51] for R. In a broader context, further work is warranted on the stability properties and biological relevance of this and other gene-selection approaches, because the multiplicity problem casts doubts on the biological interpretability of most results based on a single run of one gene-selection approach.

GENETIC MODIFICATION IN THE FORESTRY SECTOR

The first regeneration of a genetically modified (GM) forest tree was achieved in 1986 in *Populus.* Since then, the genus has become a model for genetic modification and related tree biotechnology studies. The first attempt to genetically modify a conifer (*Larix*) was reported in 1991. Introducing targeted genes into the genome of a forest tree is a way to obtain GM plants. It is also a basic research tool for a better understanding of gene functioning in woody plants.

TRANSFORMATION METHODS

Transformation methods consist of inserting into the genome of the host plant a mutated gene or a gene from another organism either by microinjection/projection (direct transformation), or through a vector such as *Agrobacterium tumefaciens* (indirect transformation).

Most successful work on genetic modification of forest trees species genomes so far has been obtained by using juvenile material, often from explants produced from early germinations tissues which have much higher regeneration capacities than older material. Successful genome modification reports of adult selected plant material are very rare, except in poplars precisely because of its greater capacity to regenerate.

Insect Resistance

The first strategy consists of using microprojectiles to insert a gene encoding an endotoxin which binds to the receptors in the intestine of Lepidoptera, Coleoptera and Diptera, lysing the organ and killing the insect. This was done in *Populus alba* × *P. grandidentata, P. tremula* × *P. tremuloides* and *Picea glauca*. GM *Populus* trees and *Pinus radiata* expressing the *Bacillus thuringiensis* endotoxin '*Bt*' have been obtained. A second transformation method is based on the introduction of a gene coding for a protease inhibitor that modifies insect digestion, causing the death of the pest. Studies used potato gene *pin*2, a protease inhibitor introduced into *P. alba* × *P. grandidentata* through *A. tumefaciens*, and the gene of a rice protease inhibitor introduced into *P. tremula* × *P. tremuloides*. A third approach focuses on simultaneous modification of two genes for enhanced resistance to insects. This was achieved in *Liquidambar styraciflua*, combining a peroxidase anionic enzyme gene involved in cell growth and wall development with a '*Bt*' gene.

Herbicide Resistance

A first method has consisted in introducing a mutated version of the gene encoding the enzyme target for various herbicides: glyphosate for *Populus alba* × *P. grandidentata, P. trichocarpa* × *P. deltoides, Eucalyptus grandis, Larix decidua* and *Pinus radiata* (the herbicide blocks the synthesis of tryptophan, tyrosine and phenylalanine), or chlorosulfuron for *Populus tremula* and *Pinus radiata* (the herbicide blocks the synthesis of leucine, isoleucine and valine). A second and more frequently used strategy consists of introducing a microbial gene encoding an enzyme for the detoxification of the herbicide, and has been applied to *Populus alba*, *P. alba* × *P. tremula*, *P. tremula* × *P. alba*, *P. trichocarpa* × *P. deltoides* and *Eucalyptus camaldulensis*.

Flowering Modification

In order to produce sterile trees and prevent possible dispersal of transgenic pollen in the environment, an approach based on genetic ablation has been tested on poplar. This technique, which consists of expressing a cytotoxic gene under the control of a very specific poplar floral promoter, resulted in more than 90 percent of transformed lines lacking floral structures. Other approaches, based on the suppression of key flowering genes, are being tested.

Quantitative and Qualitative Modification of Lignin

Modification of lignin composition or content is being actively pursued because of the expected financial gains from pulp processing improvements. Lignins, which enhance cell wall mechanical properties and hardness, are difficult to process and are a significant limitation in processing wood into paper pulp by chemical treatment. Genetic transformation to modify lignin characteristics is a key research feature on species used in the paper industry. The aim is to regulate the activity of key enzymes involved in the lignin biosynthesis pathway. Active on-going research targets the effects of lignin biosynthesis in *Populus* on soil carbon transformation and storage.

Other Traits

Other traits that have been considered for genetic modification include: resistance to diseases in *Populus*, *Betula pendula*, *Alnus* and *Juglans*.; bioremediation in *Populus*; enhanced growth through higher efficiency of nitrogen assimilation in *Picea* and *Populus*, and modification of gibberellin synthesis in *Populus*.

DATA SET ANALYSIS IN GENETIC MODIFICATION

Genetic modification activities represent 520 entries in the data set, or 19 percent of all biotechnology activities recorded worldwide. Research is reported in 35 countries and on 29 tree genera. North America is the lead region, accounting for 48 percent of genetic modification activities on forest tree species, followed by Europe (32 percent), Asia (14 percent), Oceania (5 percent) and Africa and South America (less than one percent each), while no activity is reported in the Near East. Genetic modification research at the country level is led by the United States (219 entries in the data set, or 42 percent of the world's recorded activities on GM forest trees). Studies carried out by, or in cooperation with, agencies or companies from developed countries have been identified in Brazil, Chile, China, Indonesia, South Africa and Uruguay.

Populus accounts for 47 percent of data set entries, followed by *Pinus*. (19 percent), *Eucalyptus* (7 percent) and *Liquidambar* and *Picea* (5 percent each). The pedigree of a number of hybrid poplars used in China has not been ascertained. Several countries have already established GM tree field trials.

Interestingly, while the first GM forest trees were developed some fifteen years ago, about three-quarters of the total number of species recorded in the dataset have been genetically transformed during the past five years.

The main studies underpinning genetic modification target gene expression (21 percent), *in vitro* regeneration tissue culture (18 percent), herbicide resistance (13 percent), biotic resistance (12 percent) and lignin properties (9 percent).

GLOBAL SYSTEM ON CONSERVATION AND SUSTAINABLE USE OF PLANT GENETIC RESOURCES

The salient points related to the progress of the global system on conservation and sustainable use of plant genetic resources for food and agriculture which emerged in the seventh regular session of the FAO Commission on Genetic Resources for Food and Agriculture (CGRFA) at Rome during 15-23 May, 1997 may be reviewed as follows:

- A multi-year work programme on agricultural biodiversity established as per decision III/11 of COP shall be pursued by the FAO Commission and CBD Secretariat though cooperation. Similarly, FAO's monitoring/reporting on agricultural biodiversity shall be consistent with CBD, other relevant inter-governmental bodies and world food summit' Plan of Action.
- CBD through its decision II/5 has requested GEF to give priority support to efforts for conservation and sustainable use of agro-biodiversity. Many countries stressed the need to have separate funding! FAO commission requested FAO to provide assistance and guidance to countries on request in implementation/monitoring of the Global Plan of Action (GPA). Further, the monitoring of GPA and updating/reporting on the state of world's PGR were held complementary activities which do not need separate reports (*i.e.*, need only one report each) by countries.
- No changes in the CCGCT (Code of Conduct for Germplasm Collecting and Transfer) or any further work on Code of Conduct for Biotechnology were recommended, till the revision of IUPGR. However, a decision guide on regeneration of accessions in seed collections, prepared by IPGRI, has been made available to the FAO commission for comments by members.
- The existing arrangements between FAO and the 12 CGIAR centres shall be extended, pending the revision of IUPGR.

- Statutes of the inter-governmental technical working group on AGR (Animal Genetic Resources) and PGR were presented. Both working groups have 5 members each for Asia. The AGR were specifically addressed in view of the broadened mandate of the Commission and to facilitate integrated approach to agro-biodiversity.
- The summary statement presented by IPGRI informed the Commission that 0.5 million accession were now covered under the FAO-CGIAR agreements with the designation of 50,000 new accession under this arrangement.
- The summary statement presented by the CBD Secretariat showed that at least three decisions of the COP-III, Buenos Aires, Nov., 1996, were of direct relevance to the ongoing negotiation for revision of IUPGR, namely, Decision III/11-Conservation and Sustainable use of agro-biodiversity, Decision III/15-Access to Genetic Resources, and Decision III/17-Intellectual Property Rights. The COP-III noted the narrow options for legal status of revised International Undertaking, as being, i) a voluntary agreement, ii) a binding instrument, and iii) protocol to CBD. It encouraged parties to implement GPA at national levels. Pollinators and soil-micrograms in agriculture needed special attention (as bio-indicators) and comprehensive reporting by countries to Subsidiary Body on Scientific, Technical and Technological Advice (SBSTTA)/Conference of Parties (COP) of the CBD.
- An identified group of competent people needs to resolve the PGR issues and help prioritise various PGR activities. There is need to identify crops/regions on priority.
- The dialogue between the public and private sector should continue and there is a need to create awareness about PGR.
- An information bank should be established to enhance literacy, management and utilisation of genetic wealth. Emphasis should also be given to regional languages in developing the information base.

World Information and Early Warning System

The FAO's initiative to develop a systematic information base on PGR and related technologies under the World Information and Early Warning System on Plant Genetic Resources (WIEWS) is aimed to collect, collate, maintain and disseminate facts and figures on country profiles, national PGR programmes, *ex situ* collections, database on PGR, existing crop varieties, seed production and supplying agencies, *in situ* on farm conservation, bioprospecting and exchange of seed.

The nature of such an activity requires a comprehensive information input and a grid approach for analysis and interpretation of data. The system is open and responsive to queries from all stakeholders and interest groups. In practical terms, it would be relatively simpler to attend to details of the resource availability and the custodian(s) but, considering the importance of implementing CBD and WTO provisions and the imminent revision of the International Undertaking on Plant Genetic Resources, the access to resources for the purpose of replenishment could pose certain difficulties. The access would have to be required from any one of the following sources:

- The *ex situ* seed collections;
- The seed multiplication and supplying agencies; and
- The local communities/community gene banks.

The implications, in each case, could be several and manifold. The access from *ex situ* CG system collections is being debated wherein two clear-cut options have been put forth. The pre-CBD collections are still available freely which means their replenishment would be only a matter of time actually required for the multiplication of their seed. However, in case of requirement of post CBD collections, several access-related issues shall have to be sorted prior to the multiplication of seed for replenishment. The requirement of seed from commercial/public seed multiplication and supplying agencies could be easily met but such a replenishment would not serve the purpose effectively. This being so, the agencies could provide seed of few dominant varieties, which they are currently handling, in contrast with the requirement of a wider genetic base. The replenishment of seed from farmers' varieties/land races faces two different situations:

- The issue of consent/access; and
- The issue of compensation or sharing of benefit accrued from the use of their heritage seed.

The setting up of community seed banks and the Global Biodiversity Fund have definitely been provided as suitable institutions which could help in build up at an appropriate system and come to the rescue in cases of crisis. The setting up of a separate Global Biodiversity Fund is still being debated as also the Farmers' Rights. The development of the *sui generis* systems to meet the countries' own (local) needs, has to be accelerated. Through this mechanism, the whole idea is to encourage local level replenishment through procurement of required seed from nearly/similar climate logical areas.

Some Queries Seeking Answers

Increased investments on *ex situ* conservation are unavoidable with each increment in the quantum of work and management skills. The cost/benefit ratio does not work in such cases because the conservation is for posterity. In this endeavour, the obvious questions seeking answers would be:

- What should be the estimate of funds for genetic resources conservation at the CG Centres, at the National genebanks and at community seed banks?

- What kind of activities we need to take up in the beginning which ensure fast replenishment in cases of crisis?
- What shall be the mechanism of funding-whether the Global Biodiversity Fund or private sector contribution shall be forthcoming or not?
- Which sites for *in situ*/on-farm conservation to begin with?
- How to identify genuine NGOs for their involvement in PGR Conservation?
- What could be a standard format of collecting and collating information for WIEWS on PGR *ex situ* conservation, country status, on-farm conservation and wild relatives? Do the existing proforma suffice?
- What would be a suitable indicator or an appropriate method for monitoring of genetic erosion for use under Early Warning Systems?
- What should be the suitable proforma for multicrop passport data?
- How to generate compatible country reports providing information for the State of the World's PGR and the action on GPA?
- How to minimise diversity and redundancy in use of computer software for various Database Management Activities?

STATE OF GENETIC MODIFICATION IN THE FORESTRY SECTOR

In vitro regeneration of transformed plants is still a technical limitation for many species and genotypes.

Genetically transformed lines have been regenerated for *Populus*, *Eucalyptus*, *Pinus* (including *P. taeda*) and *Picea*. Technical processes involved in regeneration are not always well controlled or even studied; especially in mature tissues where the situation is the most critical.

No GM tree is reported to be commercialized in international databases to date. Insertion of genes controlling traits of interest in a given forest species is limited by insufficient knowledge of the molecular control of these traits. Most of the traits important in wood production, such as growth rate, adaptability, and stem and wood quality, are under polygenic control, and therefore depend on more than a single gene.

Biosafety regulations governing GM trees are often a concern due to the long period of time necessary to monitor environmental effects (including gene stability, pollen and seed dispersal, and impact on other elements of the ecosystem).

Most existing regulations are both stringent and involve high development costs compared to annual crops. Collaborative genomics programmes such as those carried out on *Populus, Pinus* and *Eucalyptus*, with the on-going characterization of genes of interest (including those coding for flowering, and lignin and cellulose biosynthesis), and the identification of specific promoters of transgenic expression, are expected to boost advances in genetic modification of forest trees. Studies to assess the stability of GM tree growth under abiotic stress, and their environmental impact, have been initiated.

BIOTECHNOLOGY RESEARCH AND APPLICATIONS IN FORESTRY

Unlike work carried out on crop or farm animal species, the domestication of forest trees is, with few exceptions, very recent. However, within a few decades, knowledge transfer from agriculture to forestry has been so fast that genetic knowledge of a few tree model genera, such as poplar (*Populus*), pine (*Pinus*) and *Eucalyptus*, is about to become as important, in terms of complexity and quality, as the genetic knowledge of the main crop species. Modern biotechnologies have potentially outstanding applications in the forestry sector, mainly because of the potential genetic gains they could confer. Expected gains for tree breeders include new genetic pools and significant reduction in tree selection time. Owing to technical difficulties, high costs, and time needed to evaluate forest species, tree breeding programmes have long been limited to the mass selection of the best individuals.

Between the 1970s and 1990s, development of horticultural techniques such as grafting, propagation by cuttings and controlled pollination, as well as information technology development, contributed to the implementation of sophisticated and efficient tree improvement activities. Long-term conventional tree breeding programmes and large-scale clonal propagation techniques have resulted in, and will continue to provide, significant productivity gains, especially for fast-growing species such as *Eucalyptus* and *Populus*. While the objectives of tree breeding are clearly defined and address end-users' (tree growers') requirements (in terms of yield, wood quality, and biotic and abiotic resistance), the situation in biodiversity conservation is less clear. Several issues (including knowledge of species biology, ecosystem conservation, restoration of endangered species, and decision-making tools for ecosystem or forest management) are common to biological diversity and forestry.

However, in spite of a considerable number of studies carried out on hundreds of forest tree species, there are still very few reported biotechnology applications. Many countries are currently engaged in biotechnology-supported forest biodiversity conservation programmes. Nonetheless, there are only a few reported direct impacts at policy or technical levels. The impact of biotechnology on genetic studies has allowed the use of cutting-edge methodologies with forest tree species in spite of the late domestication process compared with crop species.

The sequencing of the *Populus* genome has significant implications for forestry. *Eucalyptus* gene sequencing, which was reported during a IUFRO (International Union of Forest Research Organizations) Tree Biotechnology meeting in 2003, has since been pursued by the Eucalypt Genome Initiative, the Eucalyptus Genome Sequencing Project Consortium and the Brazilian project Genolyptus. The Dendrome Project is a collection of forest tree genome databases and other forest genetic information resources, with a special emphasis on conifers and *Pinus*. It is in genome sequencing that research seems currently most active.

The development of large-scale vegetative propagation techniques, essentially based on micropropagation or somatic embryogenesis, makes it theoretically possible to deploy superior planting materials rapidly and effectively. This is already the case for some coniferous and broad-leaved tree species including *Eucalyptus* and teak (*Tectona grandis*). Micropropagation techniques are also necessary to regenerate GM plants.

Genetic modification has been applied to forest trees mainly during the last two decades, at the experimental stage in the laboratory, to alter various traits such as herbicide, metal, salt, or insect resistance or cold tolerance. The rationale for producing trees with reduced lignin content for the pulp and paper industry is generally based on the expected environmental and economic benefits.

Equally if not more important, genetic modification is often considered as a tool to improve knowledge of tree biology and functioning, with significant potential applications; for example, for studying cell wall properties and wood formation. Since genes regulating secondary tissue and lignin formation in the annual plant *Arabidopsis thaliana* were identified, research on the genetic basis of wood quality in forest trees has significantly increased (mainly in *Eucalyptus, Pinus* and *Populus*). With the exception of micropropagation, for which some applications are reported, most public domain information relates to research activities. Field applications appear extremely limited, according to the literature. Nevertheless, and in spite of unclear economic rationale, the use of biotechnology in forestry research has increased and provided new knowledge on tree biology and functioning. It also appears that forest biotechnology is increasingly funded by the public sector in academic research, while the private sector seems to focus investments more on specific traits of commercial interest such as wood quality.

THE GREAT DIVERSITY FOUND IN FOREST TREES

There are an estimated 60,000 tree species on earth, and approximately 30 of the 49 plant orders contain tree species. Clearly, the tree phenotype has evolved many times in plants. The diversity of plant structures, development, life history, environments occupied and so on in trees is nearly as broad as higher plants in general, but trees share the common characteristic that all are perennial and many are very long lived. Because of the sessile nature of plants, each tree must survive and reproduce in a specific environment over the seasonal cycles of its lifetime. This tight association between individual genotypes and their environment provides a powerful research setting, just as it has driven the evolution of a plethora of uniquely arboreal adaptations. Understanding these evolutionary strategies is a long-standing area of study of tree biologists, with many broader biological implications.

Completed and current genome-sequencing projects in forest trees are limited to about 25 species from just 4 of more than 100 families: Pinaceae (pines, spruces and firs), Salicaceae (poplars and willows), Myrtaceae (eucalyptus) and Fagaceae (oaks, chestnuts and beeches). Large-scale sequencing projects such as the 1000 Human Genomes, 1000 Plant Genomes (1KP) or the 5000 Insect Genome (i5k) projects have not yet been proposed for forest trees.

RAPIDLY DEVELOPING GENOMIC RESOURCES IN FOREST TREES

Genome resources are developing rapidly in forest trees in spite of the challenges associated with working with large, long-lived organisms and sometimes very large genomes [2]. Complete genome sequencing, however, has been slow to advance in forest trees owing to funding limitations and the large size of conifer genomes. Black cottonwood (*Populus trichocarpa* Torr. and Gray) was the first forest tree genome to be sequenced by the US Department of Energy Joint Genome Institute (DOE/JGI) [6] (Table 1). Black cottonwood has a relatively small genome (450 Mb) and is a target feedstock species for cellulosic ethanol production, and thus fits into the DOE/JGI priority of sequencing bioenergy feedstock species. The genus *Populus* has 30+ species (aspens and cottonwoods) with genome sizes of approximately 500 Mb. Several species are being sequenced by DOE/JGI, and other groups around the world, and it seems likely that all members of the genus will soon have a genome sequence (Table 1). The next forest tree to be sequenced was the flooded gum (*Eucalyptus grandis* BRASUZ1, which is a member of the Myrtaceae family), again by DOE/JGI. Eucalyptus species and their hybrids are important commercial species grown in their native Australia and many regions throughout the southern hemisphere. Several more eucalyptus species are being sequenced, each with relatively small genomes (500 Mb), but it will probably take many years before all 700+ members of this genus are completed. Several members of the Fagaceae family are now being sequenced (Table 1). Members of this group include the oaks, beeches and chestnuts, with genome sizes less than 1 Gb.

The gymnosperm forest trees (such as the conifers) were the last to enter the world of genome sequencing. This was entirely due to their very large genomes (10 Gb and greater) as they are extremely important economically and ecologically, and phylogenetically they represent the ancient sister lineage to that of angiosperm species. Genome resources needed to support a sequencing project were reasonably well developed, but it was not until the introduction of next-generation sequencing (NGS) technologies that sequencing conifer genomes became tractable. Currently, there are at least ten conifer (Pinaceae) genome-sequencing projects under way. Aside from reference genome sequencing in forest trees, there is significant activity in transcriptome sequencing and resequencing for polymorphism discovery (Tables 2 and 3). We have only listed the transcriptome and resequencing projects in Table 1 that are associated with a species that has an active genome-sequencing project.

THE OPPORTUNITY FOR COMPARATIVE-GENOMIC APPROACHES IN FOREST TREES

The power of comparative-genomic approaches for understanding function in an evolutionary framework is well established. Comparative genomics can be applied to sequence data (nucleotide and protein) at the level of individual genes or genome-wide. Genome-wide approaches provide insight into both chromosome evolution and the diversification of biological functions and interactions. Understanding of gene function in forest tree species is challenged by the lack of standard reverse-genetic tools routinely used in other systems - for example, standard marker stocks, facile transformation and regeneration - and by the long generation times. Thus, comparative genomics becomes the more powerful approach to understanding gene function in trees.

Comparative genomics requires not only data availability but also cyber-infrastructure to support exchange and analysis. The TreeGenes database is the most comprehensive resource for comparative-genomic analyses in forest trees. Several smaller databases have been created to facilitate collaborations, including: Fagaceae genomics web, hardwoodgenomics.org, Quercus portal, PineDB, ConiferGDB, EuroPineDB, PopulusDB, PoplarDB, EucalyptusDB and Eucanext (Tables 1, 2, and 3). These resources vary greatly in their scope, relevance and integration. Some are static and archival, whereas others focus on current sequence content for a specific species or a small number of related species. This results in overlapping and conflicting data among repositories. In addition, each database uses its own custom interfaces and back-end database technology to serve sequence to the user.

The US National Science Foundation funding for large-scale infrastructure projects, such as iPlant, is leading efforts aimed towards centralizing resources for research communities [15]. Without centralized resources, researchers are forced to employ inefficient data-mining methods through queries of independently maintained databases or inconsistently formatted supplemental files on journal web sites. Specific areas of interest for the forest tree genomic community include the ability to connect sequence, genotype and phenotype to individual, geo-referenced trees. This type of integration can only be achieved through web services that allow disparate resources to communicate in ways that are transparent to the user[16]. With the recent increase of genome sequences available for many of these species, there is a need to facilitate community-level annotation and research support.

The need for a Better-Developed Open-Access Culture

The Human Genome Project established a culture of open access and data sharing in genomics research for both humans and animal models that has been extended to many other species, including *Arabidopsis*, rat, cow, dog, rice, maize and more than 500 other eukaryotes. Beginning in the late 1990s, these large-scale projects released data very rapidly to the scientific community, often years before publication. This rapid release of data with few restrictions has allowed thousands of scientists to begin work on specific genes and gene families, and on functional studies, long before the genome papers have appeared. One of the driving motivations for this culture, and the reason that many scientists support it, is that large-scale sequencing can be done most efficiently when centers that have expertise in sequencing technology take the lead. With all the sequencing concentrated, the body of data needs to be shared freely in order to get it in the hands of the widely distributed experts. This open-access culture has dramatically accelerated scientific progress in biological research.

The Path to Success Avoids Delays

Careful inspection of Table 1 reveals that forest tree genome projects are very slow to release sequence data into the public domain. Once a project is finished and submitted for publication, a draft genome becomes available - for example, the poplar genome was released and published in 2006. However, pre-publication releases are infrequent, exceptions being the PineRefSeq project that has made three releases and the SMarTForest project that has made one (Table 1). This is unfortunate because good-quality sequence contigs and scaffolds could be made available years before publication, delivering an extremely important resource to the community. This delay can be understood from privately financed projects seeking commercial advantages, but nearly all the projects listed in Table 1 are financed by public funds whose stated mission is advancing science and development of community resources. Publication rights are easily protected by data-use policy statements such as the Ft Lauderdale [17] and Toronto agreements [18], but unfortunately these conventions are not often used and data access is restricted by password-protected web sites (Tables 1, 2, and 3). We hope the opinion offered here will lead to a discussion in the forest tree community, to a more open-access culture and thus to a more vibrant and rapidly advancing research area.

GENETIC MODIFICATION OF FOREST TREES

GMOs are defined as organisms that have been modified by the application of recombinant DNA technology (where DNA from one organism is transferred to another organism). The term "transgenic trees" is also used for GM trees, where a foreign gene (a transgene) is incorporated into the tree genome. One of the first reported trials with GM forest trees was initiated in Belgium in 1988 using poplars. A study carried out in 1999 indicated that, since then, there have been more than 100 reported trials, involving at least 24 tree species-most of which are timber-producing species.

The majority of the field trials were carried out in the USA and Canada. Whereas it is estimated that roughly 40 million hectares of transgenic agricultural crops were grown commercially in 1999 (ISAAA figures), there is no reported commercial-scale production of transgenic trees. Information on field trials of GM trees has been published by the OECD and the World Wide Fund for Nature (1999). Traits for which genetic modification can realistically be contemplated in the near future include insect and virus resistance, herbicide tolerance and lignin content.

However, insertion of any gene into a tree species with expected functional results will be a substantial undertaking and insertion of enough genes to confer *e.g.* long-term insect resistance in a perennial species even more so. Virus and insect resistance, in particular, are of major significance for crop plants. By contrast, these traits are not the most important in breeding programmes of forest tree species (poplars being an exception). Reduction of lignin is a valuable objective for species producing pulp for the paper industry; work on this aspect is underway in aspen.

A major technical factor limiting the application of genetic modification to forest trees, is the current low level of knowledge regarding the molecular control of traits which are of most interest, notably those relating to growth and stem and wood quality. Genetic modification of these traits remains a distant prospect. Investments in genetic modification technologies should be weighed against the possibilities of exploiting the large amounts of genetic variation, which are generally untapped, available within any single species in nature. Biosafety aspects of GM trees need careful consideration because of the long generation time of trees, their important role in ecosystem functioning and the potential for long distance dispersal of pollen and seed.

Forestry in Developing Countries

Forests cover approximately 30 percent of the world's total land area. They are the source of vital commodities, including raw materials and food and are essential for maintaining agricultural productivity and the environmental well-being of the planet as a whole. They protect soil and water and buffer the effects of wind and rain, thus helping to decrease soil erosion and they are an important sink for carbon dioxide. Forests are also among the most important repositories of biological diversity. Roughly 500 million rural people live in, or close to, forests. Most communities use a variety of forest products, particularly those in developing countries. Plant stems, tubers and fruits provide additional food during hungry seasons or when crops fail; wild animals are harvested for meat and hides; and the forests provide fuelwood, fodder for livestock, medicines and other products and services.

The most important trend in forestry in developing countries is the progressive reduction in the area of forests due to changes in land use. Another important trend, evident at a global level, is increasing forest degradation through unmanaged use. When forests are degraded, their productive functions and their capacity as regulators of the environment are reduced, increasing flood and erosion hazards, reducing soil fertility and contributing to the loss of forest products and overall loss of biological diversity.

While forests are being lost, there is growing demand both for environmental services and for wood and wood products which they provide. A forecast by FAO predicts that wood demand is expected to increase by 25 percent from 1996 to 2010. This demand will, increasingly, have to be met by forest plantations, and with decreasing land areas available for forestry, plantation methods will have to be increasingly intensive. This will necessitate better tree improvement programmes in which biotechnology may play a role.

THE WEALTH OF INTRAPOPULATION GENETIC VARIATION

Within populations, four forces (selection, migration, mutation and population size) determine the structure of genetic variation. The interaction of all four factors can be highly complex and their effects are determined by the mating system and the kinds of gene effects that exist. A problem in analysing even this level of genetic evolution is that the kinds of gene effects that exist depend on the effects that other forces of evolution have had on gene actions themselves. To consider how the forces can affect evolution, we can first start with each of them and then consider how their interactions can change predictions.

For selection to have an effect on speciation and the structure of forests, there must be a difference among genotypes in the traits affected by survival or reproduction events and the individuals must accordingly leave more or fewer progeny than other genotypes. The genotype is defined as the individuals with a distinctive ensemble of alleles at its gene loci that distinguishes it from other genotypes. This is a multiple gene locus definition, which is often simplified by considering every one of the loci to be independent in the occurrence of its alleles in both frequency and effect.

In this case, we can consider the genotype to be the summation of all independent loci and hence can examine gene effects and dynamics one locus at a time. This is, of course, an oversimplified view since genes are in fact linked on chromosomes and therefore have linkages that correlate their occurrences, and most genes affect processes that are combined in their net effect on traits; furthermore, most genes affect enzymatic pathways that ultimately affect several traits when measured at the level of whole organism performance. These effects are called linkage, epistasis and pleiotropy, respectively, and tie the whole organism together into an integrated set of actions and functions that can survive and reproduce.

While undoubtedly true in general, there is little evidence that individuals are so bound by such constrained genetic constitutions and physiological limitations that substantial independence among loci exists. The evidence for the existence of strongly selected, tightly linked 'coadapted gene complexes' is not strong and hence the model of independence can be usefully employed, even though it is surely wrong at least in detail.

Within a gene locus, many alleles may exist and the pairwise interactions among them are called the dominance relations. For a diploid organism, if both alleles at a locus are the same, the individual is said to be a homozygote; if the two alleles are different, the individual is said to be a heterozygote. If the heterozygote has a phenotype intermediate between the two homozygotes, the dominance effect is considered to be zero and the gene action is said to be additive. If the heterozygote is other than intermediate, it may be less than the lesser of the two homozygotes. If its phenotype lies between the homozygotes but is not exactly intermediate and if it appears to be greater than the greater of the homozygotes, the gene action is said to be underdominant, partially dominant or overdominant, respectively.

The effect of selection would depend on the kind of gene action, since directional selection for greater or lesser phenotypes would favour different homozygote and heterozygote states. Selection may also be stabilizing if it favours an intermediate phenotype or disruptive if it disfavours the intermediate and its effect on the frequency of the alleles would depend on the gene actions.

Directional selection experiments seem to indicate that most alleles have partial dominance actions, though many may be involved in multiple locus interactions. Since selection would be expected to favour survival and reproduction of certain genotypes, we expect that it would increase the population level of fitness. Selection is therefore usually considered to have a positive effect on the evolutionary potential of a population or species in small increments every generation but may not generate an optimal condition at any one time. For alleles that have partial dominance effects, the dynamics of the rates of displacement of one allele by another are such that the frequency of the allele itself affects its own rate of displacement. When alleles are at either low or high frequency, their rates of change are slower than when they are at an intermediate frequency. Except in small populations where accidental loss can occur, selection would not generally lose alleles that are favoured for adaptation.

Another force of evolution is mutation, which is usually considered to be a random change in gene structure and function that occurs indiscriminately throughout the genome. Since most species have evolved a genetic and physiological level that is close to a sufficient capacity for continued evolution, novel changes are not expected to improve performance or fitness but rather to generally decrease adaptability. However, there are always going to be mutations and while most may be disfavoured by selection, there is likely to be a balance between the rate at which new mutants are introduced and the rate of their elimination.

Most mutations occur at low rates for any one locus in any one generation and would not be expected to affect the adaptability of a population unless they are accumulated in small populations. For most conifer species where the frequency of deleterious mutants has been estimated, there are a larger number of lethal alleles carried at one or another locus in most individual trees than in most other plant or animal species. This might indicate that there are many effective loci in trees that can mutate to a harmful state, that the mutation rate is high or that the mutants are favoured in some conditions or are of some advantage in heterozygous conditions. However, independent mutations and selection can generate differences among populations, which can increase fitness in each. Hybridization among populations can then generate high levels of genetic variance in traits that may then be useful in new environments.

A third force of evolution is migration, which can occur among many different kinds of population structures. The input of pollen or seed from other populations that have different allele frequencies may bring in alleles that might otherwise be lost, and if the migration rate is high enough among all populations may induce all populations to have approximately the same frequencies.

If the immigrant alleles are favoured, then further immigration would advance the rate of its increase; if disfavoured, immigration would maintain alleles at higher frequencies than expected in a manner similar to mutation; and if neither, would tend to make frequencies homogeneous among all populations. In the presence of divergent selection among populations that exist in different environments, selection would act to eliminate migrant alleles and would maintain population divergence. The fourth force is sampling error induced by small population sizes, which tends to increase the probability that low-frequency alleles are lost by chance. If populations remain at low sizes for several generations, the chances of random loss are accumulated; even populations of more than 20-50 adults can lose alleles that may have started with moderate frequencies.

Since most forest trees seem to carry many deleterious mutants, even if at low frequency at any one locus, another problem is that small populations will suffer from inbreeding depression and may eventually suffer such debilitation from several mildly depressive mutants that the population goes extinct. Individual populations would have to be large enough to sustain viability, and to avoid environmental accidents and catastrophes if no migrants were recruited.

There are two problems faced in persistently small populations, the loss of alleles and inbreeding depression; hence, larger population sizes are generally better for avoiding extinction. However, the effects of selection, migration and mutation are confounded with the sampling errors incurred in real populations of finite size. Thus, if populations are separated and selected for different environmental optima, migration can introduce alleles in high frequency that depress fitness. On the other hand, migration can maintain alleles in populations where they might be useful, but would be lost due to sampling accidents in small populations.

It is reasonable to expect that all species are under the influence of several or all of the forces of evolution simultaneously and that the balance between the several forces is not the same for all genes in all populations. Furthermore, since the physical and biotic environments of forests are rarely at an equilibrium, the genetic and ecological dynamics of most species would have to be considered to be in other than a stable state.

The actual distribution of the multiple functional forms of alleles at each of the several tens of thousands of gene loci in each of the individuals of a population must therefore be the resultant of the mixture of forces felt at each locus, subject to historical events. The vast possibilities of mixed distributions are overwhelming, although it is also obvious that for most of the species we can study, species are not random mixtures of an infinite number of possible combinations.

DEFINITION OF TREES OUTSIDE THE FOREST

Trees outside the forest are defined by default, as all trees excluded from the definition of forest and other wooded lands. Trees outside the forest are located on "other lands", mostly on farmlands and built-up areas, both in rural and urban areas. A large number of TOF consist of planted or domesticated trees. TOF include trees in agroforestry systems, orchards and small woodlots. They may grow in meadows, pastoral areas and on farms, or along rivers, canals and roadsides, or in towns, gardens and parks. Some of the land use systems include alley cropping and shifting cultivation, permanent tree cover crops (e.g. coffee, cocoa), windbreaks, hedgerows, home gardens and fruit-tree plantations.

Classification of trees outside the forest presents certain difficulties. There are existing classifications for agroforestry, but none applicable to all trees outside the forest. For practical reasons, the FRA 2000 definition of "forest" combines aspects of both land cover and land use. This approach creates difficulties not only for classification of forest, but also for classification of TOF.

In a study on data-gathering on TOF in Latin America, where classification was primarily based on land use criteria, separating land use and land cover aspects was found to be a main source of misinterpretation. There was a possibility of confounding coffee plantations and trees in pasture with forest, given their high density.

This clearly shows some of the problems involved in establishing a simple and reliable *a posteriori* classification.In France, the National Forest Inventory (IFN) and Teruti Land Use Study-have begun to attempt coordinating classifications of trees outside the forest. The objective is eventually to use the annual Teruti data to update the IFN ten-year data, with a single national nomenclature as a possible end result.

FUNCTIONS AND CHALLENGES

In industrialized countries, farmers list shade and shelter, soil protection and improvement of the landscape and rural environment as their main reasons for growing trees. In the tropics, farmers grow woody species for food security and subsistence. Trees outside the forest are a major source of food. Livestock fodder produced by TOF can be a matter of life and death in semi-arid or mountainous areas.

Fuelwood remains the prime source of energy in developing countries, representing up to 81 per cent of the wood harvest (FAO 1999). In contrast, in the industrialized countries, fuelwood accounts for less than 10 per cent of total fuel consumption (FAO 1998). Very few studies have reported on overall fuelwood output from stands and single trees outside the forest, but agroforestry systems and orchards are known to provide a large part of the resource.

Trees outside the forest have an important ecological role. Planted trees and shrubs in fields help to check runoff and erosion and control flooding, as well as helping to purify water and protect against wind. Trees lining rivers and streams help to maintain biodiversity, providing spawning beds for fish and shellfish and shade which reduces eutrophication.

The unique role of trees in soil protection and conservation, checking wind and water erosion and maintaining soil fertility is universally acknowledged. Also important are the cumulative benefits of trees on smallholdings to soil and water conservation, in particular in the larger context of mountain watershed management; their positive impact on climate; and their role in buffering the effects of desertification and drought.

RESULTS OF SELECTED STUDIES

In spite of the limits of data at the regional or global level, a number of local initiatives have been carried out. The approaches of the various studies differ in accordance with the purpose and scale of the analysis. Few studies use methods resembling the conventional forest inventory. Many studies rely on existing literature or estimates drawn from surveys and interviews. The quantification of products is often based on different parameters, such as estimates of global output, marketed output, observed or potential productivity or economic value. Thus the reliability of the results is uncertain.

The following are the results of some national initiatives that have assessed trees outside the forest.

In Kerala, the most densely inhabited state of India, a study estimated that of the total annual production of 14.6 million cubic metres of wood in the state, about 83 per cent was from homesteads (house compounds and farmlands), 10 per cent from estates (plantations of rubber, cardamom, coffee and tea) and only about 7 per cent from forest areas 26.6 per cent of the state area is under forest cover. Trees outside the forest met about 90 per cent of the fuelwood requirements of the state. Fuel from coconut trees alone, including both wood and non-wood materials (pruned and fallen), constituted about 70 per cent of the total fuelwood supply.

A study in Haryana State in India, an intensively cultivated state with about 3.8 per cent of its area classified as forest land but only about 2 per cent under actual forest cover showed that farm forestry (trees along farm bonds and in small patches up to 0.1 ha) accounted for 41.2 per cent of the total growing stock of wood. Multiple tree rows along roads and canals accounted for 13 per cent and 9.6 per cent, respectively; village woodlots for 24 per cent; and block plantations of less than 0.1 ha for 10.6 per cent.

In Morocco, where forest cover is less than 5 per cent of the land cover and other wooded lands only 7 per cent, nearly 20 per cent of the land may be occupied by trees outside the forest, namely as wooded pasture (84 per cent) and fruit-tree plantations (12 per cent) (Rosaceae, citrus, olives trees, palm trees, walnut trees, fig trees, almond trees).

Fruit production has an important place in the national economy. It is noteworthy that even when a forest is largely destroyed, the carob is one of the few species traditionally conserved, as it is highly appreciated by farmers for multiple purposes, providing both fodder and income from the sale of its fruit for export. However, there are no reliable data on the distribution and potential of this "forest" resource, which is of interest for farmers, herders, concessionaires and the government and distributed on agricultural and forest lands.

In the Sudan, the National Forest Inventory has undertaken a national land use inventory to provide area and volume statistics for planning at the subnational and national levels. The inventory was designed to provide preliminary estimates regarding products other than the traditional fuelwood and timber, such as the amount of gum, fruit or nuts that can be collected and the distribution of non-wood species of interest.

In Costa Rica, the Tropical Agricultural Research and Higher Education Centre (CATIE), in collaboration with Freiburg University, Germany, is developing a regional methodology for Central America to assess tree resources outside the forest.

A mix of satellite remote sensing, aerial photos and ground sampling is used to address the complexity of the resource and to allow dynamic monitoring of resources at the national and regional levels data-gathering on TOF in eight Latin American countries. None of these countries had established a database and the search for information was multisectoral. The statistics on land cover and land use gave some idea of the relative importance of trees outside the forest.

In Kenya, extensive tree planting on farmlands was promoted in the 1970s and 1980s, with land tenure security as a major incentive. There is an increasing trend of tree cover and species diversification on privately owned farms. Assuming that the present rate of increase in tree planting will continue, it was estimated that farms produced about 9.4 million cubic metres of wood in 2000 and will produce about 17.8 million cubic metres in 2020.

Their share of the total wood produced in the medium- and high-potential districts was projected to increase to 80 per cent in 2020. Indeed, while natural stands of trees have declined there has been a corresponding increase in tree planting in much of the densely populated plateaus of Kenya. As natural forests are reduced or become inaccessible, agroforestry systems help people to diversify production and income and to protect themselves from shortages of fuel and wood.

In Bangladesh, natural forest formations cover less than 6 per cent of the country and the population growth rate is extremely high. An inventory of homestead/village forests in the country indicated that trees outside the forest constitute a vital resource for local populations, providing food, fodder and fuelwood. The sampling method was based on dual village/household sampling with an agro-ecological and administrative sampling base. Rural Bangladesh was divided into six major regions considered as agro-ecological strata, each subdivided into *thanas* (administrative entities, subdistricts). The households making up the sampling units were chosen at random from a number of villages.

The inventory sampled data on palm trees and cane as well as trees, bamboo and thickets. The results, expressed per stratum and per inhabitant, provide volumetric data for fuelwood and sawnwood and species data for total amounts under and over 20 cm. This inventory was apparently the first to nationwide assessment of trees outside the classified forests in Bangladesh.

METHODS AND TOOLS FOR FUTURE ASSESSMENTS

A priority challenge of future assessments is to know the state and dynamics of all tree resources both in and outside the forest. A country embarking on a planning exercise cannot confine itself solely to the trees within its forests, especially when its wood resources appear to be insufficient.

The choice of tools and methods used to describe or assess trees outside the forest depends on the scale of analysis, kind of data and degree of exactitude desired. The tools used are not generally specific or new; rather, they are combined and implemented in original ways.

The inventory in Bangladesh described above is one of the numerous examples of methods developed for gathering data on TOF. The Bangladesh study gives evidence of the adaptations needed to bring conventional forest inventory procedures in line with the specific nature of this resource.

In some aspects - e.g. structure, spatial distribution and extent of area cover - trees outside the forest are more difficult to assess than forest formations. The assessment of TOF does not lend itself to the potential cost savings associated with expanded uses of remote sensing technology. Remote sensing by satellite presents more difficulties for assessing TOF resources than for assessing attributes such as forest area. However, satellite data do allow a region to be stratified on the basis of ecological criteria and land cover, providing the basis for a good working document for more specific work in the future. The most commonly used remote sensing technology for TOF resources is aerial photography, which can be used to describe spatial distribution and to distinguish TOF cover classifications, providing the appropriate scale is chosen. However, high costs prohibit widespread use of aerial photography for TOF assessments in most countries. The new 1 m resolution satellite sensors represent a possible future alternative to aerial photography.

Some TOF field inventories are modelled on forest inventory methods and keep to biological and physical criteria; others emphasize social aspects, choosing villages as the sampling units. For measurements on the ground, sampling arrangements designed for forest stands may not be the most effective arrangements for trees. Less traditional sampling plans which would theoretically be better suited to this resource should be tested on various categories of TOF, especially those covering fairly large areas.

Studies of the social and economic benefits or impacts of TOF often rely on household surveys, interviews or standardized appraisals such as rapid or participatory rural appraisal. The integration of the last two approaches - biophysical inventory and socio-economic analysis - is not simple and calls for caution given the great variety of social situations that are only meaningful in the local context.

Environmental benefits or impacts of TOF might be indirectly assessed by linking measurable indicators, such as the number and type of trees, with environmental variables such as water quality or erosion. In an urban setting, tree cover might have direct impact on the ambient temperature. Measuring the environmental impact of tree management is an issue for all natural resource planning or management operations. Assessment of trees outside the forest requires geographical, ecological, biophysical, social and economic data.

However, this implies that an important amount of information will have to be carefully processed. The diversity of end-uses for this information, including land use planning and analysis based on inventory, will need to be considered in data assembly and processing and in the presentation of results.

It is important to know the status of trees outside the forest at any given moment, but it is even more essential to be able to trace patterns of change over time in the same area. The two most commonly used approaches have been comparison of aerial photos taken at sufficiently long intervals and surveys among villagers/managers combined with field inventories.

Some countries, such as France and the United Kingdom, have undertaken periodic inventories based on the establishment of permanent plots linked to permanent forest inventories. However, the high cost of this type of operation limits the number of countries able to adopt it. India and Bangladesh are now experimenting with options for the future.

The current trend towards decentralized authority in land use planning suggests the importance of carrying out assessments at the local level, where the geographical, historical and socio-economic context is relatively harmonious. A minimum number of common rules concerning methods and arrangements is necessary, however, if the data are to be comparable at the country level. Certainly, the technical side of assessing trees outside the forest is complex and more research is needed to better pinpoint the resource.

Bibliography

M.R.K. Jerram: *A Text-Book on Forest Management : In Agriculture, Horticulture and Forestry*, Asiatic Pub, 2006.

S.P. Singh: *Advances in Horticulture and Forestry : Volume 8*, Scientific, 2001.

S P Singh: *Advances in Horticulture and Forestry : Volume 9*, Scientific, 2006.

S.P. Singh: *Advances in Horticulture and Forestry, Volume 6*, Scientific, 1999.

S.P. Singh: *Advances in Horticulture and Forestry, Volume 7,* Scientific, 1999.

D. Joseph Bagyaraj: *Microbial Biotechnology for Sustainable Agriculture Horticulture and Forestry*, New India Publishing Agency, 2011.

Sudakshina Gupta: *A Few Aspects of Economics of Agriculture in Ancient India : Compared to Present Day India*, Manak Publications, 2011.

S.C. Panda: *A Hand Book of Agriculture*, Agrobios, 2012.

S Dutt: *A Handbook of Agriculture*, ABD, 2005.

Mahipal Singh Rana: *A Handbook of Agriculture in India*, Cyber Tech Pub, 2009.

Afzal Sharieff, Masood Ali Khan and A. Balakishan: *A New Geography of Agriculture*, Sarup, 2010.

Sandeep Bajpai: *A Survey on Indian Agriculture*, Cyber Tech Pub, 2009.

M.R.K. Jerram: *A Text-Book on Forest Management : In Agriculture, Horticulture and Forestry* , Asiatic Pub, 2006.

S.R.J. Singh: *A Textbook of Agriculture Statistics*, Oxford Book Company, 2012.

V.C. Srivastava: *Achieving Sustainability in Agriculture : Issues, Challenges and Opportunities*, Agrobios, 2010.

B B Pawde; V B Bhise and S R Takle: *Adoption and Impact of New Agricultural Technology on Tribal Agriculture*, Serials Pub, 2008.

S B Singh and O P Charasia: *Advances In Agriculture Environment and Health*, Satish Serial Pub, 2009.

S.R. Maloo and L.L. Somani: *Advances in Experimental Agriculture*, Vol. I, Agrotech Publishing Academy, 2009.

S.R. Maloo and L.L. Somani: *Advances in Experimental Agriculture: Vol. 3*, Agrotech Publishing Academy, 2013.

H.P. Singh, Anil Kumar V.A. Parthasarathy and Babita Singh: *Advances in Horticulture Biotechnology Nanotechnology in Agriculture: Vol. 6* , Westville Publishing House, 2013.

K. Jayaraman: *A Handbook on Statistical Analysis in Forestry Research*, Kerala Forest Research Institute, 2001.

L K Jha and P K Sen Sarma: *A Manual of Forestry Extension Education*, A.P.H, 2008.

Benu Singh: *A Modern Book on Forestry and Horticulture*, Vista International Pub, 2010.

Benu Singh: *A Survey of the Forestry Research*, Vista International Pub, 2011.

M.R.K. Jerram: *A Text-Book on Forest Management : In Agriculture, Horticulture and Forestry*, Asiatic Pub, 2006.

L K Jha: *Advances in Agroforestry*, APH, 2009.

Satish Kumar Sinha: Advances in Agroforestry, Random Pub, 2012.

Praveen Taank: *Advances in Forestry Research in India*, Cyber Tech Pub, 2009.

R.K Bhardwaj: *Advances in Forestry Research in India*, Oxford Book Company, 2009.

Ram Parkash: *Advances in Forestry Research in India : Vol. XXV*, International Book Distributors, 2002.

S.K. Mukherjee: *Advances in Forestry Research in India : Vol. XXVI*, International Book, 2004.

Ram Parkash: *Advances in Forestry Research in India Vol: XXIII*, International, 2001.

Index

A

Abiotic Stress 28
Absorption 38, 39, 41, 42, 43, 66, 67
Acetosyringon 114
Adaptation 143, 144, 166, 168
Adequacy 73, 76
Advantages 254, 291
Agrobacterium 113, 114, 115
Assessment 88, 93, 107

B

Benzyladenine 114
Biochemical Sequence 6
Biological 65
Bombardment 156
Brassicas 123
Breakfast Cereals 72, 73
Breeding System 133, 163

C

Canning 67
Cereals 68, 71, 72, 73
Comparative Maps 168
Comparisons 274
Cornmeal 71
Crop Improvement 90
Crop plants 88, 265

D

Definition 194
Dietary 68, 74, 75, 76
Drift 161, 162
Drosophila 9, 11
Durum Wheat 183
Durum wheat 183, 184, 187

E

Ecological 53, 56, 58, 91, 161
Elimination 89, 91, 145, 146
Engineered Plants 153
Engineering 254
Environmental 49, 50, 51, 53, 57

F

Fish 42
Flours 71
Flowering Plants 255
Food Fortification 76, 77
Food Legumes 188
Foods 48, 52, 57
Functions 255, 271, 290

G

Gametes 5, 6, 7, 8, 9, 189
Gene Mapping 117
Gene Technology 141
Genetic 5, 6, 7, 8, 9, 11, 189
Genetic Efficiency 157
Genetic Gain 187
Global System 283
Grains 66, 68, 69, 70, 71

H

Haploid Lines 133
Haploid Lines 134, 171

Health Implications 75
Heredity 5
Heterogeneity 158, 159, 162
Heterosis Potential 171
Horticultural Crops 263
Hospital 217

I

Improvement 108, 109, 272
Indolebutyric 114
Industry 53, 57, 192, 193, 197
Information Technology 287
Integration 290

L

Lethality 145
Linkage Drag 170
Linkage Groups 9

M

Measurement 250, 254, 255
Meiosis 5, 6, 7, 9, 189
Melanogaster 9
Methods 49, 193, 196
Micronutrient 66, 68, 74, 75, 76, 77, 78

N

Nondestructive 170
Novel Genes 90
Novel Technologies 88

O

Organization 46
Ornamental Plants 262

P

Packaging 64, 65, 69, 70, 73, 74
Per cent 299
Pesticides 90, 91
Phenylalanine 8
Phylogeny 160
Plant Breeding 174

Q

Quality assurance 74, 76

R

Regulation 47, 75, 77
Rhizogenes 113
Risk Assessment 58

S

Safety 50, 51
Salt 67, 71, 72, 74, 75, 76
Scenarios 277
Science 209, 210, 211
Sector 192, 196, 197, 198
Segregation 8
Spatial Scale 158

T

Taxonomic 89, 94, 99
Technological 192, 193, 284
Temporal Scale 164
The Stem 257
Thermal Processes 64
Transduction 48, 51, 58

U

Uniqueness 274, 277

V

Vectors 55
Vegetative 251, 252
Vegetative Propagation 266

W

Wide-hybridization 144

Z

Zinc 69, 71, 72